Flexural-Torsional Buckling of Structures

New Directions in Civil Engineering

Series Editor: **W.F. Chen** Purdue University

Flexural-Torsional Buckling of Structures

N.S. TRAHAIR
Challis Professor of Civil Engineering
The University of Sydney
Australia

CRC Press
Taylor & Francis Group
Boca Raton London New York

CRC Press is an imprint of the
Taylor & Francis Group, an **informa** business
A SPON PRESS BOOK

First published 1993 by E & FN Spon
First edition 1993

Published 2019 by CRC Press
Taylor & Francis Group
6000 Broken Sound Parkway NW, Suite 300
Boca Raton, FL 33487-2742

First issued in paperback 2019

No claim to original U.S. Government works

ISBN 13: 978-0-367-44983-4 (pbk)
ISBN 13: 978-0-8493-7763-1 (hbk)

Visit the Taylor & Francis Web site at
http://www.taylorandfrancis.com

and the CRC Press Web site at
http://www.crcpress.com

Typeset in 10/12pt Times by Thomson Press (India) Ltd, New Delhi

A catalogue record for this book is available from the British Library

Library of Congress Cataloging-in-Publication Data
Trahair, N.S.
    Flexural-torsional buckling of structures   N.S. Trahair.
        p.   cm. – (New directions in civil engineering)
    Includes bibliographical references and index.
    ISBN 0–8493–7763–3
    1. Buckling (Mechanics) 2. Structural stability. I. Title.
II. Series.
TA656.2.T73   1993   93-3714
624.1'76–dc20   CIP

Contents

Preface

Flexural-torsional buckling is a mode of structural failure in which one or more members of a frame suddenly deflect and twist out of the plane of loading. Because flexural-torsional buckling reduces the load-carrying capacity of the structure, designers must prevent it either by providing additional bracing, or by using larger members.

From the 1930s, the subject of flexural-torsional buckling has been dealt with in S.P. Timoshenko's widely used textbook *Theory of Elastic Stability*, last published (with J.M. Gere) in 1961, and later in F. Blelch's *Buckling Strength of Metal Structures*, published in 1952. These cover a wide range of structural stability topics, and so their coverage of flexural-torsional buckling is limited to a few chapters.

These books were both written before the advent of the electronic digital computer in the 1950s, and the subsequent explosion in published research on the subject. Consequently, more recent treatments of stability theory have been more limited either in their scope or in their depth. Strangely, there have been few, if any, recent books published which provide thorough treatments of flexural-torsional buckling.

This book is intended to provide both an up-to-date treatment of modern methods of analysing flexural-torsional buckling, and also to provide sufficiently detailed summaries of knowledge on flexural-torsional buckling that it can be used as a source book by both designers and researchers.

It may also be used for teaching purposes as a text or reference book. In advanced level undergraduate courses, the teacher will want to simplify and edit the material given. Such a course may introduce the theory of structural stability given in Chapter 2, and the hand and computer methods of analysis dealt with in Chapters 3 and 4. Subsequent material on column, beam, and beam-column buckling may be selected from Chapters 5, 7 and 11, and on design against flexural-torsional buckling from Chapter 15.

For normal level post-graduate courses, these topics may be presented in greater detail, and expanded with material on restrained buckling from Chapters 6, 8, 9, 10 and 12, and possibly with a treatment of inelastic buckling based on Chapter 14. For advanced courses, these topics would be studied more thoroughly, while additional topics may be introduced from Chapters 13 and 16.

It is not often these days that a researcher is allowed to develop a fascination for a subject and spend the amount of time on it that I have on flexural-torsional buckling, and I count myself as being privileged in this regard. My first introduction to the subject was through a Structures Honours course given to me as an undergraduate by the late J.W. Roderick in 1955. As a masters research student, a topic on the flexural-torsional buckling of beam-columns was suggested to me by one of my co-students, P.G. Lowe, and my work on this, which was supervised by Roderick, confirmed my interest in the subject. In early 1961 my curiosity on the lateral buckling of beams was stimulated by the British Standard BS 449:1959. This led me into work for the Standards Association of Australia for the development of an Australian code for the design of steel structures, and into my doctoral study of the flexural-torsional buckling of frame structures, again supervised by Roderick.

In 1968 I had the good fortune to spend a study leave at Washington University with T.V. Galambos, who inspired me with his own fascination for the subject. In the early 1970s, I learnt much with J.M. Anderson, S. Kitipornchai, P. Vacharajit-tiphan, S.T. Woolcock, T. Poowannachaikul and B.R. Mutton, an exceptional group of graduate students at the University of Sydney. My good fortune continued in 1974–75, when I spend another study leave, this time at the University of Sheffield where I collaborated with D.A. Nethercot. In the 1980s, I again had a number of outstanding students, M.A. Bradford, P.E. Cuk and J.P. Papangelis, while two of my colleagues at the University of Sydney, G.J. Hancock and N.L. Ings, collaborated with me on flexural-torsional buckling research, as did M.A. Bradford and S. Bild as post-doctoral fellows, and later Y.L. Pi.

In the early 1970s I began teaching the same Structures Honours course that Roderick taught me in 1955, and also a post-graduate course on Structural Stability. I gave a related stability course at the University of Alberta in 1985. The stimulus of preparing and developing these courses and of challenging and reacting to my students has done much to extend my own understanding of the subject, as well as to give me a broad outline for this book.

My studies of flexural-torsional buckling have not been limited to research, and my work for nearly 30 years with the Standards Association of Australia on the preparation of codes for the design of steel structures has given me a keen appreciation of the need to translate research findings on flexural-torsional buckling into forms that are easily understood and used by designers.

I have been greatly influenced in the preparation of this book by my teaching and research experiences in Australia at the University of Sydney, in the USA at Washington University, in the UK at the University of Sheffield, and in Canada at the University of Alberta. While a significant proportion of the material in this book has been developed by me and my colleagues and students, much of it is not original, but has been gathered from many sources. Unfortunately, it is very difficult or even impossible to acknowledge all individual sources, and so the references given in this book are restricted to those which the general reader may wish to consult for further information.

I would like to thank the School of Civil and Mining Engineering of the University of Sydney for the facilities that it has made available to assist in the preparation of this book. The manuscript was expertly typed by Jean Whittle and Cynthia Bautista, and the diagrams prepared by Ron Brew and Kim Pham.

Many valuable comments and suggestions were made by my colleagues and students at the University of Sydney, especially by K.J.R. Rasmussen.

Finally, I wish to acknowledge the unfailing help and support of my wife, Sally, without whom the writing of this book would not have been possible.

Nicholas Trahair

Units and conversion factors

Units

While most expressions and equations used in this book are arranged so that they are non-dimensional, there are a number of exceptions. In almost all of these, SI units are used which are derived from the basic units of kilogram (kg) for mass, metre (m) for length, and second (s) for time.

The SI unit of force is the newton (N), which is the force which causes a mass of 1 kg to have an acceleration of 1 m/s^2. The acceleration due to gravity is 9.807 m/s^2 approximately, and so the weight of a mass of 1 kg is 9.807 N.

The SI unit of stress is the pascal (Pa), which is the average stress exerted by a force of 1 N on an area of 1 m^2. The pascal is too small to be convenient in structural engineering, and it is common practice to use either the megapascal $(1 \text{ MPa} = 10^6 \text{ Pa})$ or the identical newton per square millimetre $(1 \text{ N/mm}^2 = 10^6 \text{ Pa})$. The megapascal (MPa) is used generally in this book.

Table of conversion factors

To Imperial (British) Units		To SI units	
1 kg	= 0.068 53 slug	1 slug	= 14.59 kg
1 m	= 3.281 ft	1 ft	= 0.304 8 m
	= 39.37 in.	1 in.	= 0.025 4 m
1 mm	= 0.003 281 ft	1 ft	= 304.8 mm
	= 0.039 37 in.	1 in.	= 25.4 mm
1 N	= 0.224 8 lb	1 lb	= 4.448 N
1 kN	= 0.224 8 kip	1 kip	= 4.448 kN
	= 0.100 36 ton	1 ton	= 9.964 kN
1 MPa*[†]	= 0.145 0 kip/in.²(ksi)	1 kip/in.²[†]	= 6.895 MPa
	= 0.064 75 ton/in.²	1 ton/in.²	= 15.44 MPa
1 kNm	= 0.737 6 kip ft	1 kip ft	= 1.356 kNm
	= 0.329 3 ton ft	1 ton ft	= 3.037 kNm

*1 MPa = 1 N/mm².
[†] There are a few dimensionally inconsistent equations used in this book which arise because a numerical value (in MPa or kip/in.²) is substituted for the Young's modulus of elasticity E while the yield stress F_Y remains algebraic. The value of the yield stress F_Y used in these equations should therefore be expressed in either MPa or kip/in.², whichever is appropriate. Care should be used in converting these equations from SI to Imperial units, or vice versa.

Glossary of terms

Arch A member curved in the plane of loading.

Beam A member which supports transverse loads or moments only.

Beam-column A member which supports transverse loads or moments which cause bending and axial loads which cause compression.

Beam-tie A member which supports transverse loads or moments which cause bending and axial loads which cause tension.

Brace A secondary member which prevents or restrains deflection or twist rotation of a main member.

Braced beam A beam with a number of cross-sections braced against lateral deflection and twist rotation.

Buckling A mode of failure in which there is a sudden deformation in a direction or plane normal to that of the loads or moments acting.

Cantilever A member with an end which is unrestrained against lateral deflection and twist rotation.

Capacity factor A factor used to multiply the nominal capacity to obtain the design capacity.

Column A member which supports axial compression loads.

Conservation of energy A principle describing the conditions under which a structure and its loads may deform without any change in the total energy of the system.

Continuous beam A beam which is continuous over one or more supports.

Design capacity The capacity of the structure or element to resist the design loads. Obtained as the product of the nominal capacity and the capacity factor.

Design load The combination of factored nominal loads which the structure is required to resist.

Distortion A mode of deformation in which the cross-section of a member changes shape.

Effective length The length of an equivalent simply supported member which has the same elastic buckling load as the actual member.

Elastic behaviour Deformations without yielding.

Energy method A method of buckling analysis based on the principle of conservation of energy.

Finite element analysis A computer method of numerical analysis in which a complete structure is divided into a number of elements of finite size.

First-order analysis Elastic linear analysis in which equilibrium is formulated for the undeformed position of the structure, so that the moments caused by products of the loads and deflections are ignored.

First-yield moment The value of the bending moment which nominally causes the first yield of a cross-section.

Flexural-torsional buckling A mode of buckling in which a member deflects and twists.

Frame A skeletal structure consisting of a number of members connected together at joints.

Frame buckling A mode of buckling in which all the members of a frame participate.

Geometrical imperfections Initial crookedness or twist.

Inelastic behaviour Deformations accompanied by yielding.

In-plane behaviour The behaviour of a member which deforms only in the plane of the applied loads.

Lateral buckling Flexural-torsional buckling of beams.

Limit states design A method of design in which the performance of the structure is assessed by comparison with a number of limiting conditions of usefulness. The most common conditions are the strength limit state and the serviceability limit state.

Load and resistance factor design The limit states method of design in which the factored (reduced) resistance is compared with the factored (increased) loads.

Load factor A factor used to multiply a nominal load to obtain part of the design load.

Local buckling A mode of buckling which occurs locally (rather than generally) in a thin plate element of a member.

Member One-dimensional structural element which supports transverse or longitudinal loads or moments.

Member buckling A mode of buckling involving the complete length of a member.

Nominal capacity Capacity of a member or structure computed using the formulations of a design code or specification.

Nominal load Load magnitude determined from a loading code or specification.

Non-uniform torsion The general state of torsion in which the twist of the member varies non-uniformly.

Out-of-plane buckling The buckling of a member out of the plane of loading.

Plastic analysis A method of analysis in which the ultimate strength of a structure is computed by considering the conditions for which there are sufficient plastic hinges to transform the structure into a mechanism.

Plastic hinge A fully yielded cross-section of a member which allows the member portions on either side to rotate under constant moment (the plastic moment).

Plastic moment The value of the bending moment which will cause a section to become fully yielded.

Post-buckling behaviour Behaviour after buckling.

Potential energy Energy associated with height of a gravitational load above a datum.

Pre-buckling behaviour Behaviour before buckling.

Purlin A horizontal member between main beams which supports roof sheeting.

Reduced modulus The modulus of elasticity used to predict the buckling of inelastic members under constant applied load, so called because it is reduced below the elastic modulus.

Residual stresses The stresses in an unloaded member caused by uneven cooling after rolling, flame cutting, or welding.

Resistance Capacity.

Restraint An element which restrains the deflection or twisting of a member.

Second-order analysis Non-linear analysis in which equilibrium is formulated for the deformed position of the structure, so that the moments caused by products of the loads and deflections are included.

Shear centre The point in the cross-section of a beam through which the resultant transverse force must act if the beam is not to twist.

Shear modulus The initial modulus of elasticity for shear stresses.

Squash load The value of the compressive axial load which will cause yielding throughout a short member.

Strain energy Energy associated with the straining of a structure.

Strain-hardening A stress-strain state which occurs at stresses which are greater than the yield stress.

Strength limit state The state of collapse or loss of structural integrity.

Tangent modulus The slope of the inelastic stress-strain curve which is used to predict the buckling of inelastic members under increasing load.

Total potential The sum of the strain energy of a structure and the potential energy of the gravitational loads acting on it.

Uniform torque That part of the total torque which is associated with the rate of change of the angle of twist rotation of the member.

Uniform torsion The special state of torsion in which the twist of the member varies linearly.

Virtual work A principle used to assess whether a structure is in an equilibrium position.

Warping A mode of deformation in which plane cross-sections do not remain plane.

Warping torque The other part of the total torque (than the uniform torque), which only occurs during non-uniform torsion, and which is associated with changes in the warping of the cross-sections.

Work Energy transferred during the movement of a force.

Yield stress The average stress during yielding when significant straining takes place. Usually, the minimum yield stress in tension specified for the particular steel.

Young's modulus The initial modulus of elasticity for normal stresses.

Principal notation

The following is the principal notation used in this book. Usually, only one meaning is assigned to each symbol, but in those cases where more meanings than one are possible, then the correct one will be evident from the context in which it is used.

A	Cross-sectional area
$[A_L], [A_Q]$	Matrices for linear and quadratic potential energy contributions of $\{q\}$
$[A_{LQ}], [A_{QQ}]$	Matrices for linear and quadratic potential energy contributions of $\{Q\}$
B	Flange width, or
	Bimoment
$[B_i]$	Matrix for in-plane generalized strains
$[B_L], [B_Q]$	Matrices for linear and quadratic generalized strains
$[B_u], [B_v]$	Matrices for out-of-plane generalized strains
C_{bc}	Moment gradient factor for beam-columns (equation 11.31)
C_m	Moment gradient factor for beam-columns (equation 14.55)
$[C]$	Matrix for out-of-plane nodal deformations
$[C_i]$	Matrix for in-plane nodal deformations
D	Overall depth of cross-section
$\{D\}$	Vector of restraint point shear centre deformations
$[D]$	Generalized elasticity matrix
$[D_i]$	In-plane generalized elasticity matrix
$[D_u]$	Out-of-plane generalized elasticity matrix
$[D_v]$	Generalized initial stress matrix
E	Young's modulus of elasticity
E_r	Reduced modulus of elasticity
E_s	Strain-hardening modulus of elasticity
E_t	Tangent modulus of elasticity
F_T	Translational restraint force
F_u	Ultimate tensile strength
F_Y	Yield stress
G	Shear modulus of elasticity
G_A, G_B	Relative stiffnesses of beam restraints at ends A, B
G_s	Strain-hardening shear modulus of elasticity
G_t	Tangent shear modulus of elasticity

$[G]$	Global stability matrix
$[G_e]$	Transformed element stability matrix
I_B, I_T	Second moments of area of bottom and top flanges
I_e	Second moment of area of elastic core
I_P	Polar second moment of area $= (I_x + I_y)/A$
I_{Px}	Section property $= \int_A y(x^2 + y^2)\mathrm{d}A$
I_w	Warping section constant
I_x, I_y	Second moments of area about the x, y axes
I_{yc}	Second moment of area of compression flange
J	Torsion section constant
K	Torsion parameter $= \sqrt{(\pi^2 E I_w/GJL^2)}$
$\bar{K}$	Beam parameter $= \sqrt{(\pi^2 E I_y h^2/4GJL^2)}$
$[K]$	Global out-of-plane stiffness matrix
$[K_e]$	Transformed element out-of-plane stiffness matrix
$[K_i]$	Global in-plane stiffness matrix
$[K_{it}]$	Global in-plane tangent stiffness matrix
L	Length of member
L_e	Effective length
L_R	Length of restraining segment
M	Moment, or
	Concentrated mass
M^*	Design bending moment
M_B, M_T	Bottom and top flange minor axis end moments
M_b	Nominal member moment capacity
M_d	Distortion moment
M_E	Elastic buckling moment
M_f	Flange moment
M_I	Inelastic buckling moment
M_L	Limiting moment at first yield
M_m	Maximum value of M_x
M_P	Full plastic moment
M_{Rx}, M_{Ry}, M_{Rz}	Bending and torsional restraint moments
M_s	Nominal section moment capacity
M_u	Uniform torque, or
	Ultimate moment capacity, or
	Unbraced buckling moment
M_{u0}	Ultimate moment capacity for uniform bending
M_w	Warping torque
M_x, M_y	Bending moments about x, y axes
M_Y	Moment at nominal first yield
M_{yz}	Uniform bending buckling moment $= \sqrt{\{P_y GJ(1 + K^2)\}}$
M_z	Torque about the longitudinal axis
$[M]$	Out-of-plane matrix of powers of z/L, or
	Mass matrix

$[M_i]$	In-plane matrix of powers of z/L
N	Axial tension force
$[N]$	Matrix relating $\{u, v, w, \phi\}^T$ to $\{\delta\}$
$[N_Q], [N_q]$	Matrices relating $\{\theta\}_1^T, \{\theta_2^T\}^T, \{\theta\}$ to $\{\delta\}$
$[N_\sigma]$	Matrix relating $\{\Phi\}$ to $\{\delta\}$
P	Axial compression force
P^*	Design axial compression force
P_E	Elastic buckling load
P_F	Failure load
P_I	Inelastic buckling load
P_r	Reduced modulus buckling load
P_s	Strain-hardening buckling load
P_t	Tangent modulus buckling load, or
	Nominal tension capacity
P_u	Ultimate axial force capacity
$P_x, P_y,$	Column flexural buckling loads $= \pi^2 EI_x/L^2, \pi^2 EI_y/L^2$
P_Y	Squash load
P_z	Column torsional buckling load $= (GJ + \pi^2 EI_w/L^2)/r_2^2$
Q	Concentrated load
Q_E	Elastic buckling load
$\{Q_i\}$	In-plane global nodal forces
Q_Y	Load at nominal first yield
Q_∞	Value of Q for elastic buckling with rigid torsional support restraints
R	Radius of curvature
$\{R\}$	Vector of discrete restraint actions
$[S]$	Matrix of cross-section coordinates of P
T	Flange thickness
$[T_e]$	Out-of-plane transformation matrix
$[T_{ie}]$	In-plane transformation matrix
U	Strain energy
U, V, W	Deflections in global X, Y, Z directions
U_e	Element out-of-plane strain energy
U_{fb}	Flange bending strain energy
U_i	In-plane strain energy
U_R	Discrete restraint strain energy
U_T	Total potential
U_t	Uniform torsion strain energy
V	Potential energy, or
	Shear force, or
	Volume
V_e	Element out-of-plane potential energy
V_i	In-plane potential energy
V_y	Shear force in y direction

W	Work done, or
	Wagner stress resultant
X, Y, Z	Global axes
Z_x, Z_y	Elastic section moduli about x, y axes
$\{Z\}$	Vector of four powers of z/L
$\{Z_w\}$	Vector of two powers of z/L
a	Load distance along beam
a_0	Distance from shear centre
$\{a\}$	Vector of coefficients of powers of z
b	Width of thin rectangular element, or
	Distributed bimoment per unit length
d	Depth of narrow rectangular section
$\{d\}$	Vector of shear centre deformations
f_t	Translational restraint force per unit length
$[g_e]$	Element stability matrix
h	Distance between flange centroids
k	Effective length factor
k_l	Load height effective length factor
k_{Ry}	Minor axis bending effective length factor
k_w	Warping effective length factor
$[k_e]$	Element out-of-plane stiffness matrix
$[k_{ie}]$	Element in-plane stiffness matrix
$[k_{tet}]$	Element in-plane tangent stiffness matrix
m	Moment factor
m_{rx}, m_{ry}, m_{rz}	Bending and torsional restraining moments per unit length
m_u	Uniform torque per unit length
m_w	Warping torque per unit length
n	Integer
q	Load per unit length
q_0	Value of q for an unrestrained beam
$\{q_{ie}\}$	Element in-plane distributed loads
r_x, r_y	Radii of gyration about x, y axes
r_0^2	$= (I_x + I_y)/A$
r_1^2	$= r_0^2 + y_0^2$
r_2^2	$= r_0^2 + x_0^2 + y_0^2$
$\{r\}$	Vector of continuous restraint actions
s	Distance along section mid-thickness line, or
	Distance between discrete restraints, or
	Distance along curved shear centre axis
t	Thickness of thin-walled section, or
	Thickness of web, or
	Time
t_P	Distance from mid-thickness surface
u	Shear centre deflection in X direction

u_b	Buckling component of u
u_B, u_T	Bottom and top flange deflections in X direction
u_0	Initial crookedness
u_P, v_P, w_P	Deflections of P in X, Y, Z directions
$\{u\}$	Vector of deformations
v	Shear centre deflection in Y direction
v_b	Buckling component of v
$\{v_i\}$	In-plane nodal deflections of cross section
w	$= w_0 - \omega_c \phi'$
w_b	Buckling component of w
w_c	Value of w_P at centroid
w_s	Shear centre deflection in Z direction
x, y	Principal centroidal axes
x_c, y_c	Coordinates of centre of buckling rotation
x_0, y_0	Coordinates of shear centre
x_R, y_R	Distances of discrete rotational restraints from centroid
x_r, y_r	Distances of continuous rotational restraints from centroid
x_T, y_T	Distances of discrete translational restraints from centroid
x_t, y_t	Distances of continuous translational restraints from centroid
$\bar{y}$	Distance to centroid
y_Q	Distance of concentrated load from centroid
y_q	Distance of distributed load from centroid
z	Longitudinal axis through centroid
α	Angle, or
	Beam torsional stiffness
α_c	Stiffness of critical segment
α_L	Limiting value of stiffness
α_m	Buckling factor for beams
α_R	Stiffnesses of flange minor axis rotational end restraint, or
	Stiffness of restraining segment
$\alpha_{Rx}, \alpha_{Ry}, \alpha_{Rz}$	Stiffnesses of discrete bending and torsional restraints
$\alpha_{rx}, \alpha_{ry}, \alpha_{rz}$	Stiffnesses of continuous bending and torsional restraints
α_s	Slenderness reduction factor
α_{Tx}, α_{Ty}	Stiffnesses of discrete translational restraints
α_{tx}, α_{ty}	Stiffnesses of continuous translational restraints
α_W	Stiffness of discrete warping restraint
α_w	Stiffness of continuous warping restraints
$[\alpha_B]$	Discrete restraint stiffness matrix
$[\alpha_b]$	Continuous restraint stiffness matrix
β	Ratio of end moments
β_x	Monosymmetry section constant $= I_{Px}/I_x - 2y_0$
β_1, β_2	Major and minor axis end restraint parameters
γ	Shear strain, or
	End moment coefficient
γ_F	Stiffness factor for moment distribution

γ_M	Stiffness factor for restraints at far end
γ_P	Shear strain at P
γ_y	Load height factor for continuously restrained beams (equation 8.25)
γ_α	Restraint factor for continuously restrained beams (equation 8.24)
$\delta, \delta_x, \delta_y,$	Central or end deflections
δ_0	Initial central crookedness
$\{\Delta\}$	Global out-of-plane nodal deformations
$\{\Delta_i\}$	Global in-plane nodal deformations
$\{\delta_e\}$	Element out-of-plane nodal deformations
$\{\delta_{ie}\}$	Element in-plane nodal deformations
ε	Normal strain, or
	Dimensionless distance of load from centroid
ε_P	Normal strain at P
ε_r	Residual strain $= \sigma_r/E$
ε_s	Strain-hardening strain
ε_Y	Yield strain
$\{\varepsilon\}$	Generalized strain vector
$\{\varepsilon_i\}$	Generalized in-plane strain vector
$\{\varepsilon_u\}$	Generalized stiffness strain vector
$\{\varepsilon_v\}$	Generalized stability strain vector
η	Crookedness parameter, or
	Coefficient of viscosity
θ	Rotation
θ_0	Initial central twist rotation
θ_d	Distortional twist rotation of flange
θ_f	Warping rotation of flange
$\theta_X, \theta_Y, \theta_Z$	Rotations about global X, Y, Z axes
κ	Curvature
λ	Buckling load factor
λ_c	Zero interaction buckling load factor of critical segment
λ_n	Zero interaction buckling load factor of nth segment
λ_R	Zero interaction buckling load factor of restraining segment
μ	$= \sqrt{(P/EI_x)}$
ν	Poisson's ratio
ρ	Density, or
	$= I_{yc}/I_y$
ρ_0	Perpendicular distance from shear centre
σ	Normal stress
σ_a	Allowable working stress
σ_m	Maximum normal stress
σ_r	Residual stress
σ_P	Normal stress at P

σ_w	Warping normal stress
$\{\sigma\}$	Generalized out-of-plane stress vector
$\{\sigma_i\}$	Generalized in-plane stress vector
τ	Shear stress
τ_P	Shear stress at P
Φ	Rotation about global Z axis
ϕ	Twist rotation, or
	Capacity factor
ϕ_b	Buckling component of ϕ
ϕ_0	Initial twist
Ω	Nature frequency of vibrations
ω	Section warping function

1 Introduction

1.1 General

Thin-walled structural members may fail in a flexural-torsional buckling mode, in which the member suddenly deflects laterally and twists out of the plane of loading. This form of buckling may occur in a member which has low lateral bending and torsional stiffnesses compared with its stiffness in the plane of loading.

The most common form of flexural-torsional buckling is for I-section beams which are loaded in the planes of their webs, but which buckle by deflecting laterally and twisting, as shown in Figures 1.1 and 1.2a. Flexural-torsional buckling may also occur in concentrically loaded columns. This can be regarded as a general case, of which flexural buckling without twisting is one limiting example (Figure 1.2b). Some columns may buckle torsionally without bending (Figure 1.2c), which is the other limiting example of the flexural-torsional buckling of columns. Beam-columns bent in a plane of symmetry may also buckle in a flexural-torsional mode.

Flexural-torsional buckling is not confined to individual members, but also occurs in rigid-jointed structures, where continuity of rotations between adjacent members causes them to interact during buckling.

Flexural-torsional buckling is a primary consideration in the design of steel structures, as it may reduce the load-carrying capacity. Unless it is prevented by using either sufficient bracing or members which have adequate flexural and torsional stiffnesses, then larger members must be used to avoid premature failure. The determination of these larger members will be dominated by considerations of flexural-torsional buckling.

This chapter provides an introduction to flexural-torsional buckling. An historical survey is made in section 1.2, which is followed by general reviews of structural behaviour in section 1.3, of buckling in section 1.4, and of design against buckling in section 1.5.

Chapter 2 provides a general treatment of buckling with particular reference to flexural-torsional buckling, while Chapters 3 and 4 present hand and computer methods of predicting elastic flexural-torsional buckling.

The buckling of individual columns, beams, and beam-columns is described in Chapters 5–9 and 11, while the buckling of continuous beams, frames, and arches (Figure 1.3) and rings is discussed in Chapters 10, 12, and 13.

Inelastic buckling is dealt with in Chapter 14, while the use of flexural-torsional buckling predictions in the determination of design strength is described in Chapter 15. A number of special topics are briefly discussed in Chapter 16.

Figure 1.1 Flexural-torsional buckling of a cantilever.

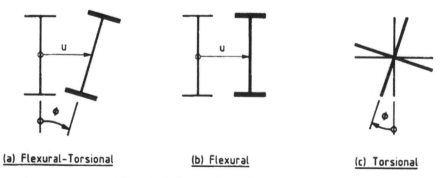

(a) Flexural-Torsional   (b) Flexural   (c) Torsional

Figure 1.2 Forms of member buckling.

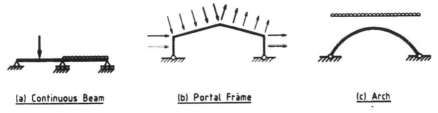

(a) Continuous Beam   (b) Portal Fràme   (c) Arch

Figure 1.3 Some structural forms.

1.2 Historical development

1.2.1 ELASTIC BUCKLING THEORY

The initial theoretical research into elastic flexural-torsional buckling was preceded by Euler's 1759 treatise [1] on column flexural buckling (Figure 1.4a), which gave the first analytical method of predicting the reduced strengths of slender columns, and by Saint-Venant's 1855 memoir [2] on uniform torsion (Figure 1.4b), which gave the first reliable description of the twisting response of members to torsion.

However, it was not until 1899 that the first treatments were published of flexural-torsional buckling by Michell [3] and Prandtl [4], who considered the lateral buckling of beams of narrow rectangular cross-section. Their work was extended in 1905 by Timoshenko [5, 6] to include the effects of warping torsion in I-section beams.

Subsequent work in 1929 by Wagner [7] and later work by others led to the development of a general theory of flexural-torsional buckling, as stated by

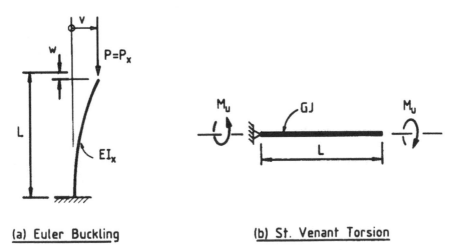

(a) Euler Buckling   (b) St. Venant Torsion

Figure 1.4 Euler buckling and St Venant torsion.

Timoshenko [8] and Vlasov [9], and incorporated in the textbooks of Timoshenko [10] and Bleich [11].

Specific studies of flexural-torsional buckling were made by many researchers, but prior to the 1960s, these were limited by the necessity to make extensive calculations by hand. Some of these are included in the 1960 survey by Lee [12].

This situation changed dramatically with the advent of the modern digital computer, and the 1960s saw an explosion in the amount of published research. As a result, the focus of research moved from the flexural-torsional buckling of isolated members under various loading conditions to the effects of end restraints exerted on a member of a rigid-jointed frame as a result of its continuity with adjacent members. Many of these studies are summarized in the 1971 survey of the Column Research Committee of Japan [13].

The extension of the general finite element method of structural analysis [14] to flexural-torsional buckling problems by Barsoum and Gallagher in 1970 [15] saw a further change, in that it was no longer necessary to publish comprehensive results of elastic flexural-torsional buckling studies, since almost any particular situation could now be analysed using a general purpose computer program. This development is similar to that which occurred in the in-plane analysis of plane rigid-jointed frames, in which the tabulations of solutions used in the 1930s were replaced by general purpose plane frame computer analysis programs.

Many of the developments of the theory of flexural-torsional buckling have been made by extensions of the previously accepted theories, as expressed either by the differential equations of elastic bending and torsion or by the energy equation for buckling. Not all of these extensions have received general acceptance, and so a number of attempts have been made through the 1980s to produce a generally acceptable theory of flexural-torsional buckling. This book includes such a general theory which is based on the use of the second-order relationships between the deformations and strains that take place during bending and torsion, the concept of the total potential, and the principles of virtual work and equilibrium, and of conservation of energy during buckling. This approach has been used, for example, to re-examine the flexural-torsional buckling of arches, early studies of which were reported by Vlasov [9] and Timoshenko [10].

1.2.2 STRENGTH AND DESIGN OF STEEL STRUCTURES

While the historical development of knowledge of flexural-torsional buckling undoubtedly was initiated by the need to prevent premature failure of steel structures in this mode, this is not well documented. It seems likely, however, that early design procedures for preventing the lateral buckling of steel beams followed and were closely related to those used for preventing the flexural failure of columns.

The need to be able to design against flexural-torsional buckling was the catalyst for the development of a theory for flexural-torsional buckling which would allow the successful prediction of failure. Early theoretical research was into the elastic buckling of perfectly straight members, some of which was verified experimentally. However, the very straight and slender members used for these experiments were unrepresentative of the real steel beams used in practice, tests of which showed that their strengths were reduced below those predicted solely by elastic buckling theory.

Theoretical research therefore extended from the elastic buckling of straight members to study the influences of crookedness, yielding, and residual stresses on the strengths of real steel beams, and to determine how to incorporate these into the procedures used in design. These developments tended to follow behind the corresponding developments from the elastic flexural buckling theory to the strengths of real steel columns. Early research on the inelastic lateral buckling of steel beams was carried out by Neal [16] and Galambos [17]. Flint [18] was one of the early researchers studying the effects of initial crookedness and twist [19] on the lateral buckling of beams and beam-columns.

Some of the early well-documented experiments on the lateral buckling of real steel beams were carried out by Hechtman, Hattrap, Styer, and Tiedmann [20]. Fukumoto and Kubo [21–23] reviewed and produced a data base of the experimental studies prior to 1977 on the lateral buckling of real steel beams.

Early rules for designing steel beams against lateral buckling were generally transpositions of rules for designing columns against flexural buckling, with perhaps the first proposal based on flexural-torsional buckling being made in 1924 by Timoshenko [24]. The first modern treatment was probably given by Kerensky, Flint and Brown [25] as the basis for the British Standard BS153–1958 [26]. More recently, most countries have or are transforming their design standards into the limit states format [27]. Current design criteria are reviewed in [28–30].

1.3 Structural behaviour

1.3.1 ELASTIC BEHAVIOUR

1.3.1.1 Linear behaviour

The simplest and most widely used model of the behaviour of a structure under static loads assumes that all of the deformations are proportional to the magnitude of the load set acting on the structure, so that the relation between load and response is linear, as shown by Curve 1 in Figure 1.5.

For this linear model to be valid, the material itself must have a linear relationship between stress and strain. Such a material is usually described as elastic. (Strictly, elastic means perfect recovery on unloading, so that an elastic material may be non-linear. However, most elastic materials are linear.) Most structural steels are linear, at least for stresses less than the yield stress F_Y, as shown in Figure 1.6, while many other structural materials are regarded as being linear over most of the range of working load.

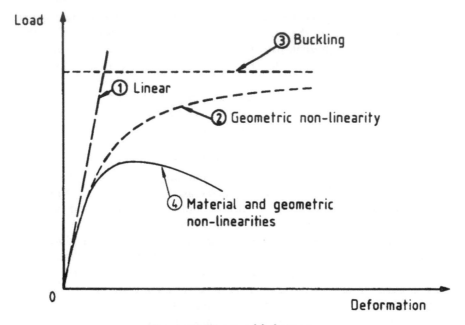

Figure 1.5 Structural behaviour.

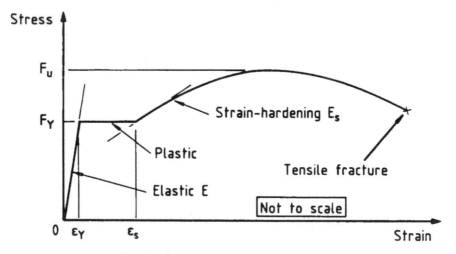

Figure 1.6 Idealized stress–strain relation for structural steel.

The structure itself must also behave linearly for the linear model to be valid. No structure is truly linear, but many are approximately so, provided the deflections are small.

The modern popularity of the linear elastic model of structural behaviour arises from the widespread availability of computer programs for linear elastic analysis [31]. These allow the deflections of the structure under load to be assessed for serviceability design under the working loads, and the member end actions to be approximated for strength design.

1.3.1.2 Non-linear behaviour

The linear elastic model of itself does not allow the strength of the structure to be assessed. For this purpose it is necessary to know of any material and structural non-linearities before the real behaviour of the structure and its maximum load-carrying capacity can be approximated.

Structural non-linearities cause the deformation response of the structure to load to become non-linear as shown by Curve 2 in Figure 1.5, even when the material remains linear. The most common structural non-linearities are associated with additional moments caused by the products of the loads and the transverse deflections of the structure or member as shown in Figure 1.7. Such effects are allowed for when equilibrium is formulated for the deformed geometry of the structure under load [32, 33], instead of the unloaded position, and so these non-linearities are usually described as being geometric, or second-order.

Geometric non-linearities may cause the load–deformation behaviour of an indefinitely elastic structure to asymptote towards a limit, as shown by Curve 2 in

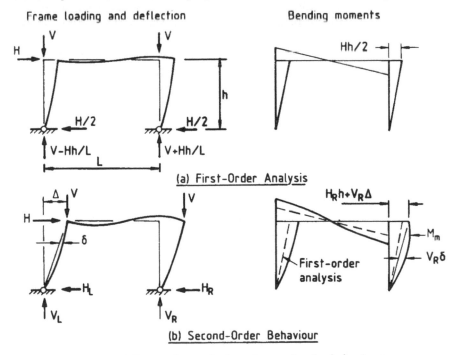

Figure 1.7 First-order analysis and second-order behaviour.

Figure 1.5. This limit (Curve 3 in Figure 1.5) is the elastic buckling load of the structure. Real structural behaviour will depart from this asymptotic behaviour when the material becomes non-linear as shown by Curve 4 in Figure 1.5, and a maximum load capacity will usually be reached, after which the load capacity will decrease.

The elastic buckling behaviour of a structure can be regarded as the limit of the elastic non-linear behaviour. In elastic buckling, the primary or pre-buckling response of the structure is in a different direction to the buckling response. For example, the pre-buckling response of the compression member shown in Figure 1.4a is due to longitudinal shortening w, while the buckling response is one of transverse bending deflections v. Thus the buckling response v remains zero until the buckling load P_x is reached, when the buckling response may initiate and continue indefinitely. For the buckling response to remain zero until the buckling load is reached, there must be no real or equivalent loads which would cause a primary response in the buckling direction.

A structure such as a concentrically loaded column or a beam loaded in the plane of the web may exhibit a real load–deformation response which differs only slightly from the idealized buckling response, in that the response in the buckling direction remains small until the buckling load is approached. Examples include straight concentrically loaded columns with small transverse loads, and columns with initial crookednesses which cause small transverse bending effects. For such members, the elastic buckling load may provide a quite accurate assessment of the strength, especially for slender members, for which small transverse loads or crookednesses are less important.

After the buckling load is reached, the post-buckling load–deformation curve may remain constant, or may rise or fall. This is caused by changes in the member stiffness that occur during buckling, which may lead to redistributions of the actions through the structure. Large deformations, for which there are gross changes in the chord lengths of some members and their rotations, may also affect the post-buckling behaviour.

1.3.2 INELASTIC BEHAVIOUR

1.3.2.1 Inelastic materials

All structural steels have a limited range over which the stress-strain behaviour is linear. Normal structural steels exhibit a horizontal yield plateau once the yield stress F_Y has been reached, as shown in Figure 1.6, followed by a slowly rising strain-hardening region. Cold-formed and stainless steels and aluminium all exhibit stress–strain curves which are rounded after a limit of proportionality is reached, as shown in Figure 1.8.

1.3.2.2 Inelastic stress distribution

The actual stress distribution at a member cross-section depends on the geometry of the section, its structural actions, the material stress–strain curve, and the

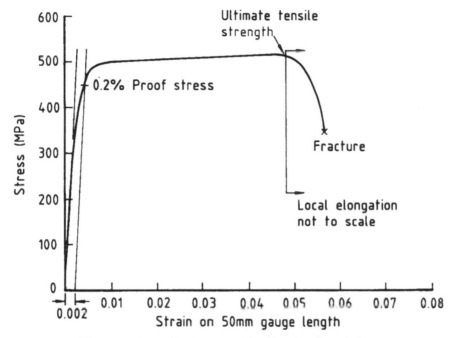

Figure 1.8 Stress–strain curve of cold reduced steel sheet.

residual stresses present before loading, such as those caused by the method of manufacture of the member.

When the stresses are low so that the member remains elastic, the stress distributions caused by axial force and bending actions are linear, as shown in Figure 1.9. Under bending actions, the maximum stresses occur at the extreme fibres, and when these reach the yield stress, a redistribution of the stresses commences. Useful structural limits (in the absence of local buckling effects) for structural steel members are provided by the moment M_P or axial force P_Y at which the cross-section becomes fully plastic.

The presence of residual stresses such as those caused by uneven cooling after hot-rolling or welding causes early initiation of yield, and generally affects the inelastic stress distribution. Because such residual stress distributions must be self-equilibrating so that they have zero axial force and bending actions in the unloaded member, they have no effect on the section full plastic capacities M_P and P_Y.

1.3.2.3 Inelastic members and structures

Inelastic effects on the behaviour of members and structures subject to buckling are best described separately in terms of their effects on the pre-buckling behaviour, on the buckling behaviour, and on the non-linear behaviour of members with geometrical imperfections such as initial crookedness or twist.

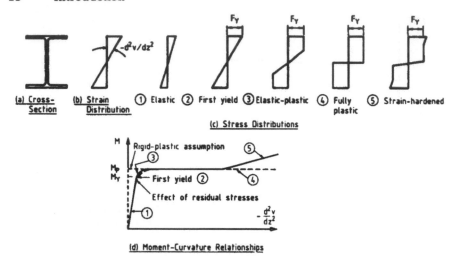

Figure 1.9 Moment–curvature relationships for steel beams.

Inelastic effects change the distributions of the bending moment, and to a lesser extent of the axial force, in an indeterminate structure before buckling. Since buckling depends on the pre-buckling distributions of these actions, these inelastic effects may be important.

Inelastic effects also increase the deflections. When there are significant geometric non-linear effects in the pre-buckling regime resulting from the pre-buckling deflections of the structure, these non-linear effects may be increased by additional deflections caused by inelastic behaviour. The advanced analysis of structures which accounts for geometric and material non-linearities including the effects of geometrical imperfections and residual stresses is described in [34].

While inelastic behaviour affects the buckling actions as described above, it also reduces the buckling resistance below the corresponding elastic resistance. Yielding causes local reductions in the cross-section stiffness which when aggregated over the complete member or structure may substantially reduce its buckling resistance. For example, one simple model of the inelastic flexural buckling of columns ignores the stiffness of any yielded regions of the column, so that its buckling resistance is based only on the regions of the column which remain elastic.

Inelastic behaviour also affects the strength of a member with small geometrical imperfections such as initial crookedness and twist. While the member remains elastic, its load–deformation behaviour asymptotes towards the elastic buckling behaviour of a perfectly straight member, as shown in Figure 1.5. The actual behaviour departs from this when the member first yields, and a maximum load is reached which is less than the elastic buckling load. This maximum load, which depends on the inelastic material properties, is sometimes approximated by the load at first yield.

1.4 Buckling

1.4.1 GENERAL

Buckling has already been described as the behaviour in which a structure or a structural element suddenly deforms in a (buckling) plane different to the original (pre-buckling) plane of loading and response. Member buckling (Figure 1.1) involves a single member, and may occur in flexural, torsional, or flexural-torsional modes. The half wave length of the buckle is of the same order as the member length.

Buckling may involve all the members of a frame, with interactions between the individual members. The buckle half wave length may be of the same order as a member length, or may be of the order of the frame size.

On the other hand, local buckling (Figure 1.10) usually takes place over a short length of a member of the same order as the cross-section width or depth. Distortional buckling (Figure 1.11) lies between member and local buckling, and is usually of a half wave length intermediate between the member and the cross-section dimensions.

These various forms of buckling are described in more detail in the following sub-sections.

Figure 1.10 Local buckling of an I-section column.

Figure 1.11 Distortional buckling of a channel section column.

1.4.2 FLEXURAL BUCKLING

Flexural buckling of a member (Figures 1.2b and 1.4a) may involve transverse displacements u or v of the member cross-sections, and is resisted by the flexural rigidity EI_y or EI_x of the member. It occurs when the second-order moments caused by the product of the axial compression force P with the displacements u or v are equal everywhere to the internal bending resistances $EI_y \mathrm{d}^2 u/\mathrm{d}z^2$ or $-EI_x \mathrm{d}^2 v/\mathrm{d}z^2$. Flexural buckling can be regarded as a limiting case of flexural-torsional buckling.

Flexural buckling may involve a single member, a group of members, or a complete frame. In braced frames, buckling is usually concentrated near one member, which is directly restrained by interactions with the adjacent members, and indirectly by the more remote members. In unbraced multi-storey structures, buckling occurs at one storey, and involves all the columns of that storey, which are restrained by the beams and columns of the adjacent storeys. Other unbraced frames may buckle in modes which directly involve many or all of the members.

1.4.3 TORSIONAL BUCKLING

Torsional buckling (Figure 1.2c) of a member involves twist rotations ϕ of the member cross-sections, and is resisted by the torsional rigidity GJ and the warping rigidity EI_w. It occurs when second-order torques $Pr_0^2 d\phi/dz$ caused by the axial compression force P and the twist $d\phi/dz$ are equal everywhere to the sum of the internal torsion resistances $GJ\, d\phi/dz$ and $-EI_w d^3\phi/dz^3$. Torsional buckling can be regarded as a limiting case of flexural-torsional buckling.

Torsional buckling may also occur in complete frames. Often the buckling resistance of these is dominated by the flexure of the individual members, as for example in tower frames whose horizontal cross-sections rotate.

1.4.4 FLEXURAL-TORSIONAL BUCKLING

Flexural-torsional buckling, which is the subject of this book, involves both displacements u, v and twist rotations ϕ, and is therefore resisted by combinations of the bending resistances $EI_y d^2u/dz$ and $-EI_x d^2v/dz^2$ and the torsional resistances $GJ\, d\phi/dz^2$ and $-EI_w d^3\phi/dz^3$.

While doubly symmetric columns whose centroidal and shear centre axes coincide buckle in either a flexural or a torsional mode, monosymmetric and asymmetric section columns may buckle in flexural-torsional modes. In these cases, the separation of the centroidal and shear centre axes causes these axes to become skew during buckling, so that the axial compression force acting along the centroidal axis has transverse components which create torques acting about the shear centre axis.

The flexural-torsional buckling of beams (Figures 1.1, 1.2a) involves lateral displacements u out of the plane of bending and twist rotations ϕ. In this case, the twist rotations ϕ cause the applied moments to have components acting out of the original plane of bending, while the lateral rotations du/dz cause the applied moments to have torque components about the axis of twist through the shear centre.

Beam-columns bent in a plane of symmetry may also buckle in flexural-torsional modes which combine those of columns and beams.

Flexural-torsional buckling may occur in frames (Figure 1.3b), where there are interactions between the adjacent members during buckling. In continuous or braced beams (Figure 1.3a), one span or segment is usually the most critical, and is restrained by the adjacent spans or segments. In three-dimensional frames, the members in each primary load-carrying plane interact during out-of-plane flexural-torsional buckling, and may be restrained by transverse members between adjacent primary frames.

Arches loaded in their plane (Figure 1.3c) may also buckle in a flexural-torsional mode by deflecting out of the plane and twisting.

1.4.5 LOCAL BUCKLING

Local buckling of a thin plate element (of thickness t) of a structural member involves deflections of the plate out of its original plane, as shown in Figure 1.10. Local buckling is resisted by the plate flexural rigidity $Et^3/12(1 - v^2)$, and occurs when the second-order actions caused by the in-plane compressions and the out-of-plane deflections are equal everywhere to the internal resistances of the plate element to bending and twisting.

Local buckling is usually concentrated near one particular cross-section of a member where the in-plane compressions of the plate elements are greatest, although multiple local buckles may occur in members whose stresses are constant along the member length. The half wave length of the buckle is of the order of the plate width. Local bucking effects may reduce the resistance of a member to flexural-torsional buckling.

Local buckling may occur in plate and shell structures, as well as in the structural members used in frame structures. Examples of plate structures include stiffened plate girders and rectangular and trapezoidal tanks, while shell structures include cylindrical and spherical containment structures. These may buckle locally in the more highly stressed regions, as well as in a more global fashion, involving larger regions of the structure.

1.4.6 DISTORTIONAL BUCKLING

Distortional buckling (Figure 1.11) describes a buckling mode intermediate between those of local and member buckling. In member buckling, the cross-section is assumed not to distort and buckling involves the whole member length, while local buckling involves relative displacements of the component plates over a short length of the member.

Distortional buckling often involves web flexure and corresponding rotations of the flanges which vary slowly along the member length, as shown in Figure 1.11. Distortional effects may reduce the flexural-torsional buckling resistances of thin-web beams.

1.5 Design against buckling

Methods of designing against flexural-torsional buckling are essentially of two types. For the first type, buckling is avoided, and the member's in-plane capacity is fully utilized. One way of achieving this is to use sections which are not susceptible to buckling. For example, closed sections have very much higher torsional rigidities GJ and higher flexural rigidities EI_y than corresponding open I-section members, and rarely buckle in a flexural-torsional mode. Less effective is the use of I-sections with comparatively wide flanges, which have higher flexural EI_y and warping EI_w rigidities than I-sections with narrow flanges.

A second way of avoiding buckling is to increase the bracing, either by reducing its spacing, or else by increasing its effectiveness, as for example when a lateral deflection brace is made more effective by using it also to restrain the twist rotation ϕ.

For the second method of designing against buckling, a reduced capacity is determined which accounts for the effects of flexural-torsional buckling, in which case the member's in-plane capacity is not fully utilized. Capacity reductions depend on the slenderness of the member, on the initial crookedness and twist, on the residual stresses, and on any loads which act above the shear centre, and are usually determined in accordance with a design code [29]. It is advantageous to take account of the effects of moment distribution and restraints which generally increase the buckling resistance, and many design codes allow this to be done.

1.6 References

1. Euler, L. (1759) Sur la force des colonnes. *Memoires Academic Royale des Sciences et Belle Lettres, Berlin*, **13**, partial translation by Van den Broek, J.A. (1947) *American Journal of Physics*, **15**, 309–18.
2. Saint-Venant, B. (1855) Memoire sur la torsion des prismes. *Memoires des Savants Etrangers*, **XIV**, 233–560.
3. Michell, A.G.M. (1899) Elastic stability of long beams under transverse forces. *Philosophical Magazine*, **48**, 298–309.
4. Prandtl, L. (1899) Kipperscheinungen, *Thesis*, Munich.
5. Timoshenko, S.P. (1953) Einige Stabilitaetsprobleme der Elastizitaetstheorie, in *Collected Papers of Stephen P. Timoshenko*, McGraw-Hill, New York, pp. 1–50.
6. Timoshenko, S.P. (1953) Sur la stabilite des systemes elastiques, in *Collected Papers of Stephen P. Timoshenko*, McGraw-Hill, New York, pp. 92–224.
7. Wagner, H. (1936) Verdrehung und Knickung von offenen Profilen (Torsion and Buckling of open sections), *NACA Technical Memorandum* No. 807.
8. Timoshenko, S.P. (1953) Theory of bending, torsion, and buckling of thin-walled members of open cross-section, in *Collected Papers of Stephen P. Timoshenko*, McGraw-Hill, New York, pp. 559–609.
9. Vlasov, V.Z. (1961) *Thin-Walled Elastic Beams*, 2nd edn, Israel Program for Scientific Translation, Jerusalem.
10. Timoshenko, S.P. and Gere, J.M. *Theory of Elastic Stability*, 2nd edn, McGraw-Hill, New York.
11. Bleich, F. (1952) *Buckling Strength of Metal Structures*, McGraw-Hill, New York.
12. Lee, G.C. (1960) A survey of literature on the lateral instability of beams, *Welding Research Council Bulletin*, No. 63, August.
13. Column Research Committee of Japan (1971) *Handbook of Structural Stability*, Corona, Tokyo.
14. Zienkiewicz, O.C. and Taylor, R.L. (1989, 1991) *The Finite Element Method, Volume 1 – Basic Formulation and Linear Problems, Volume 2 – Solid and Fluid Mechanics, Dynamics and Non-Linearity*, 4th edn, McGraw-Hill, London.

15. Barsoum, R.S. and Gallagher, R.H. (1970) Finite element analysis of torsional and torsional-flexural stability problems. *International Journal of Numerical Methods in Engineering*, 2, 335–52.
16. Neal, B.G. (1950) The lateral instability of yielded mild steel beams of rectangular cross section. *Philosophical Transactions, Royal Society of London*, A, 242 (January), 197–242.
17. Galambos, T.V. (1959) Inelastic lateral-torsional buckling of eccentrically loaded wide-flange columns, PhD Thesis, Lehigh University, Bethlehem, PA.
18. Flint, A.R. (1950) The stability and strength of slender beams. *Engineering*, 170 (December), 545–9.
19. Trahair, N.S. (1969) Deformations of geometrically imperfect beams. *Journal of the Structural Division, ASCE*, 95 (ST7), 1475–96.
20. Hechtman, R.A., Hattrap, J.S., Styer, E.F. and Tiedmann, T.L. (1955) Lateral buckling of rolled steel I-beams. *Proceedings, ASCE*, 81 (Separate N. 797), September.
21. Fukumoto, Y. and Kubo, M. (1977) A survey of tests on lateral buckling strength of beams, in *Preliminary Report, 2nd International Colloquium on Stability of Steel Structures*, ECCS-IABSE, Liege, pp. 233–40.
22. Fukumoto, Y. and Kubo, M (1977) A supplement to a survey of tests on lateral buckling strength of beams, in *Final Report, 2nd International Colloquium on Stability of Steel Structures*, ECCS-IABSE, Liege, pp. 115–7.
23. Fukumoto, Y. and Kubo, M. (1977) An experimental review of lateral buckling of beams and girders, in *Proceedings, International Colloquium on Stability of Structures Under Static and Dynamic Loads*, SSRC-ASCE, Washington, pp. 541–62.
24. Timoshenko, S.P. (1924) Beams without lateral support. *Transactions, ASCE*, 87, 1247–72.
25. Kerensky, O.A., Flint, A.R. and Brown, W.C. (1956) The basis for design of beams and plate girders in the Revised British Standard 153'. *Proceedings, Institution of Civil Engineers*, Part III, 5 (August), 396–521.
26. British Standards Institution (1958) *BS 153 Steel Girder Bridges. Part 3B: Stresses, Part 4: Design and Construction*, BSI, London.
27. Ravindra, M.K. and Galambos, T.V. (1978) Load and resistance factor design for steel. *Journal of the Structural Division, ASCE*, 104 (ST9) 1337–54.
28. Galambos T.V. (ed.) (1988) *Guide to Stability Design Criteria for Metal Structures*, 4th edn, John Wiley and Sons, New York.
29. Beedle, L.S. (ed.) (1991) *Stability of Metal Structures – A World View*, 2nd edn, Structural Stability Research Council, Bethlehem, PA.
30. Trahair, N.S. and Bradford M.A. (1991) *The Behaviour and Design of Steel Structures*, revised 2nd edn, Chapman and Hall, London.
31. Harrison, H.B. (1990) *Structural Analysis and Design – Some Microcomputer Applications, Parts 1 and 2*, 2nd edn, Pergamon Press, Oxford.
32. Harrison, H.B. (1973) *Computer Methods in Structural Analysis*, Prentice-Hall, Englewood Cliffs, NJ.
33. Hancock, G.J. (1991) Elastic method of analysis of rigid jointed frames including second order effects, in *Structural Analysis to AS4100*, School of Civil and Mining Engineering, The University of Sydney, pp. 2.1–2.32.
34. Clarke, M.J. (1991) Advanced analysis, in *Structural Analysis to AS4100*, School of Civil and Mining Engineering, The University of Sydney, pp. 6.1–6.55.

2   Equilibrium, buckling, and total potential

2.1 General

In this chapter, the conditions for the equilibrium of a structure are discussed generally, and then related to the conditions at bifurcation buckling.

Equilibrium is considered in section 2.2 for a structure of a linear elastic material. The deflection of the structure under increasing load is then discussed in relation to the changes which occur in the energy of the system used to apply the loads, in the strain energy stored in the structure, and in the potential energy of the loads.

The principle of virtual work for a structure in equilibrium is stated in section 2.4. This is then related to the first variation of the total potential defined in section 2.3, and used to derive the principle of stationary total potential.

The conditions of stable, neutral, and unstable equilibrium are discussed in section 2.5. The relationship between the condition of neutral equilibrium and the principle of conservation of energy during buckling is established. Both of these are expressed in terms of the second variation of the total potential, and restated by the theorem of minimum total potential.

The nature of bifurcation buckling, in which there is a sudden change at the bifurcation buckling load from the pre-buckling load–deflection path to a new and different buckling path, is considered in section 2.6. The neutral equilibrium conditions of bifurcation buckling under constant load are discussed, and the relationships between the virtual work principle for equilibrium of the buckled position and the conservation of energy condition for neutral equilibrium at the buckling load are established. These relationships are summarized in Figure 2.1 for the bending and flexural buckling of beams and columns.

The relationships between pre-buckling equilibrium, buckling and non-linear behaviour are exemplified in section 2.7 by considering the in-plane behaviour of beam-columns. The corresponding relationships for flexural-torsional buckling are stated in section 2.8.

2.2 Equilibrium

2.2.1 MATERIAL BEHAVIOUR

The structure shown in Figure 2.2a is composed of a material which is homogeneous, isotropic, elastic, and has the linear relationship

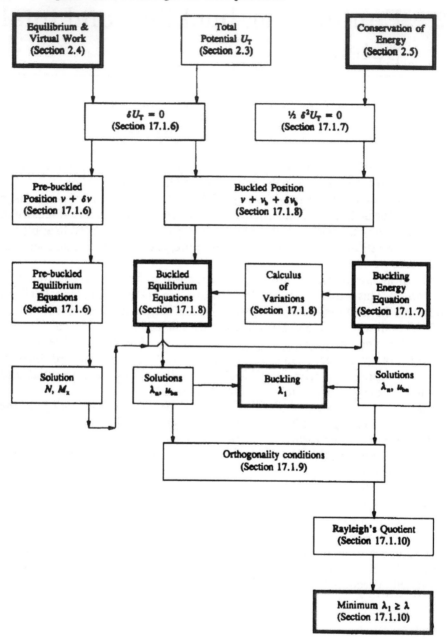

Figure 2.1 Theory of buckling.

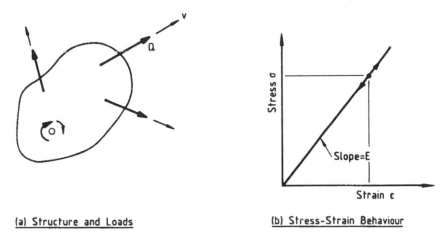

(a) Structure and Loads   (b) Stress-Strain Behaviour

Figure 2.2 Structure, loads and material behaviour.

$$\sigma = E\varepsilon \tag{2.1}$$

between the stress σ and the strain ε, as shown in Figure 2.2b, in which E is the Young's modulus of elasticity.

In the unloaded state, the material is unstressed ($\sigma = 0$) and unstrained ($\varepsilon = 0$). In the loaded state, small strains ε develop, with corresponding stresses σ.

2.2.2 STRUCTURAL BEHAVIOUR

When loads Q are added slowly (so that there are no dynamic effects) to the structure shown in Figure 2.2a, it deflects v as shown in Figure 2.3. In the deflected position, the structure is said to be in static equilibrium, in that the loads Q are exactly balanced by reaction forces and the deflections v remain constant with time.

The variations shown in Figure 2.3 of the deflections v with the loads Q define the static load–deflection behaviour. This may be linear, when the deflections v are proportional to the loads Q, or non-linear. The deflections may vary smoothly with the loads; or the rate of change of the deflection may suddenly change, in which case the load–deflection behaviour **bifurcates**; or the deflection itself may suddenly change so that the structure 'snaps' to a new position at which equilibrium is re-established.

2.2.3 STRAIN ENERGY

As the structure deflects, the strains and stresses in the structure change, and the internal strain energy of the structure changes. The increase δU in the strain

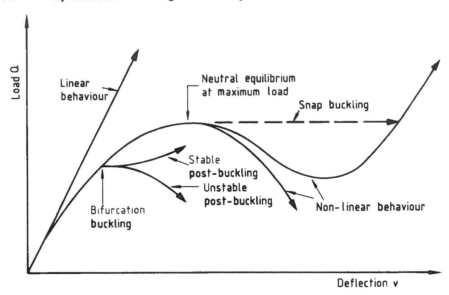

Figure 2.3 Types of load–deflection behaviour.

energy U is defined by

$$\delta U = \int_{V} \left\{ \int_{I}^{II} \sigma \, d\varepsilon \right\} dV \tag{2.2}$$

in which V is the volume of the structure, and I and II represent the initial and final strain states of the structure, as shown in Figure 2.4a. For a linear elastic material which is unstressed and unstrained at the initial state I, $\sigma = E\varepsilon$, and so

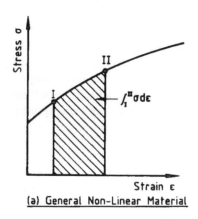

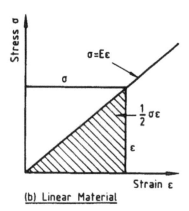

Figure 2.4 Strain energy.

the total strain energy is

$$U = \frac{1}{2} \int_V \sigma \varepsilon \, dV \qquad (2.3)$$

in which σ and ε now represent the stress and strain in the final state II, as indicated in Figure 2.4b.

Both the stress–strain and the strain–deflection relationships may be non-linear, in which case the strain energy U contains deflection terms of cubic order or higher, so that

$$U = a_2 v^2 + a_3 v^3 + a_4 v^4 + \cdots. \qquad (2.4)$$

When both the stress–strain and the strain–deflection relationships are linear, then $a_3 = a_4 = \cdots = 0$ and the strain energy U is a quadratic function of the deflection v.

2.2.4 WORK

As the loads Q on the structure and its deflections v increase, work must be done by an external power source in applying additional loads to the structure and in increasing the deflections. The work done by the external power source represents a transfer of energy from the power source to the structure and its loads.

The work W done on loads Q which deflect v in the directions of action of the loads is given by

$$W = - \sum \int_1^{11} Q \, dv \qquad (2.5)$$

in which the minus sign indicates that negative work is done if the load Q is moved in its own direction, and positive work is done if the load Q is moved in the opposite direction. Negative work done on a load when it is moved in its own direction may also be thought of as positive work done by the load, in the sense that energy equal to this positive work is transferred from the structure and its loads to the external power source.

If the load Q remains constant while it displaces v in its own direction, then the external work done on it is obtained from equation 2.5 as

$$W = - Q v \qquad (2.6)$$

in which Q and v represent the values for the final state II. For example, the lifting of a gravity load Q up through a distance h from the ground to the structure shown in Figure 2.5 requires positive work Qh to be done on the load. If this load is then added to the structure, causing it to deflect downwards by v, then a further amount of (negative) work $- Qv$ is done on the load. Thus the total work done on

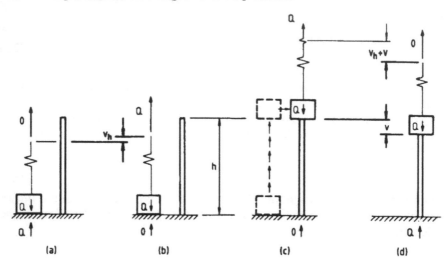

Figure 2.5 Establishment of an equilibrium position.

the load Q during these operations is

$$W = Qh - Qv. \tag{2.7}$$

If a load Q increases linearly with its deflection v, so that $Q = kv$, then the external work done on it is obtained from equation 2.5 as

$$W = -\tfrac{1}{2}Qv. \tag{2.8}$$

2.2.5 POTENTIAL ENERGY

The potential energy V of a load Q is defined as the work which must be done on the constant load to move its point of application from an initial reference position to its final position. Thus

$$V = W = -Qv \tag{2.9}$$

when the (downwards) gravity load Q is moved (downwards) though a distance v. Note that the sense of v is the same as the direction of Q, and so the potential energy decreases as the deflection increases. Equation 2.9 defining the potential energy takes the convenient but arbitrary reference position which corresponds to the zero displacement position ($v = 0$) of the load.

2.2.6 ESTABLISHMENT OF AN EQUILIBRIUM POSITION

An equilibrium position of a structure is established by steadily adding load to it until the position is reached, as indicated in Figure 2.3. It is assumed that this process is quasi-static; thus there are no dynamic effects and the kinetic energy remains zero, the structure and its supports are frictionless and elastic, there are no energy losses due to friction or plasticity, and each load continues to act in its original direction and is conservative. In this case energy is conserved, and the total energy of the system remains constant.

Care is needed in the interpretation of this statement, since the system must include the power source used to move the loads. For example, in Figure 2.5, the power source will first do work $\frac{1}{2}Qv_h$ in taking the weight of the load, where v_h is the upwards extension of the power hoist, and will then do work Qh in lifting the load into position. It will finally do negative work $-\frac{1}{2}Q(v_h + v)$ as the force Q in the hoist decreases to zero during the transfer of the load to the structure. Thus the total work done by the power source is

$$W = \tfrac{1}{2}Qv_h + Qh - \tfrac{1}{2}Q(v_h + v) \tag{2.10}$$

or

$$W = Q(h - \tfrac{1}{2}v). \tag{2.11}$$

The strain energy of the structure will increase by U while the potential energy of the load Q will increase by

$$V = Q(h - v). \tag{2.12}$$

Thus for the total energy of the system to remain constant

$$-W + U + V = 0, \tag{2.13}$$

where

$$U = W - V \tag{2.14}$$

or

$$U = \tfrac{1}{2}Qv \tag{2.15}$$

and so the structure stores strain energy equal to the negative of the work $-\frac{1}{2}Qv$ done on a force which increases from 0 to Q as its deflection increases from 0 to v.

2.3 Total potential

The total potential U_T of a structure and its loads [1,2] is defined as the sum of the strain energy U of the structure and the potential energy V of the loads, so that

$$U_T = U + V. \tag{2.16}$$

Usually, the potential energy datum is taken as the unloaded, unstrained position of the structure, in which case $U_T = 0$ for this position.

In general, the total potential depends on the load Q acting on the structure and the deflected position v assumed, so that

$$U_T = (a_2v^2 + a_3v^3 + a_4v^4 + \cdots) - Qv \tag{2.17}$$

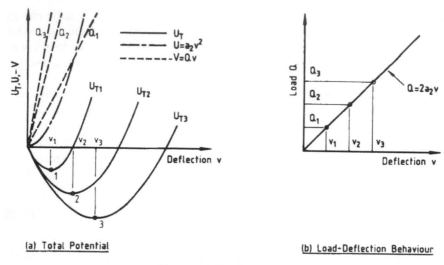

(a) Total Potential   (b) Load-Deflection Behaviour

Figure 2.6 Total potential.

after using equations 2.4 and 2.9. The particular case where $U = a_2 v^2$ is demonstrated in Figure 2.6a by the variations of U, $-V = Qv$, and U_T with Q and v. The total potential U_T has a stationary value when $dU_T/dv = 0$, in which case

$$Q = 2a_2 v. \qquad (2.18)$$

Stationary values of U_T are shown in Figure 2.6a for loads Q_1, Q_2, and Q_3. The significance of the stationary condition of U_T will be discussed in section 2.4.3 following.

Usually, only one value of v will correspond to the actual equilibrium position of the structure under the action of Q, so that all other values of v are fictitious. Thus, in general, the total potential represents a function whose value depends on the fictitious position v assumed for the structure. The total potential therefore does not of itself represent a useful concept in the consideration of the actual equilibrium condition of the structure. However, when it is combined with the principles of virtual work and conservation of energy discussed in the following sections, it becomes a powerful tool which can be used to determine the equilibrium position of the structure, and the nature of the equilibrium condition.

2.4 Virtual work

2.4.1 RIGID BODIES

The principle of virtual work is most simply applied to a rigid body which is in equilibrium under the action of a set of constant forces $Q_1, Q_2, \ldots$ and moments $M_1, M_2, \ldots$, as shown in Figure 2.7a. It is assumed that the body and its actions

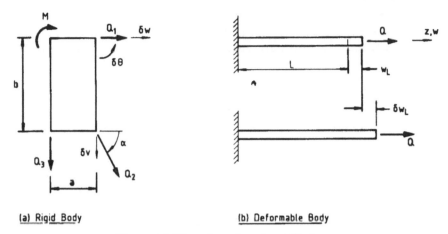

(a) Rigid Body   (b) Deformable Body

Figure 2.7 Virtual work and equilibrium.

undergo **virtual deformations** $\delta v_1, \delta v_2, \ldots, \delta\theta_1, \delta\theta_2, \ldots$. These assumed deformations are generally **fictitious**, are **infinitesimally small**, and are **consistent**, which in this case means that they correspond to a single set of rigid body displacements and rotations of the body and the points of application of its actions.

The principle of virtual work then states that **the original position of the body and its actions is one of equilibrium if the virtual work** $-\Sigma(Q\,\delta v + M\,\delta\theta)$ **done on all the actions** Q, M **is zero for every consistent set of virtual deformations** δv, $\delta\theta$. Thus

$$\delta W = -\sum (Q\,\delta v + M\,\delta\theta) = 0. \tag{2.19}$$

For example, for the virtual deformations shown in Figure 2.7a,

$$-\delta v(Q_2 \sin\alpha + Q_3) - \delta w(Q_1 + Q_2 \cos\alpha) - \delta\theta(-M + bQ_2 \cos\alpha + aQ_3) = 0 \tag{2.20}$$

and since this must hold for all sets $\delta v, \delta w, \delta\theta$, then each term must vanish independently, whence

$$Q_2 \sin\alpha + Q_3 = 0, \tag{2.21}$$

$$Q_1 + Q_2 \cos\alpha = 0, \tag{2.22}$$

$$-M + bQ_2 \cos\alpha + aQ_3 = 0, \tag{2.23}$$

which are the three equilibrium equations for the rigid body.

2.4.2 DEFORMABLE BODIES

The extension of the principle of virtual work to a deformable body requires the inclusion of the increase δU in the strain energy stored in the body as a result of its straining resulting from the virtual deformations. The principle of virtual work for a deformable body states that the original position of the body and its actions is one of equilibrium if the sum of the increase δU in the strain energy stored in the body and the virtual work δW done on its actions is zero for every consistent set of virtual deformations. Thus

$$\delta U + \delta W = 0. \tag{2.24}$$

For example, if the strain energy stored in the tension member shown in Figure 2.7b is written as

$$U = \tfrac{1}{2}(EA/L)w_L^2 \tag{2.25}$$

for real displacements $w = w_L z/L$, then for virtual displacements $\delta w = \delta w_L z/L$, the strain energy increases by

$$\delta U = (EA/L)w_L \delta w_L \tag{2.26}$$

and so the virtual work condition (equation 2.24) becomes

$$(EA/L)w_L \delta w_L - Q\delta w_L = 0 \tag{2.27}$$

whence

$$Q = EAw_L/L \tag{2.28}$$

which describes the linear relationship between the load Q and the equilibrium displacement w_L of the load Q acting on the tension member.

2.4.3 PRINCIPLE OF STATIONARY TOTAL POTENTIAL

The principle of virtual work for deformable bodies may be transformed to the principle of stationary total potential by noting that in a conservative potential energy system, the work δW done on the actions Q during virtual deformations δv is equal to the change δV in the potential energy of these actions, so that equation 2.24 becomes

$$\delta U + \delta V = 0 \tag{2.29}$$

or

$$\delta U_T = 0 \tag{2.30}$$

after using equation 2.16.

Thus the principle of virtual work is identical with the principle of stationary total potential, which states that an equilibrium position is one of stationary total potential.

Some writers describe equation 2.29 as an illustration of the principle of conservation of energy in a conservative system, in that it requires any increase δU in the strain energy to be matched by a corresponding decrease $-\delta V$ in the potential energy. However, this interpretation is an artificial one, since it refers to changes which occur when virtual deformations δv take place from the real equilibrium position v to an assumed position $(v + \delta v)$ which is generally not one of equilibrium.

When the total potential U_T varies with v and Q as in equation 2.17, then equation 2.30 leads to the equilibrium condition

$$Q = 2a_2v + 3a_3v^2 + 4a_4v^3 + \cdots \tag{2.31}$$

which describes the load–deflection behaviour of the structure.

Conditions of this type are illustrated in Figure 2.8. When the total potential U_T is quadratic $(a_3 = a_4 = \cdots = 0)$, the load–deflection behaviour of equation 2.31 is linear, as shown in Figure 2.6b.

However, when the total potential is cubic $(a_4 = \cdots = 0)$, the load–deflection behaviour is quadratic. When a_3 is positive so that d^2Q/dv^2 is positive, the structure stiffens under load, as for example is the case for tension members with bending moments. When a_3 is negative so that d^2Q/dv^2 is negative, the structure's stiffness decreases with load, as is the case for compression members with bending moments. In this case, the load Q reaches a maximum value.

When the total potential is quartic, the load–deflection behaviour is cubic. When a_3 is negative and a_4 is positive, then the early decrease in stiffness is followed by a subsequent increase in stiffness. A maximum load followed by a snap will occur if $a_2 < 3a_3^2/8a_4$, so that there is a load for which $dQ/dv = 0$.

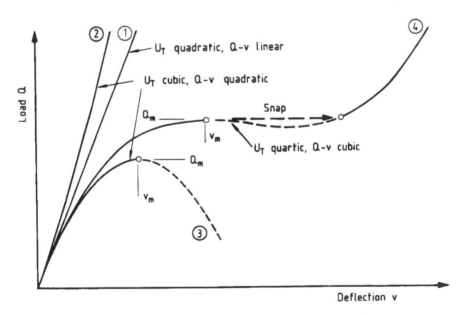

Figure 2.8 Relationship between total potential function and behaviour.

2.5 Nature of equilibrium

2.5.1 STABLE, NEUTRAL, AND UNSTABLE EQUILIBRIUM

An equilibrium position of a structure under load may be stable, neutral or unstable. One method of determining the type of equilibrium is to consider the behaviour of the structure and its loads when an infinitesimally small disturbance is first applied to displace the structure while its loads remain constant, and then removed.

If the structure returns to its original position for every disturbance, then the original equilibrium position is said to be stable. For example, consider the small sphere shown in Figure 2.9a at the low point at Position 1 of the spherical hollow, which is an equilibrium position for the sphere under the action of its own weight and the support reaction. If the sphere is displaced to Position 2 and then released, it will roll down the hollow towards Position 1, and so this equilibrium position is stable.

If, however, there is any disturbance for which the structure remains in the displaced position after the disturbance is removed, then the original equilibrium position is said to be neutral. Thus the sphere shown on a horizontal plane in Figure 2.9b will remain in the displaced Position 2. Because the structure remains in the displaced position after the disturbance is removed, this too is one of equilibrium.

Finally, if there is any disturbance for which the structure displaces further from the original equilibrium position when the disturbance is removed, then the equilibrium position is unstable. This is illustrated in Figure 2.9c for a sphere which is initially in equilibrium at the high point of Position 1 of a spherical mound. When this is displaced to Position 2 and then released, it continues to move away from the equilibrium position.

2.5.2 NEUTRAL EQUILIBRIUM AND CONSERVATION OF ENERGY

The nature of an equilibrium position may also be determined by considering the energy input to or output from a structure and its loads during an infinitesimal

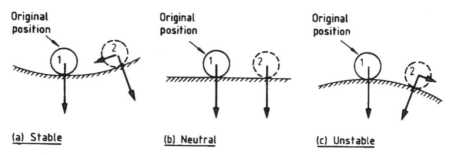

Figure 2.9 Types of equilibrium.

disturbance, in which the structure displaces δv to an adjacent equilibrium position. If a force δF in the same direction as the displacement δv is required to maintain equilibrium in the adjacent position, as shown in Figure 2.10a, then the power source providing this force does positive work

$$\tfrac{1}{2}\delta^2 W = \tfrac{1}{2}\delta F \delta v \tag{2.32}$$

on the structure and its loads. This work done causes an increase in the sum of the strain energy and the potential energy of the system, so that

$$\tfrac{1}{2}\delta^2 W = \tfrac{1}{2}(\delta^2 U + \delta^2 V) \tag{2.33}$$

(note that $\delta U + \delta V = 0$ because the original position is one of equilibrium). In this case the original position is stable.

If, however, a force δF in the opposite sense to δv is required to maintain equilibrium (δF is negative), as shown in Figure 2.10c, then work $-\tfrac{1}{2}\delta^2 W = -\tfrac{1}{2}\delta F \delta v$ is done by the structure and its loads, and energy is transferred to the power source equal to the decrease $-\tfrac{1}{2}(\delta^2 U + \delta^2 V)$ in the sum of the strain energy and potential energy of the system. In this case the original position is unstable.

The original position is one of neutral equilibrium when the work done $\tfrac{1}{2}\delta F \delta v$ is zero. In this case $\delta F = 0$, and the structure and its loads can be deflect δv under a zero disturbance from the original position, as shown in Figure 2.10b. While it does so, $\tfrac{1}{2}\delta^2 W = 0$, and so

$$\tfrac{1}{2}(\delta^2 U + \delta^2 V) = 0 \tag{2.34}$$

so that there is no change in the energy of the structure and its loads. Thus the structure and its loads obey the law of conservation of energy while they deflect

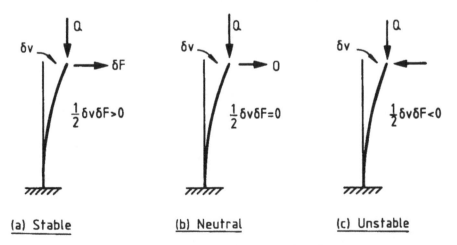

(a) Stable   (b) Neutral   (c) Unstable

Figure 2.10 Equilibrium and conservation of energy.

from the original position of neutral equilibrium to the new position, which is also one of equilibrium.

2.5.3 NEUTRAL EQUILIBRIUM AND THE TOTAL POTENTIAL

The conservation of energy condition at neutral equilibrium expressed by equation 2.34 can be related directly to the total potential U_T (see equation 2.16) by

$$\tfrac{1}{2}\delta^2 U_T = 0. \tag{2.35}$$

When the equilibrium position is unstable, $\tfrac{1}{2}\delta^2 U_T < 0$, while $\tfrac{1}{2}\delta^2 U_T > 0$ when the equilibrium position is stable. This is restated in the theorem of minimum total potential, according to which **the stationary value of U_T (for which $\delta U_T = 0$) of an equilibrium position is a minimum when the position is stable.**

This is illustrated by Curve 3 in Figure 2.8 for the case where

$$U_T = a_2 v^2 + a_3 v^3 - Qv \tag{2.36}$$

and a_3 is negative. The neutral equilibrium conditions of $\delta U_T = 0$ and $\tfrac{1}{2}\delta^2 U_T = 0$ require that

$$Q = 2a_2 v + 3a_3 v^2 \tag{2.37}$$

and

$$v_m = - a_2/3a_3, \tag{2.38}$$

so that

$$Q_m = - a_2^2/3a_3. \tag{2.39}$$

The equilibrium variation of Q with v given by equation 2.37 is shown by Curve 3 in Figure 2.8, with the neutral equilibrium values of Q_m and v_m given by equation 2.39 and 2.38. It can be seen that the neutral equilibrium condition corresponds to the maximum value Q_m of Q.

The second variation of the total potential may be expressed as

$$\tfrac{1}{2}\delta^2 U_T = a_2(1 - v/v_m)(\delta v)^2. \tag{2.40}$$

It can be seen that when $v < v_m$ then $\tfrac{1}{2}\delta^2 U_T > 0$ so that the equilibrium is stable, and that when $v > v_m$ then $\tfrac{1}{2}\delta^2 U_T < 0$ so that the equilibrium is unstable. This unstable condition can be interpreted physically by noting that in this condition $(v > v_m)$, the structure and its load require an external **supporting** force (negative δF) to keep them in equilibrium when they are disturbed δv.

Similarly, if the total potential U_T is quartic as indicated for Curve 4 of Figure 2.8, then the stiffness dQ/dv may steadily decrease as the load increases towards Q_m, and become zero at the maximum load Q_m, at which the structure is in neutral equilibrium. After this the load–deflection behaviour shows negative stiffness and is unstable until a minimum load is reached, when the load–deflection curve again starts rising. Any increase in load above Q_m is accompanied by a snap from the neutral equilibrium position v_m to the rising load–deflection curve where the structure is again stable.

2.6 Bifurcation buckling

2.6.1 LOAD–DEFLECTION PATHS

Under idealized circumstances, the load–deflection path of a structure may suddenly branch or bifurcate, as shown in Figure 2.11b and c. Thus the deflection suddenly becomes multi-valued at the bifurcation load. After the bifurcation load is passed, the original path, which is the upper path, is unstable, and any small disturbance will cause a snap to the lower path. Thus there is a change in the load–deflection behaviour from the original path to the new path. This branching behaviour is called bifurcation buckling.

Bifurcation occurs when the lower branch or buckling path v_b is independent of (or orthogonal to) the pre-buckling path v, so that there is no component in v which is of the same shape as v_b. Strictly, this idealized situation never occurs in reality, because of the presence of small imperfections in real structures. These will generally contain a small component of the same shape as v_b, which is amplified as the bifurcation load is approached. The resulting load–deflection path changes smoothly, as indicated by the dashed line in Figure 2.11c. Nevertheless, the bifurcation buckling behaviour shown by the solid line of Figure 2.11c can be regarded as the limiting behaviour of certain types of structure with vanishingly small imperfections.

The load at which bifurcation buckling takes place is called the buckling load, and sometimes the critical load, to distinguish it from the maximum load of a smoothly changing load–deflection behaviour.

The mode of bifurcation buckling considered in this book is one for which the load initially remains constant along the buckling path, as shown in Figure 2.11c.

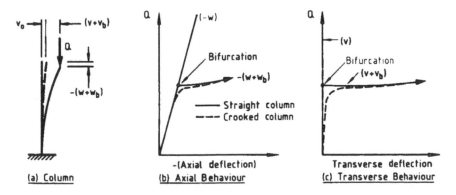

Figure 2.11 Bifurcation buckling.

It occurs when the buckling behaviour is independent of the sense (positive or negative) of the buckling path v_b.

2.6.2 PRE-BUCKLING POSITION

At the buckling load, but before the change of path takes place ($v_b = 0$), the deflected position v on the original load–deflection path is one of equilibrium. Thus, the principle of virtual work requires $\delta U_{TP} = 0$ (Equation 2.30) for every set of virtual displacements δv, as indicated in Figure 2.12a, which allows this position to be found.

2.6.3 BUCKLED POSITION

When the structure buckles under constant load from a pre-buckled position v to a buckled position ($v + v_b$), then this buckled position is also one of equilibrium, and so the principle of virtual work requires $\delta U_{TB} = 0$ for virtual displacements δv_b from this buckled position ($v + v_b$), as shown in Figure 2.12b.

Because the position v at buckling is one of neutral equilibrium, then the principle of conservation of energy requires $\frac{1}{2}\delta^2 U_{TP} = 0$ for displacements from the pre-buckled position v. This second variation from the pre-buckled position v may be considered as being equivalent to a first variation δv_b from the buckled position ($v + v_b$), which itself corresponds to a first variation v_b from the unbuckled position v. Thus for bifurcation buckling, the principle of virtual work for the equilibrium of the buckled position and the law of conservation of energy during buckling are equivalent. This equivalence is illustrated diagrammatically in Figure 2.12b.

The principle of virtual work $\delta U_{TB} = 0$ can be used to obtain the differential equations for equilibrium of the buckled position v_b. These can also be obtained by first using the law of conservation of energy $\frac{1}{2}\delta^2 U_{TP} = 0$, and then using the

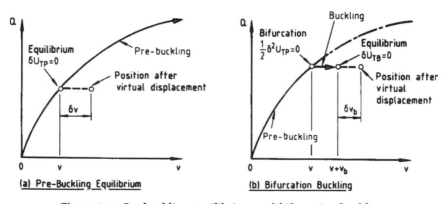

(a) Pre-Buckling Equilibrium   (b) Bifurcation Buckling

Figure 2.12 Pre-buckling equilibrium and bifurcation buckling.

calculus of variations. These methods are applied in section 2.7.2 to demonstrate the equivalence of the principle of virtual work and the law of conservation of energy for the example of the flexural buckling of a column.

2.7 In-plane behaviour of beam-columns

2.7.1 BENDING OF BEAMS

The elastic bending of doubly symmetric beams is analysed in section 17.1. For the beam shown in Figure 2.13a which has equal and opposite end moments M, the differential equation of equilibrium obtained by using the principle of virtual work is

$$M_x'' = 0 \tag{2.41}$$

in which

$$M_x = -EI_x v'' \tag{2.42}$$

and which is subject to the boundary conditions

$$(M_x)_L = (M_x)_0 = M. \tag{2.43}$$

Equation 2.41 expresses the equilibrium between the transverse distributed load effect M_x'' of the moments M and those of (zero) axial load N and (zero) distributed load q_y (see also equation 17.19).

If equation 2.41 is integrated twice and the boundary conditions of equation 2.43 substituted, then it is found that

$$M_x = M \tag{2.44}$$

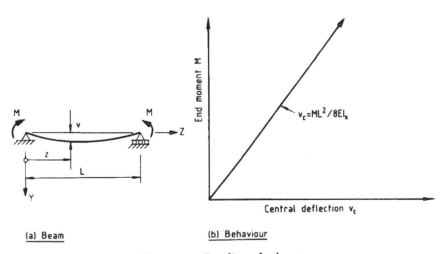

(a) Beam   (b) Behaviour

Figure 2.13 Bending of a beam.

and so

$$- EI_x v'' = M \qquad (2.45)$$

after substitution into equation 2.42. Equation 2.45 expresses the equality between the internal moment of resistance $- EI_x v''$ generated by the bending of the beam and the effect of the applied moments M.

The deflected shape is obtained by integrating equation 2.45, and when the beam is simply supported ($v_0 = v_L = 0$), then this is given by

$$EI_x v = (ML/2)(z - z^2/L) \qquad (2.46)$$

so that the central deflection v_c is

$$v_c = ML^2/8EI_x. \qquad (2.47)$$

This linear relationship between M and v_c is shown in Figure 2.13b.

2.7.2 BUCKLING OF COLUMNS

2.7.2.1 Pre-buckling behaviour

The elastic behaviour of the column shown in Figure 2.14a may be obtained from the analysis given in section 17.1. The pre-buckling deflections are given by

$$v = 0 \qquad (2.48)$$

and

$$- w = (Q/EA)z. \qquad (2.49)$$

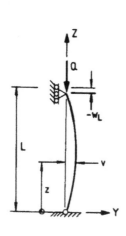

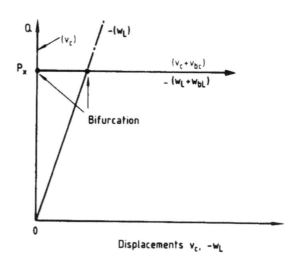

(a) Column   (b) Behaviour

Figure 2.14 Buckling of a column.

Equation 2.49 is derived from the approximate equilibrium equation

$$EAw' = -Q \tag{2.50}$$

which expresses the equality between the internal axial force resistance EAw' generated by extension of the column and the tensile axial force resultant $N = -Q$ of the applied load Q. The linear relationship between Q and $-w_L$ obtained from equation 2.49 is shown in Figure 2.14b.

2.7.2.2 Buckling behaviour

The bifurcation buckling of the column from the pre-buckled position $\{0, w\}$ to $\{v_b, w + w_b\}$ is analysed in section 17.1.7, where it is shown that the condition of neutral equilibrium leads to

$$\frac{1}{2} \int_0^L \{EI_x(v_b'')^2 + N(v_b')^2\} \, dz = 0. \tag{2.51}$$

This is the energy equation for column buckling. It expresses the principle of conservation of energy during buckling which requires the increase in the strain energy $\frac{1}{2}\int_0^L EI_x(v_b'')^2 dz$ to balance the decrease in the potential energy $-\frac{1}{2}\int_0^L N(v_b')^2 dz$ caused by the applied load $Q(= -N)$ moving downwards through a distance

$$-w_{bL} = \frac{1}{2} \int_0^L (v_b')^2 \, dz. \tag{2.52}$$

This distance is equal to the apparent axial shortening of the projection of the deflected column on the original Z axis resulting from the slope v_b' of the column (note that X, Y, Z are used to represent the original axes of the undeformed member, and x, y, z to represent the axes after deformation).

For a simply supported column ($v_0 = v_L = 0$), a possible buckled shape is given by

$$v_b = v_{bc} \sin \pi z/L \tag{2.53}$$

which leads to an estimate of the buckling load as

$$Q = P_x = \pi^2 EI_x/L^2 \tag{2.54}$$

for any magnitude v_{bc} of the buckled shape. The corresponding axial shortening is

$$-w_{bL} = \pi^2 v_{bc}^2/4L. \tag{2.55}$$

The buckling behaviour is shown in Figure 2.14b. For loads Q less than P_x, the transverse displacement v_{bc} is zero, but at $Q = P_x$, this becomes indeterminate. Similarly, at loads less than P_x, the axial displacement $-w_L$ increases linearly with Q, but becomes indeterminate at P_x. Thus the behaviour bifurcates at P_x from the prebuckling displacements $\{0, w\}$ to $\{v_b, w + w_b\}$, as shown in Figure 2.14b.

It is shown in section 17.1.8 that the differential equation for equilibrium of the buckled position can be obtained either by using the virtual work equation or by using the calculus of variations [3] on the energy equation (equation 2.51). This differential equation is

$$(EI_x v_b'')'' - (Nv_b')' = 0 \tag{2.56}$$

which is subject to

$$\begin{aligned}[(EI_x v_b'')' + Nv_b']_{0,L} &= 0, \\ -[EI_x v_b'']_{0,L} &= 0. \end{aligned} \tag{2.57}$$

Equation 2.56 expresses the equality between the resistance $(EI_x v_b'')''$ to transverse distributed load generated by the bending of the column during buckling and the transverse distributed load component $(Nv_b')'$ of the tensile axial force resultant N. The second boundary conditions of equation 2.57 express the zero moment conditions at the simply supported ends, and the first conditions express the zero shear conditions at the ends required by the symmetry of the column and its loading.

It is readily verified that for a simply supported column, these equations are satisfied by the buckled shape of equation 2.53 when the applied load $Q(= -N)$ is equal to the buckling load given by equation 2.54.

2.7.3 NON-LINEAR BENDING OF BEAM-COLUMNS

The elastic behaviour of a beam-column is analysed in section 17.1. If the beam-column has equal and opposite end loads and end moments, and is simply supported so that $v_0 = v_L = 0$ as shown in Figure 2.15a, then the differential

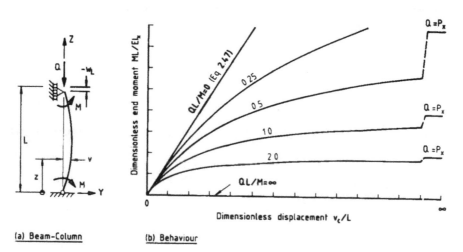

(a) Beam-Column   (b) Behaviour

Figure 2.15 Non-linear bending of a beam-column.

equilibrium equations are

$$-M_x'' - (Nv')' = 0 \tag{2.58}$$

$$-N' = 0 \tag{2.59}$$

in which

$$M_x = -EI_x v'' \tag{2.60}$$

$$N = EA(w' + v'^2/2) \tag{2.61}$$

and which are subject to certain boundary conditions. Equation 2.58 expresses the equilibrium between the transverse distributed load effect $-M_x''$ of the bending moment M_x and the corresponding effect $(Nv')'$ of the tensile axial force resultant N (see also equation 17.19). Equation 2.59 expresses the equilibrium between the longitudinal distributed load effect N' of the axial force resultant N and the (zero) distributed longitudinal load q_z (see also equation 17.20).

Equations 2.58–2.61 can be solved approximately by assuming that

$$N \approx EAw' \tag{2.62}$$

which leads to

$$N \approx -Q \tag{2.63}$$

so that $-w = (Q/EA)\, z$ as in equation 2.49, and

$$v - \left(\frac{ML^2}{8EI_x}\right)\left(\frac{8P_x}{\pi^2 Q}\right)\left\{\cos\left(\frac{\pi z}{L}\sqrt{\left(\frac{Q}{P_x}\right)}\right)\right.$$
$$\left. + \tan\left(\frac{\pi}{2}\sqrt{\left(\frac{Q}{P_x}\right)}\right)\sin\left(\frac{\pi z}{L}\sqrt{\left(\frac{Q}{P_x}\right)}\right) - 1\right\} \tag{2.64}$$

so that

$$v_c = \left(\frac{ML^2}{8EI_x}\right)\left(\frac{8P_x}{\pi^2 Q}\right)\left\{\sec\left(\frac{\pi}{2}\sqrt{\left(\frac{Q}{P_x}\right)}\right) - 1\right\} \tag{2.65}$$

at $z = L/2$. The dimensionless non-linear relationships between M and v_c given by this equation are shown in Figure 2.15b. At low loads Q, the behaviour approximates the linear beam behaviour of equation 2.47, but as the loads increase, the behaviour becomes increasingly non-linear, and asymptotes towards the column buckling behaviour as the load Q approaches the buckling load P_x of equation 2.54.

2.8 Flexural-torsional buckling

2.8.1 GENERAL

The elastic bending behaviour of members with longitudinal and transverse forces and bending moments is analysed in section 17.1, while the elastic uniform and warping torsion behaviour of members with torques and bimoments is analysed in sections 17.2 and 17.3. The approximate energy and differential equilibrium equations are derived for flexural-torsional buckling in sections 17.4 and 17.5. These approximate equations are discussed in this section for the particular cases of concentrically loaded columns, of beams bent in principal planes, and of monosymmetric beam-columns bent in planes of symmetry.

2.8.2 PRE-BUCKLING BEHAVIOUR

The pre-buckling displacements consist of longitudinal displacements w in the case of concentrically loaded columns ($u = v = \phi = 0$), in-plane transverse displacements v in the case of beams bent in the yz principal plane ($u = \phi = 0, w \approx 0$), and both v and w in the case of beam-columns bent in the yz plane of symmetry ($u = \phi = 0$).

The pre-buckling displacements induce longitudinal normal strains at a point P in the cross-section, which may be approximated by

$$\varepsilon_p = w' - yv'' \tag{2.66}$$

and corresponding stresses

$$\sigma_p = E\varepsilon_p. \tag{2.67}$$

These have stress resultants

$$N = \int_A \sigma_p \, dA, \tag{2.68}$$

$$M_x = \int_A \sigma_p y \, dA, \tag{2.69}$$

$$W = \int_A \sigma_p (x^2 + y^2) \, dA, \tag{2.70}$$

so that

$$N = EAw' \tag{2.71}$$

$$M_x = -EI_x v'' \tag{2.72}$$

$$W = -EI_P w' - EI_{Px} v'', \tag{2.73}$$

in which A, I_x, I_P, I_{Px}, are the section properties defined by

$$\left. \begin{array}{l} A = \int_A dA, \\[2mm] I_x = \int_A y^2 \, dA, \end{array} \right\} \tag{2.74}$$

$$I_P = \int_A (x^2 + y^2) \, dA,$$

$$I_{Px} = \int_A y(x^2 + y^2) \, dA.$$

These relationships allow the stress σ_P to be expressed by

$$\sigma_P = N/A + M_x y/I_x. \qquad (2.75)$$

The stress resultants of equations 2.68–2.70 consist of an axial tension force N, an in-plane bending moment M_x acting about the x axis, and a quantity W which may be referred to as a 'Wagner'. While this latter stress resultant plays no part in either the linear approximation for the pre-buckling stresses σ_P (equation 2.75), or in the determination of the approximate pre-buckling deflections v and w from equations 2.71 and 2.72, yet it does play an important role in the flexural-torsional buckling behaviour, as will be seen in section 2.8.3.

2.8.3 BIFURCATION BUCKLING OF COLUMNS

2.8.3.1 Energy equation

When the axial forces inducing compressive stress resultants $P = -N$ in an axially loaded column are large enough, the column may bifurcate from its pre-buckled position $(0, 0, w, 0)$ by deflecting $u_b, v_b, w_b,$ and twisting ϕ_b to $(u_b, v_b, w + w_b, \phi_b)$. The energy equation for the buckled position is approximated by (see section 17.4.6)

$$\frac{1}{2} \int_0^L \{ EI_y u_b''^2 + EI_x v_b''^2 + EI_w \phi_b''^2 + GJ \phi_b'^2 \} \, dz$$

$$- \frac{1}{2} \int_0^L P \{ u_b'^2 + v_b'^2 + (I_P/A + x_0^2 + y_0^2) \phi_b'^2 - 2x_0 v_b' \phi_b' + 2y_0 u_b' \phi_b' \} \, dz = 0. \qquad (2.76)$$

In this equation, P is positive for columns in compression.

The first term of this equation represents the additional strain energy stored in the column during buckling, and includes components associated with bending curvatures u_b'' and v_b'' about the y and x axes, warping 'twistatures' ϕ_b'' due to non-uniform twisting, and twists ϕ_b'. The second term in equation 2.76 corresponds to the decrease in the potential energy of the axial forces during buckling as the column shortens due to rotations u_b' and v_b' about the y and x axes and twists ϕ_b', and includes contributions which allow for the difference between the shortening of the shear centre axis (x_0, y_0) and that of the centroidal axis $(0, 0)$ along which the average axial forces act.

The components of the potential energy decrease may be interpreted physically, by considering the displacements

$$\left. \begin{array}{l} u_P = u_b - (y - y_0) \phi_b, \\ v_P = v_b + (x - x_0) \phi_b \end{array} \right\} \qquad (2.77)$$

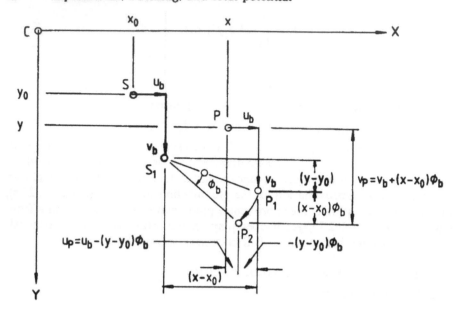

Figure 2.16 Displacements and twist rotation.

shown in Figure 2.16 of a point $P(x, y)$ in the cross-section to P_2 caused by the shear centre displacements u_b, v_b, and the twist rotation ϕ_b about the shear centre. A longitudinal element $\delta z \times \delta A$ through P rotates

$$\left. \begin{array}{l} u'_P = u'_b - (y - y_0)\,\phi''_b, \\ v'_P = v'_b + (x - x_0)\,\phi''_b \end{array} \right\} \tag{2.78}$$

and so the sorce $\sigma_P \delta A$ acting through P has components $\sigma_P \delta A\,u'_P, \sigma_P \delta A\,v'_P$ acting normal to the rotated element, as shown in figure 2.17. These components give rise to transverse forces

$$\left. \begin{array}{l} Q_{x\sigma} = -\displaystyle\int_A \sigma_P u'_P\,dA, \\[2mm] Q_{y\sigma} = -\displaystyle\int_A \sigma_P v'_P\,dA \end{array} \right\} \tag{2.79}$$

and torques about the displaced shear centre axis S_1

$$M_{z\sigma} = -\int_A \sigma_P [u'_P\{v_b - (y - y_0)\} - v'_P\{u_b - (x - x_0)\}]\,dA. \tag{2.80}$$

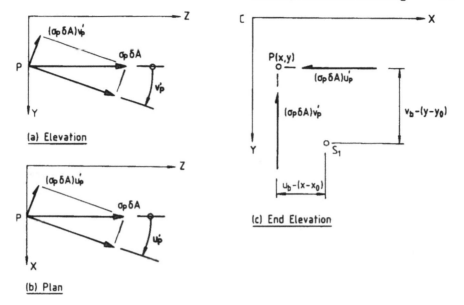

Figure 2.17 Components of axial force P.

Substituting $\sigma_P = P/A$ and integrating leads to

$$\left.\begin{array}{l} Q_{x\sigma} = P(u'_b + y_0\phi'_b), \\ Q_{y\sigma} = P(v'_b - x_0\phi'_b) \end{array}\right\} \tag{2.81}$$

and

$$M_{z\sigma} = P\{(u'_b + y_0\phi'_b)y_0 - (v'_b - x_0\phi'_b)x_0 + (I_P/A)\phi'_b\} \tag{2.82}$$

for small displacements u_b, v_b. The transverse forces undergo differential transverse displacements $u'_b\delta z$ and $v'_b\delta z$ (Figure 2.18), and so do work equivalent to potential energy decreases of

$$\left.\begin{array}{l} \dfrac{1}{2}\displaystyle\int_0^L Q_{x\sigma}u'_b\,\mathrm{d}z = \dfrac{1}{2}\int_0^L P(u_b'^2 + y_0 u'_b\phi'_b)\,\mathrm{d}z, \\[2mm] \dfrac{1}{2}\displaystyle\int_0^L Q_{y\sigma}v'_b\,\mathrm{d}z = \dfrac{1}{2}\int_0^L P(v_b'^2 - x_0 v'_b\phi'_b)\,\mathrm{d}z. \end{array}\right\} \tag{2.83}$$

The torques undergo differential twist rotations $\phi'_b\delta z$ and so do work equivalent to a potential energy decrease of

$$\frac{1}{2}\int_0^L M_{z\sigma}\phi'_b\,\mathrm{d}z = \frac{1}{2}\int_0^L P\{y_0 u'_b\phi'_b - x_0 v'_b\phi'_b + (I_P/A + x_0^2 + y_0^2)\phi_b'^2\}\,\mathrm{d}z. \tag{2.84}$$

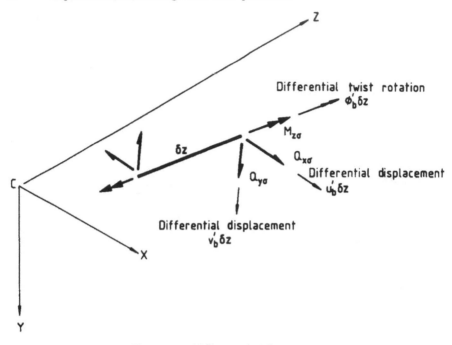

Figure 2.18 Differential deformations.

It can be seen that equations 2.83 and 2.84 provide work components which are equivalent to the potential energy decrease of equation 2.76.

Alternative physical interpretations of the potential energy decrease of equation 2.76 may be made by considering the rotations u_P', v_P' (equation 2.78) of an element δz through a point $P(x, y)$ in the cross-section. These rotations cause the projection of the element on to the Z axis to shorten by

$$\delta \Delta_P = \tfrac{1}{2}(u_P'^2 + v_P'^2)\,\delta z \qquad (2.85)$$

approximately, as demonstrated in Figure 2.19. Thus the work done by a force $\sigma_P \delta A$ acting on an element of area δA through P is

$$\delta W_P = -\tfrac{1}{2}(u_P'^2 + v_P'^2)\,\sigma_P \delta A\, \delta z. \qquad (2.86)$$

The total work done is therefore

$$W = -\int_0^L \int_A \frac{1}{2}\sigma_P(u_P'^2 + v_P'^2)\,dA\,dz. \qquad (2.87)$$

Substituting equation 2.78 and $\sigma_P = -P/A$ also leads to work terms which are equivalent to the potential energy decrease of equation 2.76.

In the special case where the shear centre and centroid coincide, the work done

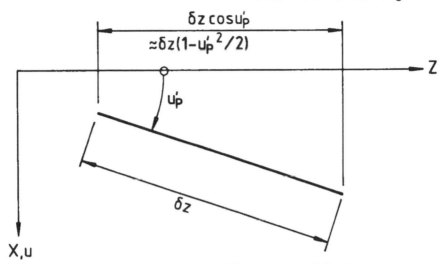

Figure 2.19 Projection of δz on to centroidal axis.

reduces to

$$W = \frac{1}{2}\int_0^L P\{u_b'^2 + v_b'^2 + (I_P/A)\phi_b'^2\}\,dz. \tag{2.88}$$

The first two terms of this can be interpreted as the work done by P due to the average shortenings $\frac{1}{2}\int_0^L u_b'^2\,dz$ and $\frac{1}{2}\int_0^L v_b'^2\,dz$ resulting from the rotations u_b' and v_b'. The third term corresponds to the work done by P due to the average shortening $\frac{1}{2}\int_0^L (I_P/A)\phi_b'^2\,dz$ resulting from the twist ϕ_b'.

More generally, the shortening due to the twist ϕ_b' can be determined by considering the rotation $a_0\phi_b'$ of a longitudinal element of area δA through a point $P(x, y)$ at a distance

$$a_0 = \sqrt{\{(x - x_0)^2 + (y - y_0)^2\}} \tag{2.89}$$

from the shear centre as shown in Figure 2.20. Thus the shortening of the element is

$$\delta\Delta_{P\phi} = \frac{1}{2}(a_0\phi_b')^2\delta z \tag{2.90}$$

and the total average shortening is

$$\frac{1}{2}\int_0^L \left\{\int_A a_0^2\,dA\right\}\phi_b'^2\,dz/A = \frac{1}{2}\int_0^L \{I_P/A + x_0^2 + y_0^2\}\phi_b'^2\,dz. \tag{2.91}$$

Thus the work done by the load P due to the twist ϕ_b' is given by

$$W_\phi = \frac{1}{2}\int_0^L P(I_P/A + x_0^2 + y_0^2)\phi_b'^2\,dz. \tag{2.92}$$

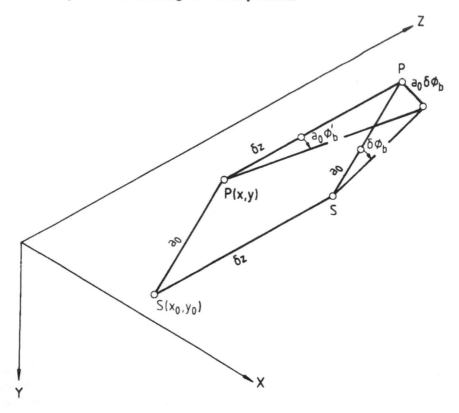

Figure 2.20 Rotation due to twist.

It may be noted that the component $\frac{1}{2}\int_0^L P(I_P/A)\phi_b'^2 dz$ in equations 2.76, 2.84, 2.88 and 2.92 may be re-expressed as $-\frac{1}{2}\int_0^L W\phi_b'^2 dz$ in which W is the 'Wagner' stress resultant obtained from equations 2.71, 2.73 and 2.75 as

$$W = -PI_P/A \tag{2.93}$$

when $M_x = 0$. Thus the work done during buckling includes a contribution from the 'Wagner' stress resultant. This contribution was first reported by Wagner [4], and is responsible for the torsional buckling ($u_b = v_b = 0$) of doubly symmetric columns ($x_0 = y_0 = 0$).

2.8.3.2 Differential equilibrium equations

The differential equilibrium equations for the flexural-torsional buckling of an axially loaded column are approximated by (see section 17.5.1)

$$(EI_y u_b'')'' + \{P(u_b' + y_0\phi_b')\}' = 0, \tag{2.94}$$

$$(EI_x v_b'')'' + \{P(v_b' - x_0 \phi_b')\}' = 0, \tag{2.95}$$

$$(EI_w \phi_b'')'' - (GJ\phi_b')' + \{Py_0 u_b'\}' - \{Px_0 v_b'\}' + \{P(I_P/A + x_0^2 + y_0^2)\phi_b'\}' = 0. \tag{2.96}$$

In these equations, P is positive for columns in compression.

The first of these equations expresses the equality between the internal resistance $(EI_y u_b'')''$ to transverse distributed load in the x direction generated by the bending about the y axis of the column during buckling and the transverse load component $-\{P(u_b' + y_0\phi_b')\}'$ of the compressive axial force resultant P (see equation 2.81) caused by the rotation $(u_b' + y_0\phi_b')$. The second equation (equation 2.95) expresses a similar equality for bending about the x axis.

The third equilibrium equation (equation 2.96) expresses the equality between the sum $(EI_w \phi_b'')'' - (GJ\phi_b')'$ of the internal warping and uniform torsion resistances to distributed torque generated by the warping and twisting of the column during buckling and the distributed torque components of the compressive axial force stress resultant P (see equation 2.82) caused by the rotations u_b', v_b' and the twist ϕ_b'.

2.8.4 BIFURCATION BUCKLING OF MONOSYMMETRIC BEAMS

2.8.4.1 Energy equation

When the transverse loads and moments inducing stress resultants M_x in a monosymmetric beam bent in the principal YZ plane are large enough, the beam may bifurcate from its pre-buckled position $(0, v, w, 0)$ by deflecting laterally u_b and twisting ϕ_b to $(u_b, v + v_b, w + w_b, \phi_b)$. The energy equation for the buckled position is approximated by (see section 17.4.7)

$$\frac{1}{2}\int_0^L \{EI_y u_b''^2 + EI_w \phi_b''^2 + GJ\phi_b'^2\}\, dz$$

$$+ \frac{1}{2}\int_0^L M_x\{2\phi_b u_b'' + (I_{Px}/I_x - 2y_0)\phi_b'^2\}\, dz = 0 \tag{2.97}$$

for shear centre loading.

The first term of this equation represents the increase in the strain energy stored in the beam during buckling due to bending curvatures u_b'' about the y axis, 'twistatures' ϕ_b'' due to non-uniform twisting, and twists ϕ_b'. The second term in equation 2.97 is equal to the increase in the potential energy of the loading system. This second term is negative, and so corresponds to a decrease in the potential energy, equal to the work done by the bending moment M_x as the beam deflects v_b due to the combined effects of the lateral deflection u_b and twist rotation ϕ_b.

A physical interpretation of the first part $-\frac{1}{2}\int_0^L 2M_x \phi_b u_b''\, dz$ of the decrease in the potential energy of the loading system given in equation 2.97 can be obtained by considering the curvature $v_b'' = \phi_b u_b''$ in the YZ plane of bending

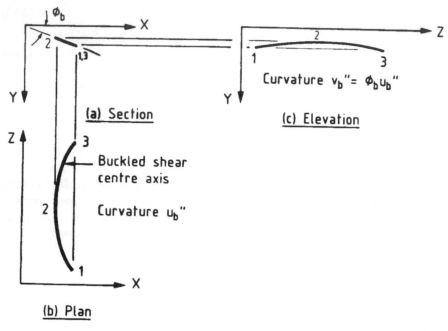

Figure 2.21 Curvature of buckled beam.

resulting from the minor axis curvature u_b'' and the twist rotation ϕ_b shown in Figure 2.21. This curvature causes the moment M_x to do work

$$-\int_0^L M_x v_b'' \, dz = -\frac{1}{2} \int_0^L 2M_x \phi_b u_b'' \, dz. \qquad (2.98)$$

An alternative interpretation of the first part $-\frac{1}{2} \int_0^L 2M_x \phi_b u_b'' \, dz$ of the decrease in the potential energy of the loading system given in equation 2.97 can be obtained by separating it into three components

$$-\frac{1}{2} \int_0^L 2M_x \phi_b u_b'' \, dz = \frac{1}{2} \int_0^L M_x \phi_b' u_b' \, dz + \frac{1}{2} \int_0^L M_x u_b' \phi_b' \, dz - \int_0^L M_x (u_b' \, \phi_b)' \, dz. \qquad (2.99)$$

The first component $\frac{1}{2} \int_0^L M_x \phi_b' u_b' \, dz$ arises from the transverse forces

$$Q_{x\sigma} = M_x \phi_b' \qquad (2.100)$$

obtained by substituting $\sigma_P = M_x y / I_x$ into equation 2.79. The transverse forces undergo differential transverse displacements $u_b' \delta z$ as shown in Figure 2.18, and so do work equivalent to a potential energy decrease of

$$\frac{1}{2} \int_0^L Q_{x\sigma} u_b' \, dz = \frac{1}{2} \int_0^L M_x \phi_b' u_b' \, dz. \qquad (2.101)$$

The second component $\frac{1}{2} \int_0^L M_x u_b' \phi_b' \, dz$ of equation 2.99 arises from the first portions of the torques

$$M_{z\sigma} = M_x u_b' - M_x(I_{Px}/I_x - 2y_0)\phi_b' \qquad (2.102)$$

obtained by substituting $\sigma_P = M_x y/I_x$ into equation 2.79. The torques undergo differential twist rotations $\phi_b'\delta z$ as shown in Figure 2.18, and so do work equivalent to a potential energy decrease of

$$\frac{1}{2}\int_0^L M_{z\sigma}\phi_b'\,dz = \frac{1}{2}\int_0^L \{M_x u_b'\phi_b' - M_x(I_{Px}/I_x - 2y_0)\phi_b'^2\}\,dz. \qquad (2.103)$$

For a doubly symmetric section, $I_{Px} = y_0 = 0$, and so

$$\frac{1}{2}\int_0^L M_{z\sigma}\phi_b'\,dz = \frac{1}{2}\int_0^L M_x u_b'\phi_b'\,dz. \qquad (2.104)$$

The third component $-\frac{1}{2}\int_0^L M_x(u_b'\phi_b)'\,dz$ of equation 2.99 arises from small rotations $-d(-yu_b'\phi_b)/dy = u_b'\phi_b$ which occur during buckling as a result of longitudinal buckling displacements $-yu_b'\phi_b$. Differential rotations $(u_b'\phi_b)'\delta z$ cause the moments M_x to do work equivalent to a potential energy decrease of $-\int_0^L M_x(u_b'\phi_b)'\,dz$.

For a monosymmetric section, the additional decrease $-\frac{1}{2}\int_0^L M_x(I_{Px}/I_x - 2y_0)\phi_b'^2\,dz$ in the potential energy of the loading system given in equation 2.97 arises from the second portions of the torques of equation 2.102. Thus the potential energy decrease resulting from the differential twist rotations $\phi_b'\delta z$ of these torques is given by equation 2.103.

It may be noted that the component $\frac{1}{2}\int_0^L M_x(I_{Px}/I_x)\phi_b'^2\,dz$ in equation 2.97 may be re-expressed as $\frac{1}{2}\int_0^L W\phi_b'^2\,dz$, in which W is the 'Wagner' stress resultant obtained from equation 2.73 as $W = M_x I_{Px}/I_x$ when $P = 0$. Thus the work done during buckling includes a contribution from the 'Wagner' stress resultant.

2.8.4.2 Differential equilibrium equations

The differential equilibrium equations for the flexural-torsional buckling of a monosymmetric beam bent in the yz principal plane are approximated by (see section 17.5.2)

$$(EI_y u_b'')'' + (M_x\phi_b)'' = 0, \qquad (2.105)$$

$$(EI_w\phi_b'')'' - (GJ\phi_b')' + M_x u_b'' - \{M_x(I_{Px}/I_x - 2y_0)\phi_b'\}' = 0. \qquad (2.106)$$

The first of these equations expresses the equality between the internal resistance $(EI_y u_b'')''$ to transverse distributed load in the x direction generated by the bending about the y axis of the beam during buckling and the transverse load component $-(M_x\phi_b)''$ of the bending moment M_x caused by the twist rotation ϕ_b.

The second equilibrium equation (equation 2.106) expresses the equality between the sum $(EI_w\phi_b'')'' - (GJ\phi_b')'$ of the internal warping and uniform torsion resistances to distributed torque generated by the warping and twisting of the beam during buckling, and the distributed torque components of the bending moment stress resultant M_x and the 'Wagner' stress resultant $W = M_x I_{Px}/I_x$.

2.8.5 BIFURCATION BUCKLING OF MONOSYMMETRIC BEAM-COLUMNS

2.8.5.1 Energy equation

When the axial loads and the transverse loads and moments bend a monosymmetric beam-column in the yz plane of symmetry ($x_0 = 0$), the beam-column may bifurcate from its prebuckled position $(0, v, w, 0)$ by deflecting laterally u_b and twisting ϕ_b to $(u_b, v + v_b, w + w_b, \phi_b)$. The energy equation for the buckled position is approximated by (see section 17.4.7)

$$\frac{1}{2} \int_0^L \{EI_y u_b''^2 + EI_w \phi_b''^2 + GJ \phi_b'^2\} \, dz$$

$$-\frac{1}{2} \int_0^L P\{u_b'^2 + (I_P/A + y_0^2)\phi_b'^2 + 2y_0 u_b' \phi_b'\} \, dz$$

$$+\frac{1}{2} \int_0^L M_x\{2\phi_b u_b'' + (I_{Px}/I_x - 2y_0)\phi_b'^2\} \, dz = 0 \qquad (2.107)$$

for shear centre loading. In this equation, P is positive for beam-columns in compression. Equation 2.107 is a combination of equation 2.76 for column buckling (with $v_b = 0$ and $x_0 = 0$), and equation 2.97 for beam buckling.

2.8.5.2 Differential equilibrium equations

The differential equilibrium equations for the flexural-torsional buckling of a monosymmetric beam-column bent in the yz plane of symmetry are approximated by (see section 17.5.2)

$$(EI_y u_b'')'' + \{P(u_b' + y_0 \phi_b')\}' + (M_x \phi_b)'' = 0, \qquad (2.108)$$

$$(EI_w \phi_b'')'' - (GJ \phi_b')' + \{Py_0 u_b'\}' + \{P(I_P/A + y_0^2)\phi_b'\}'$$
$$+ M_x u_b'' + \{M_x(I_{Px}/I_x + 2y_0)\phi_b'\}' = 0. \qquad (2.109)$$

These equations are combinations of equation 2.94 for column buckling with equation 2.105 for beam buckling, and of equation 2.96 with equation 2.106.

2.9 References

1. Langhaar, H.L. (1962) *Energy Methods in Applied Mechanics*, John Wiley, New York.
2. Rubinstein, M.F. (1970) *Structural Systems – Statistics, Dynamics, and Stability*, Prentice-Hall, Englewood Cliffs, NJ.
3. Courant, R. (1949) *Differential and Integral Calculus*, Volume II, Blackie, London.
4. Wagner, H. (1936) Verdrehung und Knickung von offenen Profilen (Torsion and Buckling of open sections). NACA Technical Memorandum, No. 807.

3 Buckling analysis of simple structures

3.1 General

Flexural-torsional buckling depends on many parameters, including those defining the structural geometry, the support and restraint conditions, the material properties, and the load arrangement. Because of these many parameters, it is not possible even in the most detailed survey to include all of the solutions that may be needed. It is therefore necessary to present methods of analysing a wide range of situations. A computer method of analysing a very wide range of flexural-torsional buckling problems which is capable of high accuracy is presented in Chapter 4.

In this chapter, the energy method [1–3] for the analysis of flexural-torsional buckling is presented in a form which is suitable for hand use. Such a method is limited to problems whose solutions do not require excessive computational effort from the analyst. For this to be so, the problems must be comparatively simple, and solutions of moderate accuracy must be acceptable.

While these conditions may seem to be restrictive, they nevertheless allow a wide range of engineering problems to be solved with an order of accuracy which is consistent with the accuracy with which many of the controlling parameters can be determined. Even when higher accuracy is required, approximate solutions usually allow the more important parameters to be identified, and may suggest methods of closely approximating more accurate solutions obtained in other ways.

The energy method is presented and demonstrated and its accuracy discussed in section 3.2. The choice of suitable buckled shapes is considered in section 3.3, five examples are worked in section 3.4, and methods of increasing the accuracy are given in section 3.5.

3.2 The energy method

3.2.1 BUCKLING AND CONSERVATION OF ENERGY

The condition of neutral equilibrium at bifurcation buckling follows from the principle of conservation of energy (section 2.5). As the structure under a fixed set of loads buckles from an unbuckled position in a quasi-static manner to an adjacent buckled position which is one of equilibrium, the increase in the strain energy $\frac{1}{2}\delta^2 U$ stored in the structure is matched by an equal decrease in the potential energy $-\frac{1}{2}\delta^2 V$ of the loads. Thus the energy equation at buckling can be expressed as

$$\frac{1}{2}(\delta^2 U + \delta^2 V) = 0. \tag{3.1}$$

For the flexural-torsional buckling of a column (section 2.8.3) these changes are given by

$$\left. \begin{array}{l} \dfrac{1}{2}\delta^2 U = \dfrac{1}{2}\displaystyle\int_0^L \{EI_y u''^2 + EI_x v''^2 + EI_w \phi''^2 + GJ\phi'^2\}\,dz, \\[3mm] \dfrac{1}{2}\delta^2 V = -\dfrac{1}{2}\displaystyle\int_0^L P\{u'^2 + v'^2 + (I_P/A + x_0^2 + y_0^2)\phi'^2 - 2x_0 v'\phi' + 2y_0 u'\phi'\}\,dz \end{array} \right\} \tag{3.2}$$

and by

$$\left. \begin{array}{l} \dfrac{1}{2}\delta^2 U = \dfrac{1}{2}\displaystyle\int_0^L \{EI_y u''^2 + EI_w \phi''^2 + GJ\phi'^2\}\,dz, \\[3mm] \dfrac{1}{2}\delta^2 V = -\dfrac{1}{2}\displaystyle\int_0^L P\{u'^2 + (I_P/A + y_0^2)\phi'^2 + 2y_0 u'\phi'\}\,dz \\[3mm] \qquad\qquad + \dfrac{1}{2}\displaystyle\int_0^L M_x\{2\phi u'' + (I_{Px}/I_x - 2y_0)\phi'^2\}\,dz \end{array} \right\} \tag{3.3}$$

for monosymmetric beam-columns with transverse loads at the shear centre (section 2.8.5), in which u, v, ϕ are the buckling deflections and twist rotation, and P is positive for members in compression.

The strain energy changes $\frac{1}{2}\delta^2 U$ in these equations include the strain energy changes due to bending about both or one of the x, y principal axes, and to warping and uniform torsion caused by the twist rotations ϕ. The potential energy changes $\frac{1}{2}\delta^2 V$ are expressed in terms of the stress resultants P and M_x induced by the applied loads.

3.2.2 DEMONSTRATION OF THE ENERGY METHOD

For demonstration purposes, the energy method will be used to find the approximate flexural compressive buckling load Q of the simply supported column ($M_x = 0$) shown in Figure 3.1. For this mode of buckling, $v = \phi = 0$, and so equations 3.1 and 3.2 simplify to

$$\frac{1}{2}\int_0^L EI_y u''^2\,dz - \frac{1}{2}\int_0^L Q u'^2\,dz = 0 \tag{3.4}$$

after replacing the compressive stress resultant P by Q.

The first step in solving this equation approximately is to guess a buckled shape which satisfies at least the kinematic boundary conditions (see section 3.3.1), which relate to the deflection and rotation constraints. In this problem, these

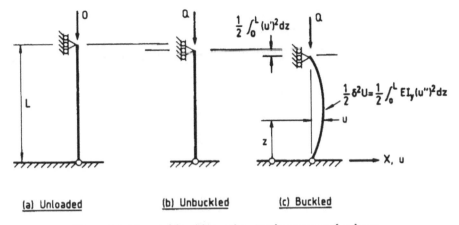

(a) Unloaded   (b) Unbuckled   (c) Buckled

Figure 3.1 Flexural buckling of a simply supported column.

boundary conditions are those of displacement prevented at the column ends, so that

$$u_0 = u_L = 0. \qquad (3.5)$$

A very simple guess for the buckled shape is

$$u = 4\delta(z/L - z^2/L^2) \qquad (3.6)$$

which satisfies the boundary conditions of equation 3.5. In this equation, δ is the magnitude of the central deflection $u_{L/2}$. The value of δ does not need to be known, since it will appear in both terms of equation 3.4 and cancel out.

The second step is to substitute the guessed buckled shape into the energy equation and integrate. In this problem, this leads to

$$\frac{1}{2}\int_0^L EI_y(4\delta)^2(-2/L^2)^2\, dz - \frac{1}{2}\int_0^L Q(4\delta)^2(1/L - 2z/L^2)^2\, dz = 0, \qquad (3.7)$$

whence

$$\tfrac{1}{2} EI_y(4\delta)^2(-2L^2)^2 L - \tfrac{1}{2}Q(4\delta)^2(L/L^2 - 2L^2/L^3 + 4L^3/3L^4) = 0, \qquad (3.8)$$

which simplifies to

$$Q = 12EI_y/L^2 \qquad (3.9)$$

which is an approximate solution for the buckling load.

This approximate solution is 22% higher than the correct solution of

$$Q = \pi^2 EI_y/L^2 \qquad (3.10)$$

which corresponds to a buckled shape

$$u = \delta\sin \pi z/L. \qquad (3.11)$$

This buckled shape also satisfies the boundary conditions of equation 3.5, and when substituted into equation 3.4 leads to

$$\tfrac{1}{2}EI_y\delta^2(-\pi^2/L^2)^2(L/2)-\tfrac{1}{2}Q\delta^2(\pi/L)^2(L/2)=0 \tag{3.12}$$

which simplifies to equation 3.10. This demonstrates that the energy method will give the correct solution for the buckling load when the correct buckled shape is assumed. However, this is not in itself a useful conclusion, since it is almost always the case that the correct buckled shape is unknown. It does, however, allow it to be inferred that the more accurate the assumed buckled shape, then the more accurate is the buckling load calculated.

3.2.3 ACCURACY OF THE ENERGY METHOD

The energy method gives a buckling load solution which is more accurate than the buckled shape guessed, provided this is reasonably close to the true buckled shape. This is because the energy method always provides an upper bound to the true buckling load when the guessed buckled shape satisfies at least the kinematic boundary conditions (see section 3.3.1), as is shown in section 17.1.10.

The upper bound nature of the calculated buckling load is demonstrated in Figure 3.2 which indicates its variation with the buckling shape guessed. If the true shape is guessed, then the true buckling load is calculated, which is equal to the minimum value of all the calculated loads.

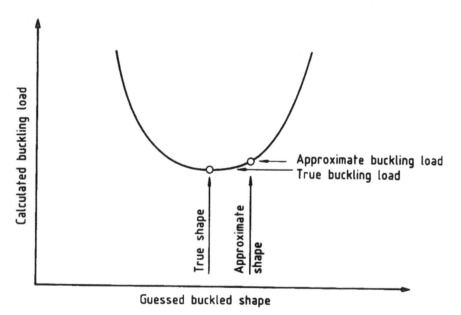

Figure 3.2 Variation of calculated load with guessed shape.

The upper bound nature of the calculated buckling load is responsible for the comparatively high accuracy of the buckling load predicted. Provided the guessed shape is reasonably close to the true shape, then the calculated load, which varies slowly with the guessed shape in this region, is quite close to the true buckling load.

3.3 Choosing the buckled shape

3.3.1 BOUNDARY CONDITIONS

The accuracy of the buckling load calculated by the energy method is related to the accuracy of the buckled shape assumed. Ideally, the buckled shape should satisfy all the boundary conditions imposed by the method of support and constraint of the structure. These boundary conditions are usually described as being either kinematic or static.

Kinematic boundary conditions relate to the geometrical constraints at the supports of the structure. These geometrical constraints prevent one or more deflections or rotations at the support points. Thus deflection prevented, rotation prevented (Figure 3.3), and warping (deflections) prevented at a support are examples of kinematic boundary conditions.

Static boundary conditions relate to the values of stress resultants such as moments, shears, bimoments and warping torques at the supports. Thus the static conditions at a frictionless hinge may be expressed by zero moment, and those at a free end by zero moment and zero shear (Figure 3.3).

The buckled shape guessed should satisfy the kinematic boundary conditions, as shapes which ignore geometrical constraints are likely to lead to buckling load predictions which are much lower than the true buckling load.

On the other hand, the guessed buckle shape need not satisfy the static boundary conditions, which are local expressions of the equilibrium equations. This is consistent with the fact that the guessed shape will not generally satisfy the differential equilibrium equations. For example, while the guessed parabolic buckled shape of equation 3.6 does satisfy the kinematic boundary conditions

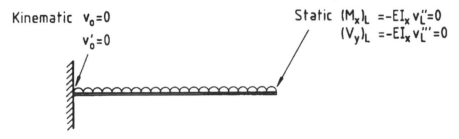

Figure 3.3 Flexural boundary conditions for a cantilever.

u		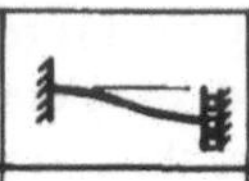					
a_0	0	0	0	0	0	0	0
a_1	0	0	0	0	1	1	1
a_2	1	3	1	1	0	-1	-3
a_3	0	-2	-1	-2	0	0	2
a_4	0	0	0	1	0	0	0

$$u = a_0 + a_1(z/L) + a_2(z/L)^2 + a_3(z/L)^3 + a_4(z/L)^4$$

Figure 3.4 Limited power series.

$u_0 = u_L = 0$, its curvature $u'' = -8\delta/L^2$ does not satisfy the static zero moment boundary conditions $EI_y u_0'' = EI_y u_L'' = 0$.

3.3.2 POWER SERIES

Perhaps the simplest buckled shapes can be obtained by using a limited power series

$$u = a_0 + a_1 z/L + a_2 z^2/L^2 + \cdots + a_n z^n/L^n \qquad (3.13)$$

in which the values of the coefficients $a_0, a_1, \ldots, a_n$ are selected to satisfy the kinematic boundary conditions. The coefficients in some examples of limited power series are shown in Figure 3.4.

The simplicity of a limited power series is due to the ease of integration of its products which arise in the energy equation.

3.3.3 TRIGONOMETRIC SERIES

Buckled shapes may also be obtained by using a limited sine series

$$u = a_1 \sin \pi z/L + a_2 \sin 2\pi z/L + \cdots + a_n \sin n\pi z/L. \qquad (3.14)$$

Each term of such a series automatically satisfies the boundary conditions $u_0 = u_L = 0$, which often allows a single term to be used.

The terms of such a series are orthogonal (section 17.1.9), which sometimes leads to simplifications when the square of the series is to be integrated as part of the solution of the energy equation. For example, if it is assumed that

$$u = a_1 \sin \pi z/L + a_3 \sin 3\pi z/L \qquad (3.15)$$

for the column buckling problem of section 3.2.2, then

$$\int_0^L (u'')^2 \, dz = (\pi^4/L^4)(L/2)\{a_1^2 + 81a_3^2\} \qquad (3.16)$$

with the coefficient of the cross-product $a_1 a_3$ equal to zero.

More generally, however, the use of trigonometric series may lead to integration difficulties, as in beam lateral buckling problems where the bending moment M_x varies as a limited power series. For example, when

$$M_x = (qL^2/2)(z/L - z^2/L^2) \qquad (3.17)$$

as it does for a simply supported beam with uniformly distributed loading q, then the work done term $\frac{1}{2}\int_0^L 2M_x\phi u'' \, dz$ of equation 3.3 will have terms of the type

$$I = \frac{1}{L}\int_0^L \left(\frac{\pi z}{L}\right)^m \sin\frac{n_1\pi z}{L} \sin\frac{n_2\pi z}{L} dz. \qquad (3.18)$$

These difficulties can be overcome by making use of the values of I given in Table 3.1 for $m = 1, 2, 3, 4$ and $n_1, n_2 = 1, 2, 3$.

Table 3.1   Definite integrals

(a) $I = \dfrac{1}{L}\displaystyle\int_0^{L/2} \left(\dfrac{\pi z}{L}\right)^m \sin\left(\dfrac{n_1\pi z}{L}\right)\sin\left(\dfrac{n_2\pi z}{L}\right) dz$

n_1	n_2	$m = 0$	1	2	3	4
1	1	0.2500	0.2759	0.3306	0.4174	0.5463
1	2	0.2122	0.1919	0.1935	0.2101	0.2404
1	3	0.0000	−0.0796	−0.1563	−0.2488	−0.3726
2	2	0.2500	0.1963	0.1744	0.1686	0.1736
2	3	0.1273	0.0472	−0.0016	−0.0411	−0.0810
3	3	0.2500	0.2052	0.2195	0.2735	0.3683

(b) $I = \dfrac{1}{L}\displaystyle\int_0^L \left(\dfrac{\pi z}{L}\right)^m \sin\left(\dfrac{n_1\pi z}{L}\right)\sin\left(\dfrac{n_2\pi z}{L}\right) dz$

n_1	n_2	$m = 0$	1	2	3	4
1	1	0.5000	0.7854	1.3949	2.6977	5.5561
1	2	0.0000	−0.2829	−0.8889	−2.3025	−5.6941
1	3	0.0000	0.0000	0.1875	0.8836	2.9980
2	2	0.5000	0.7854	1.5824	3.5813	8.5541
2	3	0.0000	−0.3056	−0.9600	−2.6171	−6.9688
3	3	0.5000	0.7854	1.6172	3.7449	9.2019

Table 3.1 (*Contd.*)

(c) $\quad I = \dfrac{1}{L}\displaystyle\int_0^{L/2}\left(\dfrac{\pi z}{L}\right)^m \cos\left(\dfrac{n_1 \pi z}{L}\right)\cos\left(\dfrac{n_2 \pi z}{L}\right) dz$

n_1	n_2	$m = 0$	1	2	3	4
1	1	0.2500	0.1168	0.0806	0.0608	0.0626
1	2	0.1061	−0.0102	−0.0447	−0.0665	−0.0879
1	3	0.0000	−0.0796	−0.0938	−0.1015	−0.1111
2	2	0.2500	0.1964	0.2369	0.3159	0.4352
2	3	0.1910	0.1345	0.1504	−0.1847	0.2336
3	3	0.2500	0.1875	0.1917	0.2110	0.2405

(d) $\quad I = \dfrac{1}{L}\displaystyle\int_0^{L}\left(\dfrac{\pi z}{L}\right)^m \cos\left(\dfrac{n_1 \pi z}{L}\right)\cos\left(\dfrac{n_2 \pi z}{L}\right) dz$

n_1	n_2	$m = 0$	1	2	3	4
1	1	0.5000	0.7854	1.8949	5.0539	13.926
1	2	0.0000	−0.3537	−1.1111	−3.3025	−9.7842
1	3	0.0000	0.0000	0.3125	1.4726	5.3716
2	2	0.5000	0.7854	1.7074	4.1703	10.928
2	3	0.0000	−0.3310	−1.0410	−2.9879	−8.5095
3	3	0.5000	0.7854	1.6727	4.0067	10.280

3.4 Worked examples

3.4.1 SIMPLE BEAM IN UNIFORM BENDING

The energy equation for flexural-torsional buckling of an elastic beam (see section 17.4.7) is given by

$$\frac{1}{2}\int_0^L \{EI_y(u'')^2 + EI_w(\phi'')^2 + GJ(\phi')^2\}\, dz + \frac{1}{2}\int_0^L M_x\{2\phi u'' + \beta_x \phi'^2\}\, dz$$

$$+\frac{1}{2}\int_0^L q_y(y_q - y_0)\phi^2\, dz + \frac{1}{2}\sum Q_y(y_Q - y_0)\phi^2 = 0. \tag{3.19}$$

For doubly symmetric beams, $\beta_x = y_0 = 0$. For a simply supported beam in uniform bending (Figure 3.5a), $M_x = M$ and $q_y = Q_y = 0$.

If it is assumed that

$$u/\delta = \phi/\theta = z/L - z^2/L^2 \tag{3.20}$$

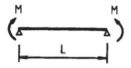

M
M
L
(a) Simple Beam in Uniform Bending
Q
L/2
L/2
(b) Central Load at Shear Centre
Q
yQ
L/2
L/2
(c) Central Load off Shear Centre
Q
L
(d) Cantilever with End Load

Figure 3.5 Worked examples.

which satisfy the kinematic boundary conditions

$$u_{0,L} = \phi_{0,L} = 0 \tag{3.21}$$

(but not the static boundary conditions $M_{y0,L} = EI_y u''_{0,L} = 0$ and $B_{0,L} = EI_w \phi''_{0,L} = 0$), then

$$\frac{1}{2}\delta^2 U = \frac{1}{2}\int_0^L \{EI_y \delta^2(-2/L^2)^2 + EI_w \theta^2(-2/L^2)^2 + GJ\theta^2(1/L - 2z/L^2)^2\}\,dz$$

$$= \frac{1}{2}\{4EI_y\delta^2/L^3 + 4EI_w\theta^2/L^3 + GJ\theta^2/3L\},$$

$$\frac{1}{2}\delta^2 V = M\theta\delta\int_0^L \{(z/L - z^2/L^2)(-2/L^2)\}\,dz = -M\theta\delta/3L.$$

Thus the energy equation

$$\tfrac{1}{2}(\delta^2 U + \delta^2 V) = 0 \tag{3.22}$$

can be written as

$$\frac{1}{2}\frac{1}{3L}\begin{Bmatrix}\delta\\ \theta\end{Bmatrix}^{\mathrm{T}}\begin{bmatrix}12EI_y/L^2 & -M\\ -M & (GJ + 12EI_w/L^2)\end{bmatrix}\begin{Bmatrix}\delta\\ \theta\end{Bmatrix} = 0. \tag{3.23}$$

This is satisfied when

$$\begin{vmatrix}12EI_y/L^2 & -M\\ -M & (GJ + 12EI_w/L^2)\end{vmatrix} = 0, \tag{3.24}$$

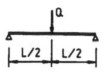

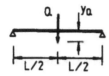

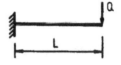

whence

$$M = \sqrt{\left\{\left(\frac{12EI_y}{L^2}\right)\left(GJ + \frac{12EI_w}{L^2}\right)\right\}} \qquad (3.25)$$

which is reasonably close to the exact solution (see Section 7.2.1)

$$M = M_{yz} = \sqrt{\left\{\left(\frac{\pi^2 EI_y}{L^2}\right)\left(GJ + \frac{\pi^2 EI_w}{L^2}\right)\right\}}. \qquad (3.26)$$

Alternatively, it may be assumed that

$$u/\delta = \phi/\theta = \sin \pi z/L \qquad (3.27)$$

which satisfies the kinematic boundary conditions of equation 3.21. In this case

$$\frac{1}{2}\delta^2 U = \frac{1}{2}\int_0^L \{[EI_y\delta^2(-\pi^2/L^2)^2 + EI_w\theta^2(-\pi^2/L^2)^2]\sin^2 \pi z/L$$

$$+ GJ\theta^2(\pi/L)^2 \cos^2 \pi z/L\} \, dz,$$

$$\frac{1}{2}\delta^2 V = \frac{1}{2}\int_0^L 2M\delta\theta(-\pi^2/L^2)\sin^2 \pi z/L \, dz.$$

Using the appropriate integrals from Table 3.1 and substituting leads to

$$\frac{1}{2}\frac{\pi^2}{2L}\begin{Bmatrix}\delta\\\theta\end{Bmatrix}^T\begin{bmatrix}\pi^2 EI_y/L^2 & -M\\ -M & (GJ + \pi^2 EI_w/L^2)\end{bmatrix}\begin{Bmatrix}\delta\\\theta\end{Bmatrix} = 0 \qquad (3.28)$$

which is satisfied by the exact solution of equation 3.26.

3.4.2 SIMPLE BEAM WITH A CENTRAL SHEAR CENTRE LOAD

For a simply supported doubly symmetric beam with a central concentrated load Q (Figure 3.5b), $M_x = Qz/2$ while $0 \leqslant z \leqslant L/2$. If the parabolic buckled shapes of equation 3.20 are assumed, and half of the symmetric beam is analysed, then

$$\frac{1}{2}\delta^2 U = \frac{1}{2}\{4EI_y\delta^2/L^3 + 4EI_w\theta^2/L^3 + GJ\theta^2/3L\}$$

as in section 3.4.1, while

$$\frac{1}{2}\delta^2 V = 2 \times \frac{1}{2}\int_0^{L/2}\{(Qz/2)(2\delta\theta)(z/L - z^2/L^2)(-2/L^2)\} \, dz = -5Q\delta\theta/96,$$

whence

$$\frac{1}{2}\frac{1}{3L}\begin{Bmatrix}\delta\\\theta\end{Bmatrix}^T\begin{bmatrix}12EI_y/L^2 & -5QL/32\\ -5QL/32 & (GJ + 12EI_w/L^2)\end{bmatrix}\begin{Bmatrix}\delta\\\theta\end{Bmatrix} = 0. \qquad (3.29)$$

so that

$$\frac{QL}{4} = 1.6 \sqrt{\left\{\left(\frac{12EI_y}{L^2}\right)\left(GJ + \frac{12EI_w}{L^2}\right)\right\}} \qquad (3.30)$$

which is reasonably close to the commonly accepted approximation

$$\frac{QL}{4} = 1.35 \sqrt{\left\{\left(\frac{\pi^2 EI_y}{L^2}\right)\left(GJ + \frac{\pi^2 EI_w}{L^2}\right)\right\}}. \qquad (3.31)$$

Alternatively, assuming the half sine wave buckled shapes of equation 3.27 leads to

$$\tfrac{1}{2}\delta^2 U = \tfrac{1}{2}(\pi^2/2L)\{\delta^2(\pi^2 EI_y/L^2) + \theta^2(GJ + \pi^2 EI_w/L^2)\}$$

as in section 3.4.1, and

$$\frac{1}{2}\delta^2 V = 2 \times \frac{1}{2}\int_0^{L/2} (Qz/2)(2\delta\theta)(-\pi^2/L^2)\sin^2 \pi z/L \, dz = -Q\delta\theta(\pi \times 0.2759)$$

after using the appropriate integral from Table 3.1. Thus the energy equation becomes

$$\frac{1}{2}\frac{\pi^2}{2L}\begin{Bmatrix}\delta\\\theta\end{Bmatrix}^T \begin{bmatrix} \pi^2 EI_y/L^2 & -(8 \times 0.2759/\pi)QL/4 \\ -(8 \times 0.2759/\pi)QL/4 & (GJ + \pi^2 EI_w/L^2) \end{bmatrix} \begin{Bmatrix}\delta\\\theta\end{Bmatrix} = 0, \quad (3.32)$$

so that

$$\frac{QL}{4} = 1.423 \sqrt{\left\{\left(\frac{\pi^2 EI_y}{L^2}\right)\left(GJ + \frac{\pi^2 EI_w}{L^2}\right)\right\}} \qquad (3.33)$$

which is within 5% of equation 3.31.

3.4.3 SIMPLE BEAM WITH AN OFF SHEAR CENTRE LOAD

If the central concentrated load of the simple beam analysed in section 3.4.2 acts at a distance y_Q below the centroidal axis as shown in Figure 3.5c, then the potential energy change during buckling must be increased to

$$\frac{1}{2}\delta^2 V = \frac{1}{2}\int_0^L 2M_x\phi u'' \, dz + \frac{1}{2}Qy_Q\phi_{L/2}^2 \qquad (3.34)$$

to account for the work done by the concentrated load Q moving upwards through the small distance $y_Q\phi_{L/2}^2/2$.

If the parabolic shapes of equation 3.20 are assumed, then this leads to

$$\frac{1}{2}\delta^2 V = \frac{-5Q\delta\theta}{96} + \frac{1}{2}Qy_Q\theta^2\left\{\frac{1}{2} - \left(\frac{1}{2}\right)^2\right\}^2,$$

whence

$$\frac{1}{2}\frac{1}{3L}\begin{Bmatrix}\delta\\\theta\end{Bmatrix}^T \begin{bmatrix} 12EI_y/L^2 & -5QL/32 \\ -5QL/32 & (GJ + 12EI_w/L^2 + 3QLy_Q/16) \end{bmatrix} \begin{Bmatrix}\delta\\\theta\end{Bmatrix} = 0 \quad (3.35)$$

so that

$$\frac{QL}{4} = 1.6\left\{\sqrt{\left[\left(\frac{12EI_y}{L^2}\right)\left(GJ + \frac{12EI_w}{L^2}\right) + \left(\frac{7.2EI_y y_Q}{L^2}\right)^2\right]} + \frac{7.2EI_y y_Q}{L^2}\right\}. \quad (3.36)$$

If the half sine wave buckled shapes of equation 3.27 are assumed, then this leads to

$$\tfrac{1}{2}\delta^2 V = - Q\delta\theta(\pi \times 0.2759) + \tfrac{1}{2}Qy_Q\theta^2,$$

whence

$$\frac{1}{2}\frac{\pi^2}{2L}\begin{Bmatrix}\delta\\\theta\end{Bmatrix}^T\begin{bmatrix}\pi^2 EI_y/L^2 & -(8 \times 0.2759/\pi)QL/4 \\ -(8 \times 0.2759/\pi)QL/4 & (GJ + \pi^2 EI_w/L^2 + 2QLy_Q/\pi^2)\end{bmatrix}\begin{Bmatrix}\delta\\\theta\end{Bmatrix} = 0. \quad (3.37)$$

The solution of this can be expressed as

$$\frac{QL}{4M_{yz}} = 1.423\left\{\sqrt{\left[1 + \left(0.577\frac{P_y y_Q}{M_{yz}}\right)^2\right]} + 0.577\frac{P_y y_Q}{M_{yz}}\right\}, \quad (3.38)$$

in which

$$M_{yz} = \sqrt{\{(\pi^2 EI_y/L^2)(GJ + \pi^2 EI_w/L^2)\}} \quad (3.39)$$

and

$$P_y = \pi^2 EI_y/L^2. \quad (3.40)$$

The approximation solution of equation 3.38 is quite close to the more accurate solution

$$\frac{QL}{4M_{yz}} = 1.35\left\{\sqrt{\left[1 + \left(\frac{0.54P_y y_Q}{M_{yz}}\right)^2\right]} + \frac{0.54P_y y_Q}{M_{yz}}\right\}, \quad (3.41)$$

obtained from section 7.6.1.

3.4.4 SIMPLE BEAM UNDER MOMENT GRADIENT

For the simply supported doubly symmetric beam with unequal end moments M and βM shown in Figure 3.6a,

$$M_x = M\{1 - (1 + \beta)z/L\}. \quad (3.42)$$

For uniform bending, $\beta = -1$, and the beam buckles with the half sine waves of equation 3.27, while for double curvature bending ($\beta = 1$), the deflected shape u is antisymmetrical, while the twist shape is symmetrical, as shown in Figure 3.6c.

This suggests that the buckled shapes may be approximated by

$$\left.\begin{matrix}\phi = \theta \sin \pi z/L, \\ u = \delta_1 \sin \pi z/L + \delta_2 \sin 2\pi z/L\end{matrix}\right\} \quad (3.43)$$

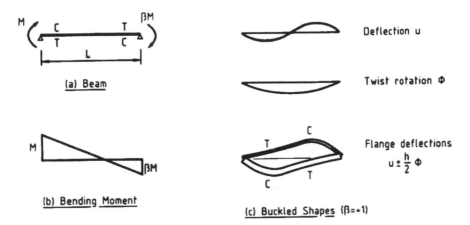

Figure 3.6 Beam under moment gradient.

so that $\delta_2 = 0$ when $\beta = -1$, and $\delta_1 = 0$ when $\beta = +1$. Equations 3.43 satisfy the kinematic boundary conditions of equation 3.21.

Using the buckled shapes of equation 3.43 leads to

$$\frac{1}{2}\int_0^L EI_y u''^2\,dz = \frac{1}{2}(\pi^4 EI_y/L^4)\int_0^L(\delta_1 \sin \pi z/L + 4\delta_2 \sin 2\pi z/L)^2\,dz$$

$$= \frac{1}{2}(\pi^2/L^2)(\pi^2 EI_y/L^2)(\delta_1^2 + 16\delta_2^2)L/2,$$

$$\frac{1}{2}\int_0^L(EI_w\phi''^2 + GJ\phi'^2)\,dz = \frac{1}{2}(\pi^2/L^2)(GJ + \pi^2 EI_w/L^2)\theta^2 L/2,$$

$$\frac{1}{2}\int_0^L 2M_x\phi u''\,dz = -\frac{1}{2}2M\theta(\pi^2/L^2)\int_0^L\left\{1 - (1+\beta)\frac{z}{L}\right\}$$

$$\times (\delta_1 \sin^2 \pi z/L + 4\delta_2 \sin \pi z/L \sin 2\pi z/L)\,dz$$

$$= -\frac{1}{2}(\pi^2/L^2)2M\theta\{0.5\delta_1 - (1+\beta)[0.7854\delta_1 - 4 \times 0.2829\delta_2]/\pi\}L,$$

so that the energy equation becomes

$$\frac{1}{2}\frac{\pi^2}{L^2}\frac{L}{2}\begin{Bmatrix}\delta_1\\\delta_2\\\theta\end{Bmatrix}^T\begin{bmatrix}\pi^2 EI_y/L^2 & 0 & -M\{1-(1+\beta)/2\}\\ 0 & 16\pi^2 EI_y/L^2 & -0.7205M(1+\beta)\\ -M\{1-(1+\beta)/2\} & -0.7205M(1+\beta) & (GJ+\pi^2 EI_w/L^2)\end{bmatrix}\begin{Bmatrix}\delta_1\\\delta_2\\\theta\end{Bmatrix} = 0. \tag{3.44}$$

This is satisfied when

$$GJ + \frac{\pi^2 EI_w}{L^2} = \frac{M^2\{1-(1+\beta)/2\}^2}{\pi^2 EI_y/L^2} + \frac{0.7205^2 M^2(1+\beta)^2}{16\pi^2 EI_y/L^2}, \tag{3.45}$$

whence

$$\frac{M^2}{M_{yz}^2} = \frac{1}{\{1-(1+\beta)/2\}^2 + \{0.1801(1+\beta)\}^2}.$$ (3.46)

This is quite close to the common approximation given by

$$\frac{M}{M_{yz}} = 1.75 + 1.05\beta + 0.3\beta^2 \leqslant 2.5$$ (3.47)

shown in Figure 7.12.

3.4.5 CANTILEVER WITH AN END LOAD

For a cantilever with an end load Q (Figure 3.5d), $M_x = -Q(L-z)$. If the parabolic buckled shapes

$$u/\delta = \phi/\theta = z^2/L^2$$ (3.48)

are assumed, which satisfy the boundary conditions

$$\left. \begin{array}{c} u_0 = u_0' = \phi_0 = 0, \\ \phi_0' = 0, \end{array} \right\}$$ (3.49a)

then the terms of the energy equation become

$$\left. \begin{array}{l} \frac{1}{2}\delta^2 U = \frac{1}{2}\int_0^L \{EI_y\delta^2(2/L^2)^2 + EI_w\theta^2(2/L^2)^2 + GJ\theta^2(2z/L^2)^2\}\,dz \\[2mm] = \frac{1}{2}\{4EI_y\delta^2/L^3 + 4EI_w\theta^2/L^3 + 4GJ\theta^2/3L\}, \\[2mm] \frac{1}{2}\delta^2 V = \frac{1}{2}\int_0^L 2(-QL)(1-z/L)\theta(z^2/L^2)\delta(2/L^2)\,dz = -Q\theta\delta/6. \end{array} \right\}$$ (3.49b)

Thus the energy equation can be written as

$$\frac{1}{2}\frac{1}{3L}\begin{Bmatrix}\delta\\\theta\end{Bmatrix}^T \begin{bmatrix} 12EI_y/L^2 & -QL/2 \\ -QL/2 & (4GJ+12EI_w/L^2) \end{bmatrix}\begin{Bmatrix}\delta\\\theta\end{Bmatrix} = 0$$ (3.50)

which is satisfied by

$$QL^2/\sqrt{(EI_yGJ)} = \sqrt{\{13.86^2 + 7.64^2(\pi^2EI_w/GJL^2)\}}.$$ (3.51)

This solution is substantially higher than the comparatively accurate approximation of

$$QL^2/\sqrt{(EI_yGJ)} = 3.95 + 3.52\sqrt{(\pi^2EI_w/GJL^2)}$$ (3.52)

obtained from section 9.3.1. This error arises because of a conflict in the boundary conditions at the support, where the kinematic condition of end warping prevented requires $\phi_0' = 0$ (equation 3.49a) for a cantilever with non-zero warping rigidity EI_w. However, for a cantilever with zero warping rigidity EI_w, such a boundary condition has no significance, and $\phi_0' \neq 0$ (a more suitable twist shape than equation 3.48 is used in section 3.5.2 to obtain a more accurate solution for $I_w = 0$). For cantilevers for which both GJ and EI_w are significant, the value of ϕ' usually increases very quickly from zero at the support, but then varies only slowly. In this case it is difficult to select suitable twist buckled shapes for hand analysis, and it is better to use a computer method to obtain solutions of reasonable accuracy.

In the case of a cantilever with $EI_w = 0$, a more accurate solution can be obtained by using the buckled shapes

$$u = \delta z^2/L^2, \tag{3.53a}$$

$$\phi = \theta z/L, \tag{3.53b}$$

for which

$$\tfrac{1}{2}\delta^2 U = \tfrac{1}{2}\{4EI_y\delta^2/L^3 + GJ\theta^2/L\}$$

and

$$\tfrac{1}{2}\delta^2 V = -Q\theta\delta/3.$$

Thus

$$\frac{1}{2L}\begin{Bmatrix}\delta\\\theta\end{Bmatrix}^{\mathrm{T}}\begin{bmatrix}4EI_y/L^2 & -QL/3\\-QL/3 & GJ\end{bmatrix}\begin{Bmatrix}\delta\\\theta\end{Bmatrix} = 0, \tag{3.54}$$

whence

$$QL^2/\sqrt{(EI_yGJ)} = 6 \tag{3.55}$$

which is much closer than equation 3.51 (with $EI_w = 0$) to equation 3.52.

3.5 Increasing the accuracy

3.5.1 USING THE MINOR AXIS BENDING EQUATION

The accuracy of the hand energy method of solving flexural-torsional beam buckling problems can often be significantly improved with comparatively little extra effort by making use of the minor axis bending differential equilibrium equation, which often takes the form (see also equation 2.105)

$$EI_y u'' = -M_x\phi \tag{3.56}$$

except when minor axis end restraints induce restraining end moments and shears. Equation 3.56 can be used to obtain

$$u'' = -M_x\phi/EI_y \tag{3.57}$$

so that a buckled shape for u is not required. In this case the accuracy of u'' is directly related to that of the guessed twist shape ϕ, and is often more accurate than the result of differentiating twice a guessed deflected shape u. This method is sometimes referred to as Timoshenko's Energy Method [1].

When equation 3.57 is substituted into the appropriate terms of the energy equation (equation 3.19), it leads to

$$\frac{1}{2}\delta^2 V = \frac{1}{2}\int_0^L 2M_x \phi u'' \, dz = -\frac{1}{2}\int_0^L 2(M_x^2 \phi^2/EI_y)\,dz \tag{3.58}$$

and

$$\frac{1}{2}\int_0^L EI_y(u'')^2 \, dz = \frac{1}{2}\int_0^L (M_x^2 \phi^2/EI_y)\,dz. \tag{3.59}$$

Equation 3.59 can be evaluated directly from equation 3.58, so that one less integration needs to be carried out.

For example, for the simple beam in uniform bending analysed in section 3.4.1, the assumption of $\phi/\theta = z/L - z^2/L^2$ leads to

$$\frac{1}{2}\int_0^L 2(M_x^2\phi^2/EI_y)\,dz = \frac{1}{2}(2M^2\theta^2/EI_y)\int_0^L (z/L - z^2/L^2)^2\,dz = \frac{1}{2}(2M^2\theta^2 L/30EI_y).$$

Thus the energy equation becomes

$$\frac{1}{2}\theta^2\{M^2 L/30EI_y + 4EI_w/L^3 + GJ/3L\} - \frac{1}{2}\theta^2\{2M^2 L/30EI_y)\} = 0, \tag{3.60}$$

whence

$$M = \sqrt{\left\{\left(\frac{10EI_z}{L^2}\right)\left(GJ + \frac{12EI_w}{L^2}\right)\right\}} \tag{3.61}$$

which is much closer than equation 3.25 to the exact solution of equation 3.26.

In the case of the simple beam with a central shear centre load analysed in section 3.4.2, the assumption of $\phi/\theta = z/L - z^2/L^2$ leads to

$$\int_0^{L/2} 2(M_x^2\phi^2/EI_y)\,dz = \int_0^{L/2} 2\{(Qz/2)^2\theta^2(z/L - z^2/L^2)^2/EI_y\}\,dz$$

$$= (Q^2\theta^2 L^3/EI_y)(29/26\,880).$$

Thus the energy equation becomes

$$\tfrac{1}{2}\theta^2\{(Q^2 L^3/EI_y)(29/53\,760) + 4EI_w/L^3 + GJ/3L\}$$
$$- \tfrac{1}{2}\theta^2\{(Q^2 L^3/EI_y)(29/26\,880)\} = 0, \tag{3.62}$$

whence

$$\frac{QL}{4} = 1.40\sqrt{\left\{\left(\frac{\pi^2 EI_z}{L^2}\right)\left(GJ + \frac{12EI_w}{L^2}\right)\right\}} \tag{3.63}$$

which is much closer than equation 3.30 to the accurate solution of equation 3.31.

3.5.2 EXTENDING THE BUCKLED SHAPES

The accuracy of the hand energy method can also be improved by extending the buckled shapes so that they satisfy the static boundary conditions as well as the kinematic boundary conditions (see section 3.3.1). This method generally increases the number of terms in the buckled shapes, and therefore increases the amount of effort required to reach a solution.

For example, the static boundary conditions for the simple beam in uniform bending analysed in section 3.4.1 are

$$u_0'' = u_L'' = 0 \qquad (3.64\text{a})$$

for zero end moments M_y, and

$$\phi_0'' = \phi_L'' = 0 \qquad (3.64\text{b})$$

for zero warping restraint. These and the kinematic boundary conditions of equation 3.2.1 are satisfied by the buckled shapes

$$u/\delta = \phi/\theta = z/L - 2z^3/L^3 + z^4/L^4 \qquad (3.65)$$

which have three terms instead of the two of equation 3.20 used in section 3.4.1 to satisfy the kinematic boundary conditions only. Substituting equation 3.65 in the energy equation leads to

$$\frac{1}{2}\begin{Bmatrix}\delta\\\theta\end{Bmatrix}^{\text{T}}\begin{bmatrix} 24EI_y/5L^3 & -17M/35L \\ -17M/35L & (17GJ/35L + 24EI_w/5L^3) \end{bmatrix}\begin{Bmatrix}\delta\\\theta\end{Bmatrix} = 0 \qquad (3.66)$$

which is satisfied when

$$M = \sqrt{\left\{\left(\frac{9.882EI_y}{L^2}\right)\left(GJ + \frac{9.882EI_w}{L^2}\right)\right\}} \qquad (3.67)$$

which is very close to the exact solution of equation 3.26.

In the case of the cantilever with $I_w = 0$ and an end load analysed in section 3.4.5, the static boundary condition of

$$\phi_L' = 0 \qquad (3.68)$$

for zero end torque M_z is satisfied by

$$\phi/\theta = 2z/L - z^2/L^2 \qquad (3.69)$$

which has two terms instead of the single term of equation 3.53b. Substituting equations 3.69 and 3.53a in the energy equation leads to

$$\frac{1}{2L}\begin{Bmatrix}\delta\\\theta\end{Bmatrix}^{\text{T}}\begin{bmatrix} 4EI_y/L^2 & -QL/2 \\ -QL/2 & 4GJ/3 \end{bmatrix}\begin{Bmatrix}\delta\\\theta\end{Bmatrix} = 0, \qquad (3.70)$$

which is satisfied when

$$QL^2/\sqrt{(EI_y GJ)} = 4.61$$

which is much closer than equation 3.55 to equation 3.52 (with $EI_w = 0$).

3.5.3 MINIMIZING THE BUCKLING LOAD

The accuracy of the hand energy method can be improved by extending the buckled shapes so that they contain an arbitrary parameter. The calculated buckling load then depends on the arbitrary parameter, and can be minimized to obtain the most accurate solution possible for all the buckled shapes defined in this way. This method requires an additional effort in the minimization process, which is generally not justified for a hand calculation.

In the case of the simple beam in uniform bending analysed in section 3.4.1, the parabolic buckled shapes of equation 3.20 can be extended to

$$u/\delta = \phi/\theta = (z/L - z^2/L^2) + \alpha(z/L - z^2/L^2)^2 \tag{3.71}$$

in which α is an arbitrary parameter. Substituting these in the energy equation leads to

$$\frac{1}{2L}\begin{Bmatrix} \delta \\ \theta \end{Bmatrix}^T \begin{bmatrix} (4 + 4\alpha^2/5)EI_y/L^2 & -M(1 + 2\alpha/5 + 2\alpha^2/35)/3 \\ -M(1 + 2\alpha/5 + 2\alpha^2/35)/3 & \{GJ(1 + 2\alpha/5 + 2\alpha^2/35)/3 + 4EI_w(1 + \alpha^2/5)L^2\} \end{bmatrix} \begin{Bmatrix} \delta \\ \theta \end{Bmatrix} = 0, \tag{3.72}$$

so that

$$\frac{M^2L^2}{EI_y} = \frac{12(1 + \alpha^2/5)\{GJ(1 + 2\alpha/5 + 2\alpha^2/35) + 12EI_w(1 + \alpha^2/5)/L^2\}}{(1 + 2\alpha/5 + 2\alpha^2/35)^2}. \tag{3.73}$$

Although this equation for M^2L^2/EI_y can be minimized formally with respect to α, much tedious manipulation can be avoided by programming equation 3.73 and finding a numerical minimum by trial and error. Such a procedure for the two limiting cases of $GJ = 0$ and $EI_w = 0$ discloses minima for $\alpha = 1.0759$, and substituting this into equation 3.73 leads to

$$M = \sqrt{\left\{\frac{9.875EI_y}{L^2}\left(GJ + \frac{9.875EI_w}{L^2}\right)\right\}} \tag{3.74}$$

which is very close to the exact solution of equation 3.26.

3.6 Problems

PROBLEM 3.1

A uniform simply supported column has a compression load Q acting at one end, and a second compression load Q acting at mid-height, which is unbraced, as shown in Figure 3.7a. Determine approximate values of PL^2/π^2EI at elastic flexural buckling,

(a) by assuming $u = 4\delta(z/L - z^2/L^2)$;
(b) by assuming $u = \delta \sin \pi z/L$;
(c) by assuming an asymmetrical buckled shape which reflects the asymmetrical loading;
(d) by assuming a buckled shape which satisfies both the static and the kinematic boundary conditions; and

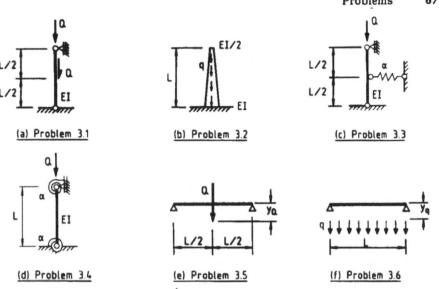

Figure 3.7 Problems 3.1–3.6.

(e) by assuming a variable buckled shape with an undetermined parameter, and minimizing the calculated buckling load.

PROBLEM 3.2

A non-uniform column is built-in at its base and free at its top, and has compression loads per unit length q uniformly distributed down its length, as shown in Figure 3.7b. The flexural rigidity varies linearly from EI at the bottom to $EI/2$ at the top. Determine an approximate value of $4qL^3/\pi^2 EI$ at elastic flexural buckling.

PROBLEM 3.3

A uniform simply supported column has a compression load Q at one end and an elastic translational restraint at mid-height which exerts a restraining force $\alpha u_{L/2}$, as shown in Figure 3.7c. Determine the variation of $QL^2/\pi^2 EI$ at flexural buckling with the dimensionless brace stiffness $\alpha L^3/EI$.

PROBLEM 3.4

A simply supported column has a compression load Q and elastic rotational restraints at both ends which exert restraining moments $\alpha u'_{0,L}$, as shown in Figure 3.7d. Determine the variation of $QL^2/\pi^2 EI$ at flexural buckling with the dimensionless brace stiffness $\alpha L/EI$.

PROBLEM 3.5

A simply supported monosymmetric beam has a central concentrated load Q acting a distance y_Q below the centroid at mid-span, which is unbraced, as shown in Figure 3.7e. Determine the variation of $QL/4M_{yz}$ at elastic flexural-torsional buckling with $y_Q P_y/M_{yz}$ and $\beta_x P_y/M_{yz}$,

  (a) by assuming parabolic shapes for u and ϕ;
  (b) by assuming half sine waves for u and ϕ;
  (c) by assuming a parabolic shape for ϕ and using the minor axis bending equation; and
  (d) by assuming shapes which satisfy both the static and the kinematic boundary conditions.

PROBLEM 3.6

A simply supported doubly symmetric beam ($\beta_x = 0$) has a uniformly distributed load per unit length q acting at a distance y_q below the centroid, as shown in Figure 3.7f. Determine the variation of $qL^2/8M_{yz}$ at elastic flexural-torsional buckling with $y_q P_y/M_{yz}$,

  (a) by assuming parabolic shapes for u and ϕ;
  (b) by assuming half sine waves for u and ϕ;
  (c) by assuming a parabolic shape for ϕ and using the minor axis bending equation; and
  (d) by assuming shapes which satisfy both the static and the kinematic boundary conditions.

3.7 References

1. Timoshenko, S.P. and Gere, J.M. (1961) *Theory of Elastic Stability*, 2nd edition, McGraw-Hill, New York.
2. Temple, G. and Bickley, W.G. (1956) *Rayleigh's Principle and Its Applications to Engineering*, Dover Publications, New York.
3. Bleich, F. (1952) *Buckling Strength of Metal Structures*, McGraw-Hill, New York.

4   Finite element buckling analysis

4.1 General

In Chapter 3, the energy method of analysing flexural-torsional buckling was presented in a form which is suitable for hand use. The need to avoid excessive computational effort by the analyst limits the use of this hand method to problems which are comparatively simple and for which solutions of moderate accuracy are acceptable.

This chapter presents a computer method of analysing flexural-torsional buckling which is capable of very high accuracy, and which can be applied to a very wide range of structures. The method is the same energy method as that discussed in section 2.5 and used in Chapter 3, but is presented in the form of the very widely used finite element method [1–5].

The application of the finite element method to elastic buckling problems [6] is discussed generally in section 4.2, and then particularized to the flexural-torsional buckling of columns in section 4.3, and to the flexural-torsional buckling of monosymmetric beam-columns in section 4.4.

The results of many finite element analyses of the flexural-torsional buckling of columns, beams, cantilevers, continuous beams, and beam-columns are included in Chapters 5–11. Extensions of the finite element method of elastic buckling analysis are discussed in Chapter 12 on frame buckling, in Chapter 13 on the buckling of arches and rings, and in Chapter 14 on inelastic buckling.

4.2 The finite element method for buckling analysis

The application of the finite element method to buckling problems in general [1–6], and to flexural-torsional buckling problems in particular [7–10] involves the replacement of the quantities in the energy equation for the complete structure

$$\tfrac{1}{2}(\delta^2 U + \delta^2 V) = 0 \tag{4.1}$$

by the sums of the approximations $\tfrac{1}{2}\delta^2 U_e, \tfrac{1}{2}\lambda\delta^2 V_e$ for the strain energies stored in and work done on each of a finite number of elements into which the structure is divided, so that

$$\tfrac{1}{2}\sum(\delta^2 U_e + \lambda\delta^2 V_e) = 0 \tag{4.2}$$

in which $\tfrac{1}{2}\delta^2 V_e$ represents the work that would be done on an initial load set, and λ is the buckling load factor by which the initial load set must be multiplied to obtain the buckling load set.

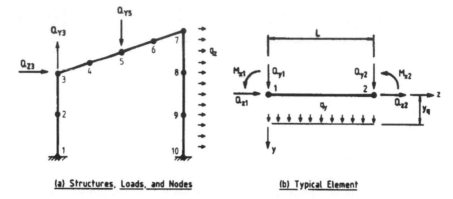

(a) Structures, Loads, and Nodes   (b) Typical Element

Figure 4.1 Division of a structure into elements.

The first step is therefore to divide the complete structure into a number of finite elements connected at nodes, as for example in Figure 4.1. Approximations are then made for the variations of the buckling deformations u, v, ϕ over the length of a typical element, which are then related to the element values $\{\delta_e\}$ of the buckling nodal deformations. This allows the element strain energy stored to be written as

$$\tfrac{1}{2}\delta^2 U_e = \tfrac{1}{2}\{\delta_e\}^{\mathrm{T}}[k_e]\{\delta_e\} \tag{4.3}$$

and the element work done as

$$\tfrac{1}{2}\lambda\delta^2 V_e = \tfrac{1}{2}\lambda\{\delta_e\}^{\mathrm{T}}[g_e]\{\delta_e\} \tag{4.4}$$

in which $[k_e]$ is the element stiffness matrix, and $[g_e]$ is the element stability matrix associated with the initial load set.

To facilitate the summations of equation 4.2, the element nodal buckling deformations $\{\delta_e\}$, which are defined in relation to the local element axis system, are transformed to the global nodal buckling deformations $\{\Delta\}$ at all the nodes defined in relation to the structure axis system. These transformations take the form of

$$\{\delta_e\} = [T_e]\{\Delta\} \tag{4.5}$$

in which $[T_e]$ is a transformation matrix.

This allows the element local stiffness and stability matrices to be transformed to the element global matrices

$$\left.\begin{aligned}
[K_e] &= \{T_e\}^{\mathrm{T}}[k_e]\{T_e\}, \\
[G_e] &= \{T_e\}^{\mathrm{T}}[g_e]\{T_e\},
\end{aligned}\right\} \tag{4.6}$$

so that

$$\left.\begin{aligned}
\tfrac{1}{2}\delta^2 U_e &= \tfrac{1}{2}\{\Delta\}^{\mathrm{T}}[K_e]\{\Delta\}, \\
\tfrac{1}{2}\lambda\delta^2 V_e &= \tfrac{1}{2}\lambda\{\Delta\}^{\mathrm{T}}[G_e]\{\Delta\}.
\end{aligned}\right\} \tag{4.7}$$

These element global matrices can then be summed to form the structure stiffness and stability matrices

$$\left.\begin{array}{l}[K] = \sum[K_e],\\ [G] = \sum[G_e],\end{array}\right\} \tag{4.8}$$

so that the energy equation (equation 4.2) becomes

$$\tfrac{1}{2}\{\Delta\}^{\mathrm{T}}([K] + \lambda[G])\{\Delta\} = 0. \tag{4.9}$$

This summation may be carried out in stages after each element matrix $[k_e]$ or $[g_e]$ is transformed.

The boundary (support) conditions of the structure will require that some of the global nodal buckling deformations $\{\Delta\}$ are zero. In this case, the global stiffness and stability matrices $[K]$, $[G]$ are reduced by eliminating the rows and columns corresponding to the zero buckling deformations. These reductions may be done in stages at each of the nodal deformation transformations of equation 4.5.

Equation 4.9 is described as a 'generalized linear eigen-problem' [4]. When the matrices are of order n, then it has n eigenvalues λ_n and n non-zero eigenvectors $\{\Delta\}_n$ which satisfy it. The lowest eigenvalue λ_1 defines the load set at which the structure first buckles, and the corresponding eigenvector $\{\Delta\}_1$ defines the corresponding buckled shape or buckling mode of the structure. Numerical procedures for determining the lowest eigenvalue λ_1 and the corresponding eigenvector $\{\Delta\}_1$ are discussed in [2, 4, 5].

Alternatively, the eigenvalues λ_n can be found by finding the roots of the determinantal equation

$$|K + \lambda G| = 0. \tag{4.10}$$

Methods of finding the lowest root are discussed in [11, 12].

4.3 Flexural-torsional buckling of columns

4.3.1 BUCKLING DEFORMATION FIELDS

A wide range of approximate deformation fields may be assumed for the shear centre buckling deformations u, v, ϕ which occur during the flexural-torsional buckling of a column. It is most common to use cubic deformation fields of the form

$$u = a_0 + a_1(z/L) + a_2(z^2/L^2) + a_3(z^3/L^3) \tag{4.11}$$

in which L is the length of the element (Figure 4.1b) and z is the distance along the element. Equation 4.11 can be written as

$$u = \{Z\}^{\mathrm{T}}\{a\} \tag{4.12}$$

in which

$$\{Z\}^{\mathsf{T}} = \{1, z/L, z^2/L^2, z^3/L^3\} \tag{4.13}$$

and

$$\{a\} = \{a_0, a_1, a_2, a_3\}^{\mathsf{T}}. \tag{4.14}$$

This simple field has four undetermined constants $a_0 - a_3$. These can be replaced by the four nodal deformations

$$\{\delta_u\} = \{u_1, u_2, u_1', u_2'\}^{\mathsf{T}} \tag{4.15}$$

at the ends 1, 2 of the element by modifying the field formulation. This is done by making appropriate differentiations and substitutions so that

$$\{\delta_u\} = [C_u]\{a\} \tag{4.16}$$

in which

$$[C_u] = \begin{bmatrix} 1 & 0 & 0 & 0 \\ 1 & 1 & 1 & 1 \\ 0 & 1/L & 0 & 0 \\ 0 & 1/L & 2/L & 3/L \end{bmatrix}. \tag{4.17}$$

Equation 4.16 can be inverted as

$$\{a\} = [C_u]^{-1}\{\delta_u\} \tag{4.18}$$

in which

$$[C_u]^{-1} = \begin{bmatrix} 1 & 0 & 0 & 0 \\ 0 & 0 & L & 0 \\ -3 & 3 & -2L & -L \\ 2 & -2 & L & L \end{bmatrix}. \tag{4.19}$$

Thus the deformation field can be rewritten as

$$u = \{Z\}^{\mathsf{T}}[C_u]^{-1}\{\delta_u\}. \tag{4.20}$$

Similar cubics can be used to represent the other buckling deformations v, ϕ, so that

$$\{u\} = [M][C]^{-1}\{\delta_e\}, \tag{4.21}$$

in which

$$\{u\} = \{u, v, \phi\}^{\mathsf{T}}, \tag{4.22}$$

$$\{\delta_e\} = \{u_1, u_2, u_1', u_2', v_1, v_2, v_1', v_2', \phi_1, \phi_2, \phi_1', \phi_2'\}^{\mathsf{T}}, \tag{4.23}$$

$$[M] = \begin{bmatrix} \{Z\}^{\mathsf{T}} & \{0\}^{\mathsf{T}} & \{0\}^{\mathsf{T}} \\ \{0\}^{\mathsf{T}} & \{Z\}^{\mathsf{T}} & \{0\}^{\mathsf{T}} \\ \{0\}^{\mathsf{T}} & \{0\}^{\mathsf{T}} & \{Z\}^{\mathsf{T}} \end{bmatrix} \tag{4.24}$$

and

$$[C]^{-1} = \begin{bmatrix} [C_u]^{-1} & [0] & [0] \\ [0] & [C_u]^{-1} & [0] \\ [0] & [0] & [C_u]^{-1} \end{bmatrix}. \tag{4.25}$$

4.3.2 ELEMENT STIFFNESS MATRIX

The increase in the strain energy stored in an element length L during column flexural-torsional buckling (section 2.8.3.1)

$$\tfrac{1}{2}\delta^2 U_e = \tfrac{1}{2} \int_0^L \{ EI_y u''^2 + EI_x v''^2 + EI_w \phi''^2 + GJ\phi'^2 \} dz \tag{4.26}$$

can be expressed in the form of Equation 4.3 by first transforming it to

$$\tfrac{1}{2}\delta^2 U_e = \tfrac{1}{2} \int_0^L \{\varepsilon_u\}^T [D_u]\{\varepsilon_u\} dz \tag{4.27}$$

in which $\{\varepsilon_u\}$ is a generalized strain vector

$$\{\varepsilon_u\} = \{u'', v'', \phi', -\phi''\}^T \tag{4.28}$$

and $[D_u]$ is a generalized elasticity matrix

$$[D_u] = \begin{bmatrix} EI_y & 0 & 0 & 0 \\ 0 & EI_x & 0 & 0 \\ 0 & 0 & GJ & 0 \\ 0 & 0 & 0 & EI_w \end{bmatrix}. \tag{4.29}$$

The strain vector $\{\varepsilon_u\}$ can be obtained by appropriate differentiation of the deformation fields of Equation 4.21 as

$$\{\varepsilon_u\} = [B_u][C]^{-1}\{\delta_e\}, \tag{4.30}$$

in which

$$[B_u] = \begin{bmatrix} \{Z''\}^T & \{0\}^T & \{0\}^T \\ \{0\}^T & \{Z''\}^T & \{0\}^T \\ \{0\}^T & \{0\}^T & \{Z'\}^T \\ \{0\}^T & \{0\}^T & -\{Z''\}^T \end{bmatrix}, \tag{4.31}$$

and

$$\{Z'\}^T = \{0, 1/L, 2z/L^2, 3z^2/L^3\} \tag{4.32}$$
$$\{Z''\}^T = \{0, 0, 2/L^2, 6z/L^3\}. \tag{4.33}$$

Substituting equation 4.30 into equation 4.27 leads to

$$\tfrac{1}{2}\delta^2 U_e = \tfrac{1}{2}\{\delta_e\}^T [C]^{-T} \left(\int_0^L [B_u]^T [D_u][B_u] dz \right) [C]^{-1}\{\delta_e\} \tag{4.34}$$

and this is identical with equation 4.3 if the element stiffness matrix is written as

$$[k_e] = [C]^{-\mathrm{T}} \left(\int_0^L [B_u]^{\mathrm{T}} [D_u] [B_u] \, dz \right) [C]^{-1}. \tag{4.35}$$

4.3.3 ELEMENT STABILITY MATRIX

The work done on an axial force acting on an element of length L during flexural-torsional buckling (section 2.8.3.1) can be expressed as

$$\frac{1}{2} \lambda \delta^2 V_e = -\frac{1}{2} \lambda \int_0^L P \{ u'^2 + v'^2 + (I_P/A + x_0^2 + y_0^2) \phi'^2 - 2x_0 v' \phi' + 2y_0 u' \phi' \} \, dz \tag{4.36}$$

in which P is the initial axial compression force and λ is the load factor by which the initial force must be multiplied to cause column buckling.

Equation 4.36 can be transformed to

$$\frac{1}{2} \lambda \delta^2 V_e = \frac{1}{2} \lambda \int_0^L \{ \varepsilon_v \}^{\mathrm{T}} [D_v] \{ \varepsilon_v \} \, dz \tag{4.37}$$

in which $\{ \varepsilon_v \}$ is another generalized strain vector

$$\{ \varepsilon_v \} = \{ u', v', \phi' \}^{\mathrm{T}} \tag{4.38}$$

and $[D_v]$ is a generalized initial stress matrix

$$[D_v] = - \begin{bmatrix} P & 0 & Py_0 \\ 0 & P & -Px_0 \\ Py_0 & -Px_0 & P(I_P/A + x_0^2 + y_0^2) \end{bmatrix}. \tag{4.39}$$

The strain vector $\{ \varepsilon_v \}$ can be obtained by appropriate differentiation of the deformation fields of equation 4.21 as

$$\{ \varepsilon_v \} = [B_v][C]^{-1} \{ \delta_e \} \tag{4.40}$$

in which

$$[B_v] = \begin{bmatrix} \{Z'\}^{\mathrm{T}} & \{0\}^{\mathrm{T}} & \{0\}^{\mathrm{T}} \\ \{0\}^{\mathrm{T}} & \{Z'\}^{\mathrm{T}} & \{0\}^{\mathrm{T}} \\ \{0\}^{\mathrm{T}} & \{0\}^{\mathrm{T}} & \{Z'\}^{\mathrm{T}} \end{bmatrix}. \tag{4.41}$$

Substituting equation 4.40 into equation 4.37 leads to

$$\frac{1}{2} \lambda \delta^2 V_e = \frac{1}{2} \lambda \{ \delta_e \}^{\mathrm{T}} [C]^{-\mathrm{T}} \left(\int_0^L [B_v]^{\mathrm{T}} [D_v] [B_v] \, dz \right) [C]^{-1} \{ \delta_e \} \tag{4.42}$$

and this is identical with equation 4.4 if the element stability matrix is written as

$$[g_e] = [C]^{-\mathrm{T}} \left(\int_0^L [B_v]^{\mathrm{T}} [D_v] [B_v] \, dz \right) [C]^{-1}. \tag{4.43}$$

4.3.4 DEFORMATION TRANSFORMATIONS

Before summing the strain energy changes in and work done on all the elements of a structure, the element nodal buckling deformations $\{\delta_e\}$ are transformed from the local element axis system to a global system which is common for all the structure.

Usually, the elements of a structure subject to column flexural-torsional buckling have collinear z axes (structures whose elements are not collinear usually have bending moments), although their local x, y axes may be inclined at angles α to the global X, Y axes, as shown in Figure 4.2. In this case, the element nodal buckling deformations

$$\{\delta_{en}\} = \{u_n, u'_n, v_n, v'_n, \phi_n, \phi'_n\}^{\mathrm{T}} \tag{4.44}$$

at node n are related to the global buckling nodal deformations

$$\{\Delta\} = \{\{\Delta_1\}^{\mathrm{T}}, \{\Delta_2\}^{\mathrm{T}}, \cdots \{\Delta_n\}^{\mathrm{T}}, \cdots \}^{\mathrm{T}} \tag{4.45}$$

in which

$$\{\Delta_n\} = \{U_n, \theta_{Yn}, V_n, -\theta_{Xn}, \theta_{Zn}, \theta_{Zn}'\}^{\mathrm{T}} \tag{4.46}$$

by

$$\{\delta_{en}\} = [T_{en}]\{\Delta_n\} \tag{4.47}$$

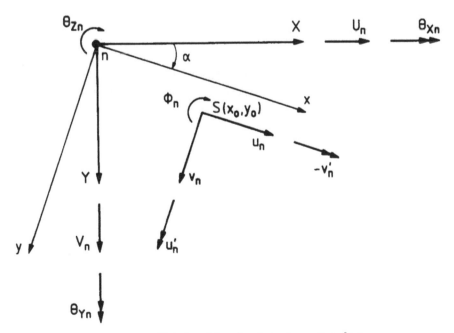

Figure 4.2 Global and local axis systems at node n.

in which

$$[T_{en}] = \begin{bmatrix} \cos\alpha & 0 & \sin\alpha & 0 & -y_0 & 0 \\ 0 & \cos\alpha & 0 & \sin\alpha & 0 & -y_0 \\ -\sin\alpha & 0 & \cos\alpha & 0 & x_0 & 0 \\ 0 & -\sin\alpha & 0 & \cos\alpha & 0 & x_0 \\ 0 & 0 & 0 & 0 & 1 & 0 \\ 0 & 0 & 0 & 0 & 0 & 1 \end{bmatrix}. \tag{4.48}$$

Most of the deformation transformations of equations 4.44–4.48 are obvious, except for the transformation

$$\phi'_n = \theta'_{zn}. \tag{4.49}$$

Physically, this can only be interpreted in terms of compatibility of the end warping displacements $w = \omega\phi'$ of two elements which have a common node. Equation 4.49 can only represent this warping compatibility if there is no conflict between the warping functions ω of the two elements in those regions of the XY plane where the cross-sections of the two elements coincide. This condition is most commonly satisfied when the two elements have identical and coinciding cross-sections. When this is not the case, then continuity between the non-coinciding cross-sections would normally only be met by the provision of a rigid node element that prevents the warping ($\phi' = 0$) of each element.

4.3.5 BOUNDARY CONDITIONS

The boundary conditions for the flexural-torsional buckling of a column usually require one or more of the global nodal deformations $\{\Delta\}$ (see equation 4.46) to be zero, corresponding to the prevention of the deformation. There should always be a sufficient number of boundary conditions to prevent rigid body deflections (U, V) or rotations $(\theta_Y, \theta_X, \theta_Z)$ of the column, which would occur under zero loads. For example, at one end of a cantilever, the boundary conditions

$$\{U_0, \theta_{Y0}, V_0, \theta_{X0}, \theta_{Z0}\}^T = \{0\} \tag{4.50}$$

are sufficient to prevent rigid body buckling motions.

The effects of the boundary conditions are allowed for by eliminating the rows and columns of the global stiffness and stability matrices corresponding to the zero nodal buckling deformations, usually as part of the nodal deformation transformations described in section 4.3.4.

Sometimes, boundary conditions are specified for an off-axis point $O_B(X_B, Y_B)$ and directions α_B which do not coincide with those of the global XYZ axis system, as shown in Figure 4.3. In such a case, the global nodal deformations $\{\Delta_n\}$ of equation 4.46 can be transformed to the corresponding deformations

$$\{\Delta_{nB}\} = \{U_B, \theta_{YB}, V_B, -\theta_{XB}, \theta_{ZB}, \theta_{ZB}'\}^T \tag{4.51}$$

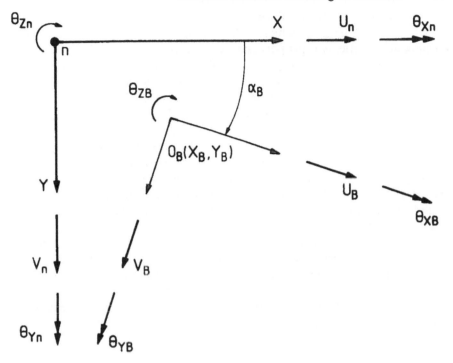

Figure 4.3 Off-axis boundary conditions.

by

$$\{\Delta_n\} = [T_{nB}] \, \{\Delta_{nB}\} \tag{4.52}$$

in which

$$[T_{nB}] = \begin{bmatrix} \cos\alpha_B & 0 & -\sin\alpha_B & 0 & Y_B\cos\alpha_B + X_B\sin\alpha_B & 0 \\ 0 & \cos\alpha_B & 0 & -\sin\alpha_B & 0 & Y_B\cos\alpha_B + X_B\sin\alpha_B \\ \sin\alpha_B & 0 & \cos\alpha_B & 0 & Y_B\sin\alpha_B - X_B\cos\alpha_B & 0 \\ 0 & \sin\alpha_B & 0 & \cos\alpha_B & 0 & -Y_B\sin\alpha_B + X_B\cos\alpha_B \\ 0 & 0 & 0 & 0 & 1 & 0 \\ 0 & 0 & 0 & 0 & 0 & 1 \end{bmatrix}. \tag{4.53}$$

4.3.6 INTERNAL HINGES

Some structures may have internal hinges which allow independent element nodal deformations at a common node. For example, flexural hinges ensure that the element moments M_x or M_y at the hinge are zero, so that the rotations v', u' of each element at the hinge are unrestrained. Other examples of hinges include torsional hinges, which ensure zero values of the torques M_z at the hinge and allow the twist rotations ϕ of each element to be unrestrained, and warping hinges, which ensure zero values of the bimoments B at the hinge and allow warping displacements w proportional to ϕ' to be unrestrained.

It should be noted that the physical conditions in a practical structure which allow a particular type of a hinge often imply other types as well. For example a torsional hinge ($M_z = 0$) often implies flexural hinges ($M_x, M_y = 0$), while flexural hinges ($M_x, M_y = 0$) often imply a warping hinge ($B = 0$), and vice versa.

In static problems without stability effects ($[G] = [0]$), internal hinges are usually allowed for by condensation techniques [2, 4], in which the free nodal deformation associated with the hinge is eliminated from the element stiffness matrix $[k_e]$. The use of such a technique in a stability problem will convert the linear eigen-problem of equation 4.9 into a non-linear eigen-problem, which will invalidate the numerical procedures commonly used to extract the eigenvalue and the eigenvector [2, 4, 5].

Internal hinges in stability problems are therefore best handled by increasing the global nodal deformation vector $\{\Delta\}$ by the element nodal deformations δ_e allowed by the hinges, and by modifying the deformation transformation matrix $[T_{en}]$ (see equation 4.48) appropriately. However, care must be taken not to create an unconstrained global nodal deformation, as this will lead to a corresponding rigid body deformation under zero load.

4.4 Flexural-torsional buckling of monosymmetric beam-columns

4.4.1 BUCKLING DEFORMATION FIELDS

Cubic buckling deformation fields (see section 4.3.1) may also be assumed for the buckling deformations u, ϕ of a monosymmetric beam-column ($x_0 = 0$). These are given by equation 4.21,

$$\{u\} = [M][C]^{-1}\{\delta_e\}$$

in which

$$\{u\} = \{u, \phi\}^T, \tag{4.54}$$

$$\{\delta_e\} = \{u_1, u_2, u'_1, u'_2, \phi_1, \phi_2, \phi'_1, \phi'_2\}^T \tag{4.55}$$

$$[M] = \begin{bmatrix} \{Z\}^T & \{0\}^T \\ \{0\}^T & \{Z\}^T \end{bmatrix} \tag{4.56}$$

and

$$[C]^{-1} = \begin{bmatrix} [C_u]^{-1} & [0] \\ [0] & [C_u]^{-1} \end{bmatrix} \tag{4.57}$$

in which $\{Z\}^T$ is given by equation 4.13, and $[C_u]^{-1}$ by equation 4.19.

4.4.2 ELEMENT STIFFNESS MATRIX

The increase in the strain energy stored in an element of length L during beam-column flexural-torsional buckling (section 2.8.5.1)

$$\frac{1}{2}\delta^2 U_e = \frac{1}{2}\int_0^L \{EI_y u''^2 + EI_w \phi''^2 + GJ\phi'^2\}\,dz \qquad (4.58)$$

can be expressed in the form of equation 4.3 by using

$$\{\varepsilon_u\} = \{u'', \phi', -\phi''\}^T, \qquad (4.59)$$

$$\{D_u\} = \begin{bmatrix} EI_y & 0 & 0 \\ 0 & GJ & 0 \\ 0 & 0 & EI_w \end{bmatrix}, \qquad (4.60)$$

so that

$$\{\varepsilon_u\} = [B_u][C]^{-1}\{\delta_e\}$$

in which

$$[B_u] = \begin{bmatrix} \{Z''\}^T & \{0\}^T & \{0\}^T \\ \{0\}^T & \{Z'\}^T & \{0\}^T \\ \{0\}^T & \{0\}^T & -\{Z''\}^T \end{bmatrix}, \qquad (4.61)$$

which leads to equation 4.35,

$$[k_e] = [C]^{-T}\left(\int_0^L [B_u]^T [D_u][B_u]\,dz\right)[C]^{-1}$$

in the same way as for the flexural-torsional buckling of columns.

4.4.3 ELEMENT STABILITY MATRIX

The work done on the actions acting on an element of length L during flexural-torsional buckling (sections 2.8.5.1 and 17.4.7) can be expressed as

$$\frac{1}{2}\lambda\delta^2 V_e = -\frac{1}{2}\lambda\int_0^L P\{u'^2 + (I_P/A + y_0^2)\phi'^2 + 2y_0 u'\phi'\}\,dz$$
$$+ \frac{1}{2}\lambda\int_0^L M_x\{2\phi u'' + (I_{Px}/I_x - 2y_0)\phi'^2\}\,dz + \frac{1}{2}\lambda\int_0^L q(y_q - y_0)\phi^2\,dz$$
$$(4.62)$$

in which P, M_x, and q are the initial axial compression force, bending moment, and distributed load, and λ is the load factor by which the initial force, moment, and distributed load must be multiplied to cause beam-column buckling.

Equation 4.62 can be expressed in the form of equation 4.4 by using

$$\{\varepsilon_v\} = \{u', u'', \phi, \phi'\}^T, \tag{4.63}$$

$$[D_u] = \begin{bmatrix} -P & 0 & 0 & -Py_0 \\ 0 & 0 & M_x & 0 \\ 0 & M_x & q(y_q - y_0) & 0 \\ -Py_0 & 0 & 0 & -P(I_P/A + y_0^2) + M_x(I_{Px}/I_x - 2y_0) \end{bmatrix}, \tag{4.64}$$

$$\{\varepsilon_v\} = [B_v][C]^{-1}\{\delta_e\},$$

$$[B_v] = \begin{bmatrix} \{Z'\}^T & \{0\}^T & \{0\}^T & \{0\}^T \\ \{0\}^T & \{Z''\}^T & \{0\}^T & \{0\}^T \\ \{0\}^T & \{0\}^T & \{Z\}^T & \{0\}^T \\ \{0\}^T & \{0\}^T & \{0\}^T & \{Z'\}^T \end{bmatrix}, \tag{4.65}$$

which leads to equation 4.43,

$$[g_e] = [C]^{-T} \left(\int_0^L [B_v]^T [D_v][B_v]\,dz \right)[C]^{-1}$$

in the same way as for the flexural-torsional buckling of columns.

4.4.4 DEFORMATION TRANSFORMATIONS

In many cases, the elements of a structure subjected to beam-column flexural-torsional buckling have collinear z axes and parallel y axes (structures with elements whose y axes are not parallel have biaxial bending actions). In these cases the element nodal buckling deformations

$$\{\delta_{en}\} = \{u_n, u'_n, \phi_n, \phi'_n\}^T \tag{4.66}$$

are identical with the element global buckling deformations

$$\{\Delta_n\} = \{U_n, \theta_{Yn}, \theta_{Zn}, \theta_{Zn}'\}^T \tag{4.67}$$

and the deformation transformations of equation 4.47

$$\{\delta_{en}\} = [T_{en}]\{\Delta_n\}$$

to the global nodal deformations of equation 4.45

$$\{\Delta\} = \{\{\Delta_1\}^T, \{\Delta_2\}^T, \cdots \{\Delta_n\}^T, \cdots \}^T$$

become very simple.

The transformation of equation 4.49

$$\phi_n' = \theta_{Zn}'$$

again can only be interpreted in terms of the compatibility of the warping displacements $w = \omega\phi'$ at a node common to two elements, and can only represent this warping compatibility if there is no conflict between the warping functions ω for the two elements.

4.4.5 OFF SHEAR CENTRE NODAL LOADS

When off shear centre loads λQ act at distances y_Q below the centroidal axis, additional work

$$\frac{1}{2}\lambda\sum\delta^2 V_{yQ} = \frac{1}{2}\lambda\sum Q_n(y_Q - y_0)_n\phi_n^2 \qquad (4.68)$$

is done which must be included in the energy equation (equation 4.2). If there are nodes at all of the load points, then the element twist rotation ϕ_n at each load point corresponds directly to the global nodal deformation θ_{zn} at the load point. Thus the effects of off shear centre loads λQ can be included by adding the terms $Q_n(y_Q - y_0)_n$ to the diagonal elements of the global stability matrix $[G]$ of equation 4.9 which correspond to ϕ_n.

4.4.6 BOUNDARY CONDITIONS AND INTERNAL HINGES

The discussion of section 4.3.5 of the boundary conditions for the flexural-torsional buckling of columns, including off-axis boundary conditions, also applies to the flexural-torsional buckling of monosymmetric beam-columns, with some simplifications that result from the elimination of V, θ_x from the global buckling deformations.

Similarly, the discussion in section 4.3.6 of internal hinges also applies to monosymmetric beam-columns.

4.5 Pre-buckling analysis

4.5.1 GENERAL

For the flexural-torsional buckling analysis of a structure, the distributions of the axial compressive forces P and in-plane bending moments M_x must be known. For statically determinate structures, these can be determined by statics.

However, for statically indeterminate structures, such as propped cantilevers, continuous beams, portal frames and the like, the in-plane pre-buckling deflections of the structure under load must be analysed. It is appropriate to use a computer method for this in-plane analysis when a computer method is being used for the out-of-plane buckling analysis. Usually it will be sufficient to carry out a first-order linear elastic in-plane analysis [13], for which in-plane stability effects are ignored. When these need to be considered, then a second-order non-linear elastic in-plane analysis [14] should be made. If inelastic buckling is being considered, then an inelastic in-plane analysis should be made [15].

When a finite element method is used to analyse flexural-torsional buckling, it is logical to use a finite element method for the in-plane pre-buckling analysis. A first-order linear finite element method of in-plane analysis is described below.

For the first-order in-plane analysis of an elastic structure, the conditions of equilibrium can be expressed using the principle of stationary total potential derived from the principle of virtual work (see section 2.4) as

$$(\delta U_i + \delta V_i)_V = 0 \tag{4.69}$$

in which

$$(\delta U_i)_V = \sum_e \delta U_{ie} \tag{4.70}$$

is the first variation of the sum of the in-plane strain energies in the elements of the structure, and

$$(\delta V_i)_V = \sum_e \delta V_{ie} + \sum_n \delta V_{in} \tag{4.71}$$

is the first variation of the sum of the potential energies of the distributed element loads $\{q_{ie}\}$ and the concentrated nodal loads $\{Q_{in}\}$.

The first variation of the element strain energy can be expressed as

$$\delta U_{ie} = \{\delta \delta_{ie}\}^T [k_{ie}] \{\delta_{ie}\} \tag{4.72}$$

in which $\{\delta_{ie}\}$ are the in-plane element nodal deformations, $\{\delta \delta_{ie}\}$ are the virtual nodal deformations, and $[k_{ie}]$ is the element in-plane stiffness matrix. These can be transformed from the local element system to the global sytstem by using the in-plane transformation matrix $[T_{ie}]$ in

$$\{\delta_{ie}\} = [T_{ie}]\{\Delta_i\} \tag{4.73}$$

in which $\{\Delta_i\}$, are the in-plane global nodal deformations. The transformed first variations of the element strain energies can be summed to obtain

$$\{\delta U_i\}_V = \{\delta \Delta_i\}^T [K_i] \{\Delta_i\} \tag{4.74}$$

in which

$$[K_i] = \sum_e [T_{ie}]^T [k_{ie}] [T_{ie}] \tag{4.75}$$

The first variation of the element potential energy can be expressed as

$$\delta V_{ie} = -\{\delta \delta_{ie}\}^T [A_e] \{q_{ie}\} \tag{4.76}$$

which can transformed and summed to obtain

$$\sum_e \delta V_{ie} = -\{\delta \Delta_i\}^T \{Q_{ie}\} \tag{4.77}$$

in which

$$\{Q_{ie}\} = \sum_e [T_{ie}]^T [A_e]\{q_{ie}\} \qquad (4.78)$$

are the nodal loads equivalent to the distributed loads $\{q_{ie}\}$.

Thus the virtual work condition of equation 4.69 can be expressed as

$$\{\delta\Delta_i\} [K_i] \{\Delta_i\} = \{\delta\Delta_i\} \{Q_i\} \qquad (4.79)$$

in which

$$\{Q_i] = \{Q_{ie}\} + \{Q_{in}\}. \qquad (4.80)$$

Since equation 4.79 must hold for all admissible sets of virtual in-plane displacements $\{\delta\Delta_i\}$, then

$$[K_i]\{\Delta_i\} = \{Q_i\} \qquad (4.81)$$

which are the in-plane equilibrium equations.

4.5.2 IN-PLANE DEFORMATION FIELDS

The in-plane deformations v, w may be approximated by cubic and linear displacement fields (see section 4.3.1) so that

$$\{v_i\} = [M_i][C_i]^{-1}\{\delta_{ie}\} \qquad (4.82)$$

in which

$$\{v_i\} = \{v, w\}^T, \qquad (4.83)$$

$$\{\delta_{ie}\} = \{v_1, v_2, v'_1, v'_2, w_1, w_2\}^T, \qquad (4.84)$$

$$[M_i] = \begin{bmatrix} \{Z\}^T & \{0\}^T \\ \{0\}^T & \{Z_w\}^T \end{bmatrix}, \qquad (4.85)$$

$$\{Z_w\}^T = \{1, z/L\}, \qquad (4.86)$$

$$[C_i]^{-1} = \begin{bmatrix} \{C_u\}^{-1} & \{0\} \\ \{0\} & \{C_w\}^{-1} \end{bmatrix}, \qquad (4.87)$$

$$[C_w]^{-1} = \begin{bmatrix} 1 & 0 \\ -1 & 1 \end{bmatrix}, \qquad (4.88)$$

and $\{Z\}^T$ and $[C_u]^{-1}$ are given by equations 4.13 and 4.19.

4.5.3 ELEMENT STIFFNESS MATRIX

The first variation of the strain energy stored in an element of length L is given by

$$\delta U_{ie} = \int_0^L \int_A \delta\varepsilon_p \sigma_p \, dA \, dz \qquad (4.89)$$

in which the strain ε_P and the stress σ_P at a point $P(x, y)$ are approximated by

$$\varepsilon_P = w' - yv'', \tag{4.90}$$

$$\sigma_P = E\varepsilon_P. \tag{4.91}$$

Equation 4.89 can be written as

$$\delta U_{ie} = \int_0^L \{\delta\varepsilon_i\}^T \{\sigma_i\} dz \tag{4.92}$$

in which

$$\{\varepsilon_i\} = \{-v'', w'\}^T, \tag{4.93}$$

$$\{\sigma_i\} = [D_i]\{\varepsilon_i\} \tag{4.94}$$

are the generalized in-plane strains and stresses, and

$$[D_i] = \begin{bmatrix} EI_x & 0 \\ 0 & EA \end{bmatrix}, \tag{4.95}$$

is the generalized in-plane elasticity matrix, whence

$$\{\sigma_i\} = \{M_x, -P\}^T \tag{4.96}$$

in which

$$-P = \int_A \sigma_P dA, \tag{4.97}$$

$$M_x = \int_A \sigma_P y dA \tag{4.98}$$

are the in-plane stress resultants.

The strain vector $\{\varepsilon_i\}$ can be obtained by appropriate differentiation of the deformation fields of equation 4.82 as

$$\{\varepsilon_i\} = [B_i][C_i]^{-1}\{\delta_{ie}\} \tag{4.99}$$

in which

$$[B_i] = \begin{bmatrix} \{-Z''\}^T & \{0\}^T \\ \{0\}^T & \{Z_w'\}^T \end{bmatrix}. \tag{4.100}$$

Substituting equations 4.94 and 4.99 into equation 4.92 leads to

$$\delta U_{ie} = \{\delta\delta_{ie}\}^T[C_i]^{-T}\left(\int_0^L [B_i]^T[D_i][B_i]dz\right)[C_i]^{-1}\{\delta_{ie}\} \tag{4.101}$$

and this is identical with equation 4.72 if the element stiffness matrix is written as

$$[k_{ie}] = [C_i]^{-T} \left(\int_0^L [B_i]^T [D_i][B_i]dz \right) [C_i]^{-1}. \tag{4.102}$$

4.5.4 EQUIVALENT NODAL LOADS

The first variation of the potential energy of constant distributed loads $\{q_{ie}\} = \{q_y, q_z\}^T$ acting on an element of length L

$$\delta V_{ie} = - \int_0^L \{\delta v_i\}^T \{q_{ie}\} dz \tag{4.103}$$

can be written as

$$\delta V_{ie} = - \{\delta \delta_{ie}\}^T [C_i]^{-T} \left(\int_0^L [M_i]^T dz \right) \{q_{ie}\} \tag{4.104}$$

which is of the form of equation 4.76 with

$$[A_e] = [C_i]^{-T} \left(\int_0^L [M_i]^T dz \right). \tag{4.105}$$

Transformation and summation leads to

$$(\delta V_{ie})_V = - \{\delta \Delta_i\}^T \sum_e [T_{ie}]^T [C_i]^{-T} \left(\int_0^L [M_i]^T dz \right) \{q_{ie}\}. \tag{4.106}$$

This is equivalent to

$$(\delta V_{ie})_V = - \{\delta \Delta_i\}^T \{Q_{ie}\} \tag{4.107}$$

when

$$\{Q_{ie}\} = \sum_e [T_{ie}]^T [C_i]^{-T} \left(\int_0^L [M_i]^T dz \right) \{q_{ie}\} \tag{4.108}$$

which are the nodal loads equivalent to the distributed loads $\{q_{ie}\}$.

4.5.5 SOLUTION

After substitution of the boundary conditions, the linear in-plane equilibrium equations (equation 4.81) can be solved for the global nodal displacements $\{\Delta_i\}$ caused by the load set $\{q_{ie}\}$, $\{Q_{in}\}$. The element nodal displacements $\{\delta_{ie}\}$ can be obtained from equation 4.73, and these can be used to determine the strains ε_P from equations 4.99 and 4.90, the stresses σ_P from equation 4.91, and the stress resultants M_x, $-P$ from equations 4.94–4.96.

4.6 Computational considerations

4.6.1 INTEGRATION

When cubic fields are used for the buckling displacements u, v, ϕ, the integrations that must be performed to determine the element stiffness and stability matrices $[k_e]$, $[g_e]$ (see equations 4.35 and 4.43) are simple enough to be carried out formally [10]. However, it is often considered to be more reliable to use a general numerical integration method which does not need to be modified for different displacement fields. Gaussian numerical integration methods [1] can be of very high accuracy if sufficient sampling (Gauss) points are taken along the element. For example, only n sampling points are required for the exact integration of a polynomial of degree $(2n - 1)$.

4.6.2 NODE SPACING AND ACCURACY

While the finite element method always predicts an approximate buckling factor which is greater than the true value, the accuracy can be improved by increasing the number of elements into which the structure is divided. Generally, at least two elements should be used to represent each member of the structure when cubic deformation fields are used. This is because a cubic field can have only one inflexion point, and often the most critical member of the structure will buckle as if elastically restrained at both ends, so that it has two inflextion points. Studies in [10] suggest that using two elements per member will often lead to errors of less than 1%.

However, experience with cantilevers [16] indicates a lower rate of convergence, and suggests that at least four elements per cantilever should be used.

4.6.3 HIGHER ORDER ELEMENTS

An alternative method of increasing the accuracy to that of increasing the number of elements is to use higher order elements, based on quintic or even higher order polynomials, than the cubics discussed in section 4.31. The use of higher order elements is discussed in [17, 18].

4.7 Problems

In Problems 4.1–4.6, a finite element computer program is to be developed for analysing the elastic flexural buckling of continuous columns. The program is to use a data file which defines the member nodes and constraints, the element properties and connections, and the initial axial forces. The solutions for the lowest buckling load factor and the buckled shape are to be written into a solution file.

PROBLEM 4.1

Write a computer routine for calculating the element stiffness matrix from the element data.

PROBLEM 4.2

Write a computer routine for calculating the element stability matrix from the element data and the initial axial forces.

PROBLEM 4.3

Write a computer routine for transforming from the element to the structure coordinate system.

PROBLEM 4.4

Write a computer routine for summing the transformed element stiffness and stability matrices.

PROBLEM 4.5

Write a computer routine for using the boundary constraints to reduce the structure stiffness and stability matrices.

PROBLEM 4.6

Write a computer routine for finding the lowest buckling load factor and the corresponding buckled shape.

PROBLEM 4.7

Extend the solutions of Problems 4.1–4.6 to develop a program for analysing the elastic in-plane flexural buckling of plane rigid-jointed frames.

PROBLEM 4.8

Write a computer program for the elastic first-order analysis of the in-plane bending of plane rigid-jointed frames.

PROBLEM 4.9

Integrate the computer programs developed for Problems 4.7 and 4.8 so that the in-plane bending program determines the initial axial forces to be used in the buckling program.

PROBLEM 4.10

Extend the solution of Problem 4.9 to allow for continuous elastic restraints (see section 6.3).

PROBLEM 4.11

Extend the solution of Problem 4.10 to allow for concentrated elastic restraints (see section 6.3).

PROBLEM 4.12

Adapt the solution of Problem 4.9 so as to produce a computer program for analysing the elastic flexural-torsional buckling of continuous monosymmetric beam-columns.

PROBLEM 4.13

Adapt the solution of Problem 4.9 so as to produce a computer program for analysing the elastic flexural-torsional buckling of continuous asymmetric columns.

4.8 References

1. Zienkiewicz, O.C. and Taylor, R.L. (1989) *The Finite Element Method, Volume 1 – Basic Formulation and Linear Problems*; (1991) *Volume 2 – Solid and Fluid Mechanics, Dynamics and Non-Linearity*, McGraw-Hill, New York.
2. Cook, R.D., Malkus, D.S. and Plesha, M.E. (1989) *Concepts and Applications of Finite Element Analysis*, 3rd edn, John Wiley, New York.
3. Gallagher, R.H. (1975) *Finite Element Analysis Fundamentals*, Prentice-Hall, Englewood Cliffs, N.T.
4. Bathe, K.-J. and Wilson, E.L. (1976) *Numerical Methods in Finite Element Analysis*, Prentice-Hall, Englewood Cliffs, N.T.
5. Irons, B.M. and Ahmad, S. (1980) *Techniques of Finite Elements*, Ellis Horwood, Chichester.
6. Galambos, T.V. (ed.) (1988) *Guide to Stability Design Criteria for Metal Structures*, 4th edn, John Wiley, New York.
7. Barsoum, R.S. and Gallagher, R.H. (1970) Finite element analysis of torsional and lateral stability problems. *International Journal of Numerical Methods in Engineering*, 2(3), 335–52.
8. Powell, G. and Klingner, R. (1970) Elastic lateral buckling of steel beams. *Journal of the Structural Division, ASCE*, 96(ST9), 1919–32.
9. Nethercot, D.A. and Rockey, K.C. (1971) Finite element solutions for the buckling of columns and beams. *International Journal of Mechanical Sciences*, 13, 945–9.

10. Hancock, G.J., and Trahair, N.S. (1978) Finite element analysis of the lateral buckling of continuously restrained beam-columns. *Civil Engineering Transactions*, Institution of Engineers, Australia, CE20(2) 120–7.

11. Gallagher, R.H. (1973) The finite element method in shell stability analysis. *Computers and Structures*, 3, 543–7.

12. Cedolin, L. and Gallagher, R.H. (1978) Frontal-based solver for frequency analysis. *International Journal of Numerical Methods in Engineering*, 12, 1659–66.

13. Harrison, H.B. (1990) *Structural Analysis and Design – Some Micro Computer Applications*, Parts 1 and 2, 2nd edn, Pergamon Press, Oxford.

14. Harrison, H.B. (1973) *Computer Methods in Structural Analysis*, Prentice-Hall, Englewood Cliffs NJ.

15. Clarke, M.J. (1991) Advanced analysis. Lecture 6 of *Structural Analysis to AS4100*, School of Civil and Mining Engineering, University of Sydney, November.

16. Trahair, N.S. (1983) Lateral buckling of overhanging beams, in *Instability and Plastic Collapse of Steel Structures*, (ed. L.J. Morris), Granada, London, pp. 503–18.

17. Papangelis, J.P and Trahair, N.S. (1986) In-plane finite element analysis of arches, in *Proceedings*, 1st Pacific Structural Steel Conference, Auckland, August, Vol. 4, pp. 333–50.

18. Papangelis, J.P. and Trahair, N.S. (1987) Finite element analysis of arch lateral buckling. *Civil Engineering Transactions*, Institution of Engineers, Australia, CE29(1) 34–9.

5 Simply supported columns

5.1 General

A concentrically loaded column may buckle flexurally by deflecting u or v, or may buckle torsionally by twisting ϕ as shown in Figure 5.1. These buckling modes are independent for columns of doubly symmetric cross-section, and the column will buckle at the lowest of the loads associated with each of these three modes.

However, for columns of asymmetric cross-section, the modes are inter-dependent, and the lowest buckling load is lower than each of the loads associated with the independent modes described above. In this case, the column buckles in a flexural-torsional mode.

The resistance to buckling depends on the column's resistances to bending and torsion. When one resistance is very low compared with the other two, then the buckling load and mode will be strongly associated with this low resistance. Thus

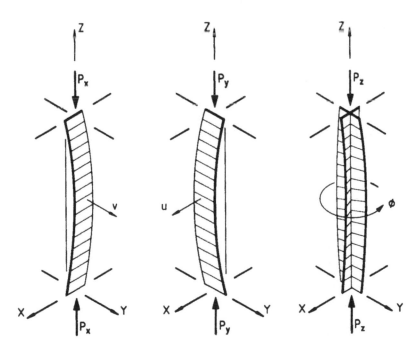

(a) Flexure About X Axis   (b) Flexure About Y Axis   (c) Torsion About Z Axis

Figure 5.1 Flexural and torsional buckling modes.

very thin-walled open sections which have very low torsional and warping rigidities GJ and EI_w are likely to buckle in a predominantly torsional mode, while sections with very low minor axis flexural rigidity EI_y are likely to buckle in a predominantly flexural mode.

The treatment given in this chapter of the elastic flexural-torsional buckling of simply supported columns discusses the effects of cross-section type on the mode of buckling. It is assumed that the columns are perfectly straight, and that the axial forces are applied concentrically.

The beneficial effects of end and intermediate restraints are discussed in Chapter 6, and the effects of end moments caused by eccentric axial forces in Chapter 11.

5.2 Elastic buckling analysis

The ends $z = 0$, L of the axially loaded columns shown in Figure 5.1 are simply supported, so that

$$0 = u_{0,L} = v_{0,L} = \phi_{0,L} \tag{5.1}$$

and

$$0 = u_{0,L}'' = v_{0,L}'' = \phi_{0,L}'' \tag{5.2}$$

If the column buckles laterally u, v and twists ϕ as shown in Figure 5.2a into an equilibrium buckled position, then the differential equilibrium equations (section 2.8.3.2) are

$$(EI_x v'')'' = -(Pv')' + (Px_0\phi')', \tag{5.3}$$

$$(EI_y u'')'' = -(Pu')' + (Py_0\phi')', \tag{5.4}$$

$$(EI_w \phi'')'' - (GJ\phi')' = -(Pr_2^2\phi')' + (Px_0 v')' - (Py_0 u')', \tag{5.5}$$

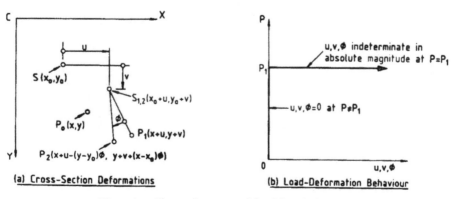

(a) Cross-Section Deformations   (b) Load-Deformation Behaviour

Figure 5.2 Flexural–torsional buckling behaviour.

in which

$$r_2^2 = (I_x + I_y)/A + x_0^2 + y_0^2 \tag{5.6}$$

and P is taken as positive when it acts in compression. Equation 5.3 represents the equality between the bending actions $-(Pv')'$ and $(Px_0\phi')'$ acting about the x axis and the corresponding flexural resistance $(EI_xv'')''$, while equation 5.4 provides a corresponding representation for bending about the y axis. Equation 5.5 represents the equality between the torsion actions $-(Pr_2^2\phi')'$, $(Px_0v')'$, and $-(Py_0u')'$ and the warping and torsional resistances $(EI_w\phi'')''$ and $-(GJ\phi')'$.

It can be verified by substitution that equations 5.1–5.5 are satisfied by the buckled shapes

$$u/\delta_x = v/\delta_y = \phi/\theta = \sin \pi z/L \tag{5.7}$$

where δ_x, δ_y, and θ are the values of u, v, and ϕ at mid-height, provided that the axial force P satisfies

$$\begin{vmatrix} (P_x - P) & 0 & Px_0 \\ 0 & (P_y - P) & -Py_0 \\ Px_0 & -Py_0 & r_2^2(P_z - P) \end{vmatrix} = 0 \tag{5.8}$$

in which

$$P_x = \pi^2 EI_x/L^2, \tag{5.9}$$

$$P_y = \pi^2 EI_y/L^2, \tag{5.10}$$

$$P_z = (GJ + \pi^2 EI_w/L^2)/r_2^2 \tag{5.11}$$

are three reference buckling loads associated with flexural buckling about the x axis (P_x), flexural buckling about the y axis (P_y), and torsional buckling (P_z).

Equation 5.8 can also be obtained by substituting the buckled shapes of equation 5.7 into the energy equation (section 2.8.3.1).

$$\frac{1}{2}\int_0^L \{EI_y(u'')^2 + EI_x(v'')^2 + EI_w(\phi'')^2 + GJ(\phi')^2\}\,dz$$

$$= \frac{1}{2}\int_0^L P\{(u')^2 + (v')^2 + r_2^2(\phi')^2 - 2x_0v'\phi' + 2y_0u'\phi'\}\,dz \tag{5.12}$$

which represents the equality at buckling between the flexural, warping, and torsional strain energy stored and the work done by the axial force.

Equation 5.8 can be written as the cubic equation

$$f(P) = P^3\{r_2^2 - x_0^2 - y_0^2\} - P^2\{(P_x + P_y + P_z)r_2^2 - P_yx_0^2 - P_xy_0^2\}$$

$$+ Pr_2^2\{P_xP_y + P_yP_z + P_zP_x\} - \{P_xP_yP_zr_2^2\} = 0 \tag{5.13}$$

which has three solutions P_1, P_2, P_3. It can be shown that these are real, and that the lowest solution is always less than or equal to the lowest of the three reference

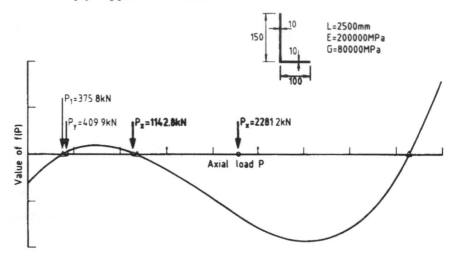

Figure 5.3 Solution of cubic equation for buckling load.

buckling loads P_x, P_y, P_z. For example, consider the case where $P_y < P_z < P_x$, as shown in Figure 5.3 for a 150 mm × 100 mm × 10 mm angle section column. It can be shown that $f(0) = -P_x P_y P_z r_2^2 < 0$ and $f(P_y) = (P_x - P_y)P_y^2 y_0^2 > 0$. Thus a solution (at which $f(P) = 0$) must occur between 0 and P_y, so that $0 < P_1 < P_y$.

The relative magnitudes of the buckled shapes u, v, ϕ can be obtained by substituting the solution of equation 5.13 into equations 5.3–5.5, whence

$$\delta_x/\theta = Py_0/(P_y - P) \tag{5.14}$$

$$\delta_y/\theta = -Px_0/(P_x - P). \tag{5.15}$$

The relative magnitudes of the buckled shapes determine an apparent axis of rotation which can be found by noting that during buckling, a general point $P(x, y)$ of the cross-section moves to a final position $P(x + u - (y - y_0)\phi, y + v + (x - x_0)\phi)$ as shown in Figure 5.2a. However, the apparent axis of rotation through $R(x_c, y_c)$ does not move, and so its coordinates x_c, y_c must satisfy $u - (y_c - y_0)\phi = 0$ and $v + (x_c - x_0)\phi = 0$, so that

$$x_c = P_x x_0/(P_x - P), \tag{5.16}$$

$$y_c = P_y y_0/(P_y - P). \tag{5.17}$$

The apparent centre of rotation of a 150 mm × 100 mm × 10 mm steel angle section column of length $L = 2000$ mm is shown in Figure 5.4.

At the axial force P defined by the solution P_1 of equation 5.13, the buckling deformations are defined in shape but are indeterminate in their absolute magnitudes, as indicated in Figure 5.2b. When the axial force does not satisfy equation 5.13, then the only solution of equations 5.1–5.5 is

$$u = v = \phi = 0 \tag{5.18}$$

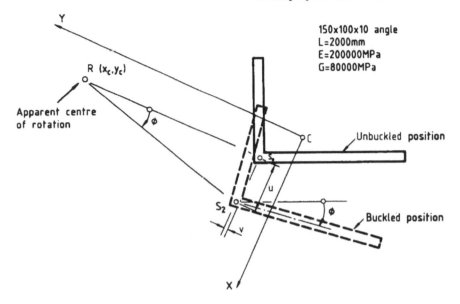

150x100x10 angle
L=2000mm
E=200000MPa
G=80000MPa

Figure 5.4 Apparent centre of rotation of an angle section column.

indicating that the column remains straight and untwisted until the lowest buckling load P_1 is reached, as shown in Figure 5.2b. Thus the state of equilibrium bifurcates at the lowest buckling load P_1 from the stable position given by equation 5.18 to neutral equilibrium positions given by equation 5.7.

5.3 Doubly symmetric sections

For doubly symmetric sections, $x_0 = y_0 = 0$, and the off-diagonal terms of equation 5.8 vanish. In this case equation 5.13 simplifies to

$$(P_x - P)(P_y - P)(P_z - P) = 0 \qquad (5.19)$$

and so the three solutions are

$$P_{1,2,3} = P_x, P_y, P_z. \qquad (5.20)$$

The column therefore buckles at the lowest of these three reference loads and in a corresponding mode. Thus buckling occurs either by flexure about the x axis at P_x, or by flexure about the y axis at P_y, or by torsion about the z axis at P_z.

The variations with length of the reference buckling loads P_x, P_y, P_z of an I-section are shown in Figure 5.5a. It can be seen that in this case the lowest reference load is always P_y, so that this section always buckles in flexure about the

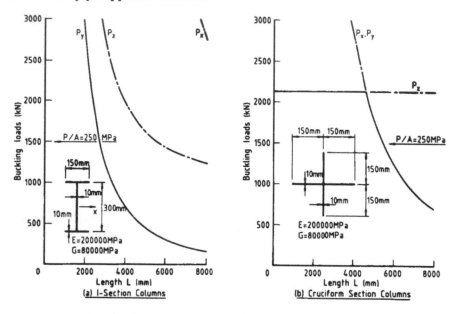

Figure 5.5 Buckling of doubly symmetric section columns.

y axis. This is generally the case for simply supported I-section columns, for which $P_x > P_y$, while $h > 0.58B$, and $I_w/r_2^2 > I_y$, while $h^3 t > B^3 T$. The exceptions are likely to have low values of h/B (which decrease the ratios of P_x/P_y and P_z/P_y), low values of t/T (which decrease the ratio of P_z/P_y), and low values of t/h and T/B (which decrease the contribution made by the torsional rigidity GJ to P_z).

The variations of the reference buckling loads of a cruciform section are shown in Figure 5.5b. It can be seen that P_z is often the lowest reference load, so that the column frequently buckles in a torsional mode. A cruciform section has zero warping rigidity EI_w, and so depends entirely on its torsional rigidity GJ for its resistance to torsional buckling, which is then independent of the column length L. Because the torsion section constant $J(\approx \Sigma bt^3/3)$ varies as the cube of the thickness t, the resistance of a cruciform column to torsional buckling reduces rapidly with the thickness t. Thus torsional buckling is common in thin-walled cruciform section columns, and it is only for long cruciform columns that the flexural buckling loads P_x, P_y may decrease below the torsional buckling load P_z, and the mode may change to flexural buckling.

On the other hand, the torsional rigidity GJ is always higher than the flexural rigidity EI for a circular hollow section, and the two are comparable for most rectangular hollow sections. Because of this, the torsional reference buckling loads P_z of practical hollow section columns are always higher than the corresponding flexural loads P_x, P_y, and so hollow sections buckle in a flexural mode at the lower of P_x and P_y.

5.4 Monosymmetric sections

For monosymmetric section columns which are symmetric about the x axis so that $y_0 = 0$, equation 5.13 simplifies to

$$(P_y - P)\{(r_0^2 + x_0^2)(P_x - P)(P_z - P) - (Px_0)^2\} = 0 \qquad (5.21)$$

which has the solutions

$$P_1 = P_y, \qquad (5.22)$$

$$P_{2,3} = \frac{(P_x + P_z) \pm \sqrt{\{(P_x + P_z)^2 - 4P_x P_z r_0^2/(r_0^2 + x_0^2)\}}}{2r_0^2/(r_0^2 + x_0^2)}, \qquad (5.23)$$

in which $r_0^2 = (I_x + I_y)/A$. Such a column therefore buckles at the lowest of the three loads P_y, P_2, or P_3, either in a flexural mode about the y axis at P_y, or in a flexural-torsional mode by flexure about the x axis and torsion at the lower of P_2, P_3.

The solutions of equation 5.13 for monosymmetric section columns which are symmetric about the y axis so that $x_0 = 0$ can be obtained by interchanging x and y in equations 5.21–5.23.

The lower flexural-torsional buckling load P_2 of equation 5.23 can be expressed non-dimensionally in terms of a modified harmonic mean buckling load $\sqrt{\{P_x P_z(1 + x_0^2/r_0^2)\}}$. The dimensionless buckling load $P_2/\sqrt{\{P_x P_z(1 + x_0^2/r_0^2)\}}$ then varies as shown in Figure 5.6a with the parameter $\frac{1}{2}(P_x + P_z)(1 + x_0^2/r_0^2)/\sqrt{\{P_x P_z(1 + x_0^2/r_0^2)\}}$ which may be thought of as the ratio of a modified arithmetic mean buckling load $\frac{1}{2}(P_x + P_z)(1 + x_0^2/r_0^2)$ to the modified harmonic mean.

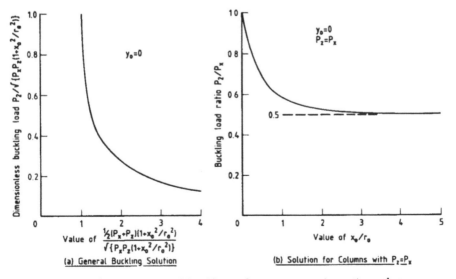

Figure 5.6 Flexural-torsional buckling of monosymmetric section columns.

It can be seen that the dimensionless buckling load decreases rapidly as the ratio of the modified arithmetic and harmonic means increases above 1.

For the special case where $P_z = P_x$, the lower flexural-torsional buckling load can be expressed more simply by the relationship shown in Figure 5.6b between the buckling load ratio P_2/P_x and x_0/r_0. The value of P_2/P_x decreases from 1.0 as x_0/r_0 increases from zero and approaches 0.5 asymptotically.

Torsional effects are often important in thin-walled monosymmetric tee-section columns, since $I_w = 0$ for these sections so that their torsional resistances depend on the torsional rigidities J which decrease rapidly with the wall thickness. The variations with length of the lowest buckling load P_{min} and the reference buckling loads, P_x, P_y, P_z of a monosymmetric tee-section are shown in Figure 5.7a. For this tee-section, P_y is greater than P_x, and so the lowest buckling mode is always flexural-torsional. For very short columns, P_x is much greater than P_z and so the mode is predominantly torsional, with P_{min} approaching P_z. For very long columns, P_z is much greater than P_x, and so the mode is predominantly flexural about the x axis, with P_{min} approaching P_x. For intermediate length columns, the buckling load is significantly less than both P_x and P_z.

Torsional effects are less important in monosymmetric unlipped channel section columns with P_x much larger than P_y, as is demonstrated in Figure 5.7b. For this section, P_y is less than P_z for long columns, and the buckling mode is flexural about the y axis, while P_2 is less than P_y for short columns, and the

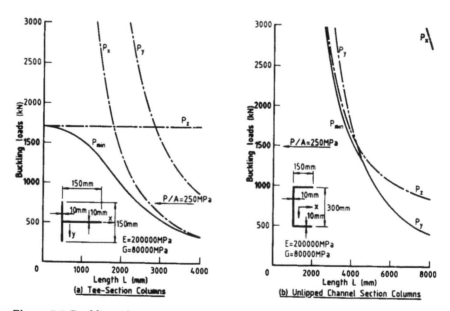

Figure 5.7 Buckling of monosymmetric tee and unlipped channel section columns.

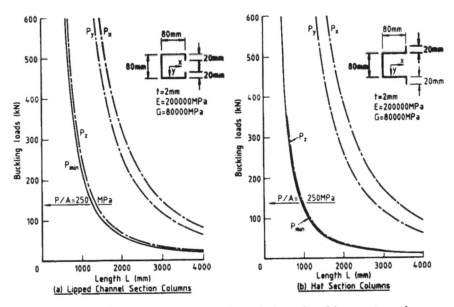

Figure 5.8 Buckling of monosymmetric lipped channel and hat section columns.

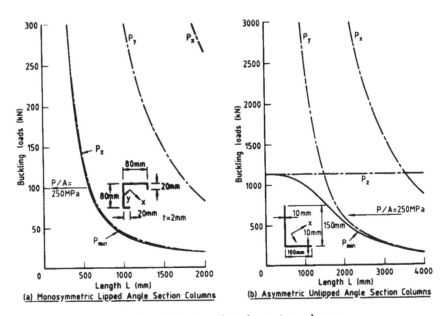

Figure 5.9 Buckling of angle section columns.

flexural-torsional mode is predominantly torsional. The mode changes at $L = 4565$ mm.

On the other hand, torsional effects may dominate in thin-walled mono-symmetric lipped channel and hat section columns, as is demonstrated in Figure 5.8. The geometries of these sections are such that in both cases P_z is much less than P_x and P_y, so that the columns buckle in flexural-torsional modes which are predominantly torsional at loads P_{min} which are a little less than the torsional reference loads P_z. Similar behaviour is shown by the thin-walled monosymmetric lipped angle section columns shown in Figure 5.9a.

5.5 Asymmetric sections

For asymmetric sections, the cubic equation 5.13 cannot be simplified, and so the lowest buckling load must be obtained numerically. This can easily be done approximately by interpolating linearly using the values of $f(P)$ for $P = 0$ and for the lowest value P_1 of P_x, P_y, P_z. Thus

$$P_{min} \approx \frac{P_1 f(0)}{f(0) - f(P_1)}. \tag{5.24}$$

The accuracy of this approximation can then be improved by again interpolating linearly, this time using $f(P_{min})$ calculated for the approximate value of P_{min} and $f(P_1)$, and this can be repeated until a sufficiently accurate estimate of P_{min} is obtained. This P_{min} is always less than the lowest of P_x, P_y, P_z, and corresponds to a flexural-torsional mode involving flexure about both the x and y axes and torsion about the z axis.

Torsional effects are often important in thin-walled angle section columns, because $I_w = 0$ and J decreases rapidly with the wall thickness. The variations with length of the lowest buckling load P_{min} and the three reference buckling loads P_x, P_y, P_z of an asymmetric angle section are shown in Figure 5.9b. For this section P_x is much greater than P_y, and so the buckling mode changes from predominantly flexural (about the y axis) for long columns to predominantly torsional for short columns. For intermediate length columns, the buckling load is significantly less than both P_y and P_z.

5.6 Problems

PROBLEM 5.1

A simply supported steel ($E = 200\,000$ MPa, $G = 80\,000$ MPa) cruciform section column is fabricated from four plates 120 mm × 8 mm × L (Figure 5.10a).

(a) Determine the variations of the elastic flexural and torsional buckling loads P_x, P_y, P_z with the length L.

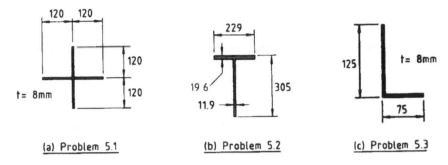

(a) Problem 5.1   (b) Problem 5.2   (c) Problem 5.3

Figure 5.10 Problems 5.1–5.3.

(b) Find the length L at which $P_z = P_y$.

(c) Compare the torsional buckling load P_z with a value obtained from the plate elastic local buckling stress

$$\sigma = \frac{\pi^2 E}{12(1 - v^2)\,(b/t)^2}\cdot k$$

in which $k = 0.425$ for a plate outstand b of thickness t, and $v = 0.25$.

PROBLEM 5.2

A simply supported tee-section column (Figure 5.10b) is formed from half of the I-section shown in Figure 7.23 by cutting it along the web centre line. The depth of the tee-section is 305 mm, and its properties are $y_0 = 65.3$ mm, $A = 7970$ mm^2, $I_x = 68.5 \times 10^6$ mm^4, $I_y = 19.7 \times 10^6$ mm^4, $J = 771 \times 10^3$ mm^4.

(a) Determine the variations of P_x, P_y, P_z with the length L.

(b) Determine the variation of the minimum elastic buckling load with the length L.

PROBLEM 5.3

A simply supported steel ($E = 200\,000$ MPa, $G = 80\,000$ MPa) angle section column has overall leg widths of 125 mm and 75 mm and a thickness of 8 mm (Figure 5.10c). Its properties are $y_0 = 33.3$ mm, $x_0 = -24.2$ mm, $A = 1500$ mm^2, $I_x = 2.68 \times 10^6$ mm^4, $I_y = 0.399 \times 10^6$ mm^4, $J = 31.7 \times 10^3$ mm^4.

(a) Determine the variations of P_x, P_y, P_z with the length L.

(b) Determine the variation of the minimum elastic buckling load with the length L.

5.7 References

1. Timoshenko, S.P. and Gere, J.M. (1961) *Theory of Elastic Stability*, 2nd edn, McGraw-Hill, New York.
2. Vlasov, V.Z. (1961) *Thin Walled Elastic Beams*, 2nd edn, Israel Program for Scientific Translations, Jerusalem.
3. Bleich, F. (1952) *Buckling Strength of Metal Structures*, McGraw-Hill, New York.
4. Chajes, A. and Winter, G. (1965) Torsional-flexural buckling of thin-walled members. *Journal of the Structural Division, ASCE*, 91 (ST4), 103–24.

6 Restrained columns

6.1 General

A column is often connected to other elements which participate in the buckling action, and significantly influence the buckling resistance. Braces are provided specifically for the purpose of increasing the buckling resistance (Figure 6.1), but many other elements, such as sheeting, which are intended primarily for other purposes, may also have important restraining actions.

These elements induce restraining actions (Figure 6.2) which restrict the buckled shapes of the column, and increase its buckling resistance. Discrete restraints act at points where braces or other restraining elements are connected to the column and induce actions which resist the buckling deflections, rotations, and warping displacements. These restraints are usually assumed to be elastic, in which case they may be characterized by their elastic stiffnesses.

In some cases the discrete restraints may be assumed to be rigid, so that they prevent one or more of the buckling deformations. In the special case where rigid restraints prevent the lateral deflections and twist rotations of the cross-sections at which they act, then the column may be described as a braced column which consists of column elements between the restraint points.

Continuous restraints are usually considered to be uniform along the length of a column, and are often used to approximate the actions of restraining elements which are connected to the column at closely spaced intervals, as in the case of wall sheeting (Figure 6.3). Continuous restraints are similar to discrete restraints, in that they induce actions which restrain the buckling deformations (Figure 6.2). They also are usually assumed to be elastic, and so may be characterized by their elastic stiffnesses. When continuous restraints acting in both principal planes can

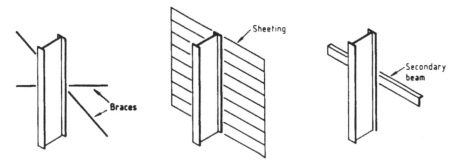

Figure 6.1 Column restraints.

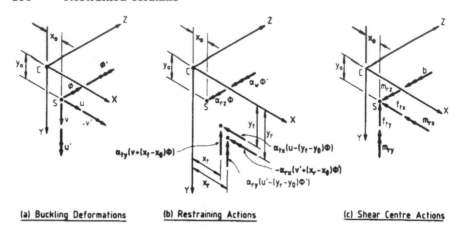

(a) Buckling Deformations   (b) Restraining Actions   (c) Shear Centre Actions

Figure 6.2 Continuous restraint actions.

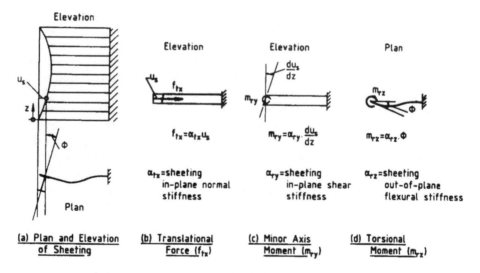

Elevation

Elevation   Elevation   Plan

$f_{tx} = \alpha_{tx} u_s$ $m_{ry} = \alpha_{ry} \cdot \dfrac{du_s}{dz}$ $m_{rz} = \alpha_{rz} \cdot \Phi$

α_{tx} =sheeting in-plane normal stiffness α_{ry} =sheeting in-plane shear stiffness α_{rz} =sheeting out-of-plane flexural stiffness

Plan

(a) Plan and Elevation of Sheeting   (b) Translational Force (f_{tx}) (c) Minor Axis Moment (m_{ry}) (d) Torsional Moment (m_{rz})

Figure 6.3 Restraint actions of continuous sheeting.

be assumed to be rigid, then they enforce a longitudinal axis about which the column cross-sections rotate during buckling.

This chapter is concerned with the influence of restraints on the elastic buckling of columns. The buckling of restrained beams is treated in Chapter 8, of restrained cantilevers in Chapter 9, of continuous beams in Chapter 10, and of restrained beam-columns in section 11.5.

6.2 Restraint stiffnesses

6.2.1 CONTINUOUS RESTRAINTS

A column may be continuously restrained as shown in Figure 6.2b by translational restraints of stiffness α_{tx}, α_{ty} which act at distances y_t, x_t from the centroid, by rotational restraints of stiffness α_{rx}, α_{ry} which act at distances y_r, x_r from the centroid, by torsional restraints of stiffness α_{rz}, and by warping restraints of stiffness α_w.

The actions exerted by these restraints can be replaced by the shear centre actions

$$\{r\} = \{f_{tx}, m_{ry}, f_{ty}, -m_{rx}, m_{rz}, b\}^T \tag{6.1}$$

shown in Figure 6.2c, in which b is the bimoment per unit length. For these actions to be statically equivalent to those of the restraints, they must be related to the shear centre deformations (Figure 6.2a)

$$\{d\} = \{u, u', v, v', \phi, \phi'\}^T \tag{6.2}$$

by

$$\{r\} = [\alpha_b]\{d\} \tag{6.3}$$

in which the continuous restraint stiffness matrix is

$$[\alpha_b] = \begin{bmatrix} \alpha_{tx} & 0 & 0 & 0 & -\alpha_{xz} & 0 \\ 0 & \alpha_{ry} & 0 & 0 & 0 & -\alpha_{yw} \\ 0 & 0 & \alpha_{ty} & 0 & \alpha_{yz} & 0 \\ 0 & 0 & 0 & \alpha_{rx} & 0 & \alpha_{xw} \\ -\alpha_{xz} & 0 & \alpha_{yz} & 0 & \alpha_{zz} & 0 \\ 0 & -\alpha_{yw} & 0 & \alpha_{xw} & 0 & \alpha_{ww} \end{bmatrix} \tag{6.4}$$

in which

$$\left.\begin{aligned} \alpha_{zz} &= \alpha_{tx}(y_t - y_0)^2 + \alpha_{ty}(x_t - x_0)^2 + \alpha_{rz}, \\ \alpha_{ww} &= \alpha_{ry}(y_r - y_0)^2 + \alpha_{rx}(x_r - x_0)^2 + \alpha_w, \\ \alpha_{xz} &= \alpha_{tx}(y_t - y_0), \\ \alpha_{yz} &= \alpha_{ty}(x_t - x_0), \\ \alpha_{xw} &= \alpha_{rx}(x_r - x_0), \\ \alpha_{yw} &= \alpha_{ry}(y_r - y_0). \end{aligned}\right\} \tag{6.5}$$

6.2.2 DISCRETE RESTRAINTS

A column may have discrete restraints which have translational stiffnesses α_{Tx}, α_{Ty} (at distances y_T, x_T from the centroid), rotational stiffnesses α_{Rx}, α_{Ry} (at distances y_R, x_R from the centroid), torsional stiffness α_{Rz}, and warping stiffness α_W.

The actions exerted by these restraints can be replaced by the shear centre actions

$$\{R\} = \{F_{Tx}, M_{Ry}, F_{Ty}, -M_{Rx}, M_{Rz}, B\}^T \tag{6.6}$$

in which B is the bimoment. For these actions to be statically equivalent to those of the restraints, they must be related to the shear centre deformations

$$\{D\} = \{u, u', v, v', \phi, \phi'\}^T \tag{6.7}$$

by

$$\{R\} = [\alpha_B]\{D\} \tag{6.8}$$

in which the discrete restraint stiffness matrix is

$$[\alpha_B] = \begin{bmatrix} \alpha_{Tx} & 0 & 0 & 0 & -\alpha_{xZ} & 0 \\ 0 & \alpha_{Ry} & 0 & 0 & 0 & -\alpha_{yW} \\ 0 & 0 & \alpha_{Ty} & 0 & \alpha_{yZ} & 0 \\ 0 & 0 & 0 & \alpha_{Rx} & 0 & \alpha_{xW} \\ -\alpha_{xZ} & 0 & \alpha_{yZ} & 0 & \alpha_{zz} & 0 \\ 0 & -\alpha_{yW} & 0 & \alpha_{xW} & 0 & \alpha_{WW} \end{bmatrix} \tag{6.9}$$

in which

$$\left.\begin{aligned} \alpha_{zz} &= \alpha_{Tx}(y_T - y_0)^2 + \alpha_{Ty}(x_T - x_0)^2 + \alpha_{Rz}, \\ \alpha_{WW} &= \alpha_{Ry}(y_R - y_0)^2 + \alpha_{Rx}(x_R - x_0)^2 + \alpha_W, \\ \alpha_{xZ} &= \alpha_{Tx}(y_T - y_0), \\ \alpha_{yZ} &= \alpha_{Ty}(x_T - x_0), \\ \alpha_{xW} &= \alpha_{Rx}(x_R - x_0), \\ \alpha_{yW} &= \alpha_{Ry}(y_R - y_0). \end{aligned}\right\} \tag{6.10}$$

6.3 Buckling analysis

In the general case of restrained columns, exact solutions for the buckling loads cannot be obtained, and a numerical method must be used to obtain approximate solutions. If the approximate energy method is used (Chapter 3), then the strain energy stored in an element during buckling should be increased to

$$\frac{1}{2}\delta^2 U = \frac{1}{2}\int_0^L \{EI_y u''^2 + EI_x v''^2 + EI_w \phi''^2 + GJ\phi'^2\}\,dz$$

$$+ \frac{1}{2}\sum \{D\}^T[\alpha_B]\{D\} + \frac{1}{2}\int_0^L \{d\}^T[\alpha_b]\{d\}\,dz \tag{6.11}$$

to account for the strain energy stored in the discrete and continuous restraints acting on the element.

If the finite element method of computer analysis is used (Chapter 4), then each element stiffness matrix should be augmented by transforming the term $\frac{1}{2}\int_0^L \{d\}^T[\alpha_b]\{d\}\,dz$ which accounts for continuous restraints, and the column stiffness matrix should be augmented by including a transformation of the term $\frac{1}{2}\{D\}^T[\alpha_B]\{D\}$ for each discrete restraint.

6.4 Continuous restraints

6.4.1 GENERAL SOLUTION

A simply supported column ($u_{0,L} = v_{0,L} = \phi_{0,L} = 0, u_{0,L}'' = v_{0,L}'' = \phi_{0,L}'' = 0$) with uniform continuous restraints buckles into n half sine waves so that

$$u/\delta_x = v/\delta_y = \phi/\theta = \sin n\pi z/L. \tag{6.12}$$

The load which causes elastic buckling can be obtained [1, 2] by substituting these buckled shapes into the energy equation

$$\frac{1}{2}(\delta^2 U + \delta^2 V) = 0 \tag{6.13}$$

in which the strain energy $\frac{1}{2}\delta^2 U$ stored during buckling is given by equation 6.11 and the work $\frac{1}{2}\delta^2 V$ done on the applied compression force P is given by

$$\frac{1}{2}\delta^2 V = -\frac{1}{2}\int_0^L P\{u'^2 + v'^2 + r_2^2\phi'^2 - 2x_0 v'\phi' + 2y_0 u'\phi'\}\,dz. \tag{6.14}$$

This leads to an equation of the form

$$\begin{vmatrix} a_{11} & 0 & a_{13} \\ 0 & a_{22} & a_{23} \\ a_{13} & a_{23} & a_{33} \end{vmatrix} = 0 \tag{6.15}$$

in which

$$\left.\begin{aligned} a_{11} &= \alpha_{ry} + \alpha_{tx}L^2/n^2\pi^2 + n^2 P_y - P, \\ a_{22} &= \alpha_{rx} + \alpha_{ty}L^2/n^2\pi^2 + n^2 P_x - P, \\ a_{33} &= \alpha_{ww} + \alpha_{zz}L^2/n^2\pi^2 + GJ + n^2\pi^2 EI_w/L^2 - Pr_2^2, \\ a_{13} &= -\alpha_{yw} - \alpha_{xz}L^2/n^2\pi^2 - Py_0, \\ a_{23} &= \alpha_{xw} + \alpha_{yz}L^2/n^2\pi^2 + Px_0. \end{aligned}\right\} \tag{6.16}$$

In general, a number of trials must be made before the integer value of n which leads to the lowest value of the buckling load P can be determined.

6.4.2 DOUBLY SYMMETRIC COLUMNS WITH SHEAR CENTRE RESTRAINTS

For doubly symmetric columns $(x_0 = y_0 = 0)$ with shear centre restraints $(\alpha_{xz} = \alpha_{yz} = \alpha_{xw} = \alpha_{yw} = 0)$, equations 6.15 and 6.16 reduce to

$$(P_{yr} - P)(P_{xr} - P)(P_{zr} - P)r_2^2 = 0 \tag{6.17}$$

in which

$$P_{yr} = n^2 P_y + \alpha_{ry} + \alpha_{tx}L^2/n^2\pi^2, \tag{6.18}$$

$$P_{xr} = n^2 P_x + \alpha_{rx} + \alpha_{ty}L^2/n^2\pi^2, \tag{6.19}$$

$$P_{zr} = (GJ + n^2\pi^2 EI_w/L^2 + \alpha_w + \alpha_{rz}L^2/n^2\pi^2)/r_2^2. \tag{6.20}$$

Thus there are three possible buckling modes; flexure about the y axis at P_{yr}, flexure about the x axis at P_{xr}, and torsion about the z axis at P_{zr}. The actual buckling load will be the lowest of P_{yr}, P_{xr}, P_{zr}.

These buckling loads depend on the number n of half waves in the buckled shapes. Each of equations 6.18–6.20 is of the form

$$P_r = n^2 P_a + P_b + P_c/n^2 \tag{6.21}$$

and the lowest value of P_r can be closely approximated by

$$P_{min} = 2\sqrt{(P_a P_c)} + P_b \tag{6.22}$$

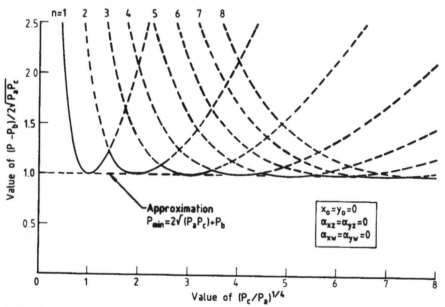

Figure 6.4 Buckling of doubly symmetric columns with continuous shear centre restraints.

except when $P_e/P_a < 1$, in which case this approximation is conservative, as shown in Figure 6.4. An accurate solution may then be obtained by using $n = 1$ in equations 6.18–6.20 or 6.21.

6.4.3 OTHER COLUMNS

For other columns with more general restraints, equations 6.15 and 6.16 cannot be so simplified, and the actual buckling load must be determined by finding the value of n which gives the lowest solution of P. In general, this solution will correspond to a flexural-torsional buckling mode in which the column simultaneously deflects u, v, and twists ϕ.

6.4.4 BUCKLING WITH AN ENFORCED CENTRE OF ROTATION

When rigid translational or rotational restraints act along a line (x_c, y_c), then the column buckles with this line as an enforced centre of rotation, so that

$$\delta_x = (y_c - y_0)\theta, \tag{6.23}$$

$$\delta_y = -(x_c - x_0)\theta. \tag{6.24}$$

In this case the only restraint strain energy arises from the torsional and warping restraints α_{rz}, α_w, and so the energy equation (equations 6.11, 6.13, 6.14) leads to

$$\frac{1}{2}\begin{Bmatrix} \delta_x \\ \delta_y \\ \theta \end{Bmatrix}^T \begin{bmatrix} (n^2 P_y - P) & 0 & -Py_0 \\ 0 & (n^2 P_x - P) & Px_0 \\ -Py_0 & Px_0 & \left(GJ + \dfrac{n^2\pi^2 EI_w}{L^2} + \alpha_w + \dfrac{\alpha_{rs} L^2}{n^2\pi^2} - Pr_2^2 \right) \end{bmatrix} \begin{Bmatrix} \delta_x \\ \delta_y \\ \theta \end{Bmatrix} = 0. \tag{6.25}$$

When equations 6.23, 6.24 are substituted, then equation 6.25 can be rearranged as

$$P = n^2 P_a + P_b + P_c/n^2 \tag{6.26}$$

in which

$$\left. \begin{aligned} P_a &= \{P_y(y_c - y_0)^2 + P_x(x_c - x_0)^2 + \pi^2 EI_w/L^2\}/r_c^2, \\ P_b &= (GJ + \alpha_w)/r_c^2, \\ P_c &= \alpha_{rz} L^2/\pi^2 r_c^2, \end{aligned} \right\} \tag{6.27}$$

and

$$r_c^2 = r_0^2 + x_c^2 + y_c^2. \tag{6.28}$$

Equation 6.26 is of the same form as equation 6.21, and can therefore be solved in the same way. When $P_c < P_a$, then the minimum solution is given by

$$P_{min} = P_a + P_b + P_c.$$ (6.29)

When $P_c > P_a$, then the minimum solution may be closely approximated by

$$P_{min} = 2\sqrt{(P_a P_c)} + P_b.$$ (6.30)

6.5 Discrete restraints

6.5.1 END RESTRAINTS

Restraints acting at the supported ends ($u = v = \phi = 0$) of a column may restrain end rotations u', v' [3] and warping displacements proportional to ϕ' [4]. The particular case of a doubly symmetric column ($x_0 = y_0 = 0$) whose ends have equal shear centre rotational and warping restraint stiffnesses of α_{Ry}, α_{Rx}, and α_W is shown in Figure 6.5a.

In this case, the buckled shapes are defined by three equations of the type

$$\frac{u}{\delta_x} = \frac{v}{\delta_y} = \frac{\phi}{\theta} = \frac{\cos(\pi z/kL - \pi/2k) - \cos(\pi/2k)}{(1 - \cos(\pi/2k))}$$ (6.31)

where k satisfies equations of the type

$$\frac{\alpha L}{EI} = -\frac{\pi}{2k}\cot\frac{\pi}{2k}.$$ (6.32)

In these equations, the values k_y, k_x, k_r, of k are associated with u, v, ϕ and

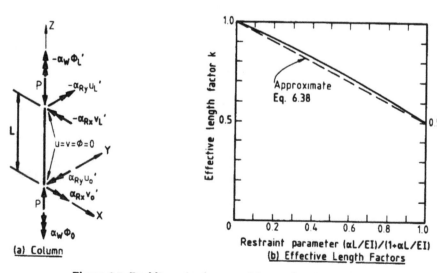

Figure 6.5 Buckling of columns with equal end restraints.

$\alpha_{Ry} L/EI_y, \alpha_{Rx} L/EI_x, \alpha_W L/EI_w$, respectively. These define three column effective lengths $k_y L, k_x L, k_z L$, equal to the distances between the inflexion points of the buckled shapes for u, v, ϕ.

Substitution of equations 6.31, 6.32 into the energy equation

$$\frac{1}{2} \int_0^L \{EI_y u''^2 + EI_x v''^2 + EI_w \phi''^2 + GJ\phi'^2\} \, dz + \frac{1}{2} \sum \{D\}^T [\alpha_B] \{D\}$$

$$- \frac{1}{2} \int_0^L P\{u'^2 + v'^2 + r_0^2 \phi'^2\} \, dz = 0 \tag{6.33}$$

leads to

$$(P_{yR} - P)(P_{xR} - P)(P_{zR} - P)r_0^2 = 0 \tag{6.34}$$

in which

$$P_{yR} = \pi^2 EI_y / k_y^2 L^2 \tag{6.35}$$

$$P_{xR} = \pi^2 EI_x / k_x^2 L^2 \tag{6.36}$$

$$P_{zR} = GJ + \pi^2 EI_w / k_z^2 L^2 \tag{6.37}$$

which can also be obtained from equations 5.9–5.11 for an unrestrained column by substituting the appropriate effective lengths kL for the actual length L.

The relationship obtained from equation 6.32 between the effective length factor k and the dimensionless end restraint stiffness $\alpha L/EI$ is shown in Figure 6.5b. It can be seen that k varies almost linearly from 1.0 to 0.5 as $(\alpha L/EI)/(1 + \alpha L/EI)$ increases from zero (for an unrestrained column) to unity (for a rigidly restrained column). This suggests that k may be closely approximated by

$$k \approx \frac{2 + \alpha L/EI}{2 + 2\alpha L/EI} \tag{6.38}$$

as shown in Figure 6.5b.

The effect of unequal minor axis flexural restraints $\alpha_{RyA}, \alpha_{RyB}$ at the ends A, B of a column is to increase the elastic minor axis flexural buckling load P_y to the value of P_{yR} given by equation 6.35 with the effective length factor k_y being the solution [3] of

$$\frac{G_A G_B}{4} \left(\frac{\pi}{k_y}\right)^2 + \left(\frac{G_A + G_B}{2}\right)\left(1 - \frac{\pi}{k_y} \cot \frac{\pi}{k_y}\right) + \frac{\tan(\pi/2k_y)}{\pi/2k_y} = 1 \tag{6.39}$$

in which

$$\left.\begin{array}{l} G_A = 2EI_y/\alpha_{RyA} L, \\ G_B = 2EI_y/\alpha_{RyB} L. \end{array}\right\} \tag{6.40}$$

These solutions are shown in Figure 6.6b.

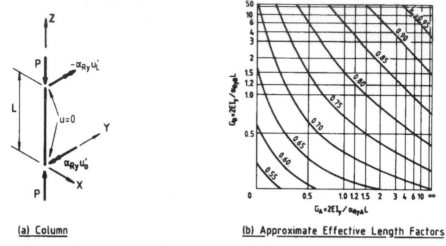

(a) Column   (b) Approximate Effective Length Factors

Figure 6.6 Buckling of columns with unequal end restraints.

These solutions may also be used for the effective length factors k_x of columns with unequal major axis end flexural restraints $\alpha_{RxA}, \alpha_{RxB}$, by substituting x for y. Similarly, solutions may also be obtained for the effective length factors k_z of columns with unequal end warping restraints α_{WA}, α_{WB} by substituting α_W for α_{Ry}, I_w for I_y, and k_z for k_y.

6.5.2 INTERMEDIATE RESTRAINTS

Intermediate restraints acting between the ends of a column may restrain the lateral deflections u, v [3] and twist rotations ϕ [4]. The particular case of a doubly symmetric column with central centroidal restraints is shown in Figure 6.7a. When the restraints are rigid so that $u = v = \phi = 0$ at the restraint point, the column buckles antisymmetrically into two half waves, each having an effective length of $L/2$, as shown in Figure 6.7b. Thus the elastic buckling resistances are given by P_{yR}, P_{xR}, P_{zR} (equations 6.35–6.37) with $k = 0.5$.

When the restraints are elastic, the column may buckle in a symmetrical mode as shown in Figure 6.7a. The buckled shapes are given by the three equations [3]

$$\frac{u}{\delta_x} = \frac{v}{\delta_y} = \frac{\phi}{\theta} = \frac{z}{L} - \frac{\sin(\pi z/kL)}{(\pi/k)\cos(\pi/2k)}. \tag{6.41}$$

while $0 \leqslant z \leqslant L/2$, where k satisfies equations of the type

$$\frac{\alpha L^3}{16EI} = \frac{(\pi/2k)^3 \cot(\pi/2k)}{(\pi/2k)\cot(\pi/2k) - 1}. \tag{6.42}$$

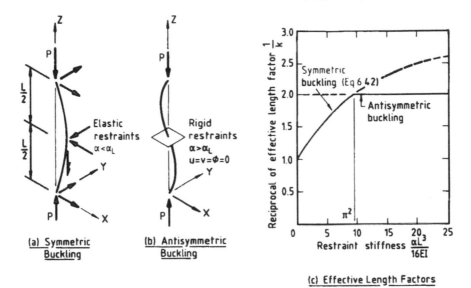

Figure 6.7 Column with central restraints.

In these equations, the three values k_y, k_u, k_z are associated with u, v, ϕ and $\alpha_{Tx}, \alpha_{Ty}, \alpha_{Rz}$ and I_y, I_x, I_w respectively.

The relationship obtained from equation 6.42 between the reciprocal of the effective length factor k and the dimensionless central restraint stiffness $\alpha L^3/(16EI)$ is shown in Figure 6.7c. It can be seen that $1/k$ increases from 1.0 towards 2.86 approximately as the restraint stiffness increases from zero towards infinity.

Also shown in Figure 6.7c is the solution $1/k = 2$ for the antisymmetric buckling mode of a column with a rigid restraint, which intersects the symmetric mode solution at a limiting restraint stiffness given by

$$\alpha_L L^3/(16EI) = \pi^2. \tag{6.43}$$

Since buckling always occurs in the mode which has the lowest buckling load, a column with $\alpha > \alpha_L$ always buckles antisymmetrically in two half waves, and at the lowest load given by P_{yR}, P_{xR}, P_{zR} with $k = 0.5$. On the other hand, a column with $\alpha < \alpha_L$ buckles symmetrically in a single half wave, and at a load obtained by using the solution of equation 6.42. This solution may be closely approximated by

$$k = \frac{1}{1 + 1.4\alpha/\alpha_L - 0.4(\alpha/\alpha_L)^2} \tag{6.44}$$

in the range $0 < \alpha < \alpha_L$.

The influence of n_r equal stiffness centroidal restraints spaced at equal intervals $s = L/(n_r + 1)$ along a doubly symmetric column has been studied [5, 6]. For high restraint values, the column may buckle between restraints into $(n_r + 1)$ half waves at a load given by the lowest of P_{yR}, P_{xR}, P_{zR} with

$$k = s/L. \tag{6.45}$$

For smaller restraint values, the column will buckle in a complex mode into a lesser number of half waves at a reduced buckling load. Solutions [7] for columns with translational restraints α_T are shown in Figure 6.8, and compared with the corresponding solutions for columns with continuous restraints. Approximate solutions may also be obtained by 'smearing' the discrete translational restraints into an equivalent uniform continuous translational restraint of stiffness

$$\alpha_t = n_r \alpha_T / L \tag{6.46}$$

and using this value in equations 6.17–6.22.

The limiting restraint stiffnesses required to ensure that a column buckles between discrete restraints in $(n_r + 1)$ half waves is approximated by

$$\frac{\alpha_{TL} L^3}{\pi^4 EI} = 0.38(n_r + 1)^3 \tag{6.47}$$

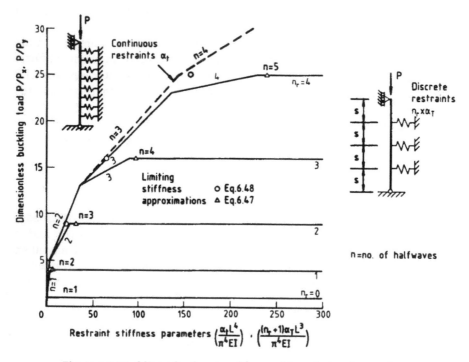

Figure 6.8 Buckling of columns with equal translational restraints.

which is somewhat higher than the corresponding approximation

$$\frac{\alpha_{tL} L^4}{\pi^4 EI} = 0.25 n^4 \qquad (6.48)$$

for a column with continuous restraints which buckles in n half waves. These approximations are also shown in Figure 6.8.

6.6 Problems

PROBLEM 6.1

Adapt the solution of Problem 4.13 so as to produce a computer program for analysing the elastic flexural-torsional buckling of continuous columns with continuous and concentrated restraints.

PROBLEM 6.2

A simply supported I-section column (Figure 6.9a) whose properties are given in Figure 7.23 is continuously restrained along its centre line so that $u = v = 0$. Determine the variations of its modified dimensionless torsional buckling load

$$P_{zr}^* = \frac{P_{zr} r_2^2 - (GJ + \alpha_w)}{\pi^2 EI_w / L^2}$$

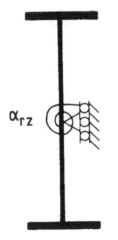

α_{rz}

(a) Problem 6.2   (b) Problem 6.3

Figure 6.9 Problems 6.2 and 6.3.

and the number n of buckle half waves with the dimensionless torsional restraint stiffness $\alpha_{rz}^* = \alpha_{rz} L^4/\pi^4 EI_w$.

PROBLEM 6.3

The tee-section column of Problem 5.2 has an enforced centre of rotation about the tip of its stem (Figure 6.9b). Compare its elastic buckling load with half of the torsional buckling load of a simply supported I-section column whose properties are given in Figure 7.23.

6.7 References

1. Trahair, N.S. (1979) Elastic lateral buckling of continuously restrained beam-columns. *The Profession of a Civil Engineer* (eds Campbell-Allen, D. and Davis, E.H.) Sydney University Press, Sydney, pp. 61–73.
2. Trahair, N.S. and Nethercot, D.A. (1984) Bracing requirements in thin-walled structures, in *Developments in Thin-Walled Structures–2*, (eds Rhodes, J. and Walker, A.C.), Elsevier Applied Science Publishers, pp. 93–130.
3. Trahair, N.S. and Bradford, M.A. (1991) *The Behaviour and Design of Steel Structures*, revised 2nd edition, Chapman and Hall, London.
4. Svensson, S.E. and Plum, C.M. (1983) Stiffener effects on torsional buckling of columns. *Journal of Structural Engineering, ASCE*, **109** (3), 758–72.
5. Horne, M.R. and Ajmani, J.L. (1969) Stability of columns supported laterally by side-rails. *International Journal of Mechanical Sciences*, **11**, 159–74.
6. Medland, I.C. (1979) Flexural-torsional buckling of interbraced columns. *Engineering Structures*, **1** (April), 131–38.
7. Winter, G. (1958) Lateral bracing of columns and beams. *Journal of the Structural Division, ASCE*, **84** (ST2) 1561.1–22.

7 Simply supported beams

7.1 General

A beam which is bent in its stiffer principal plane may buckle out of that plane by deflecting laterally u and twisting ϕ, as shown in Figures 7.1 and 7.2b. These deformations are interdependent. For example, a twist rotation ϕ of the beam cross-section will cause the in-plane bending moment M_x to have an out-of-plane component $M_x\phi$ as shown in Figure 7.3a, which will cause lateral deflections u. Conversely, lateral deflections u will cause the moment M_x to have a torque component $M_x u'$ as shown in Figure 7.3b, which will cause twist rotations ϕ.

The resistance to out-of-plane buckling depends on the resistance to lateral bending and torsion. Thus slender beams with low values of EI_y/L^2, GJ, and EI_w/L^2 are likely to buckle in the elastic range under quite low loads, as indicated in Figure 7.4. Steel beams with moderate resistances to lateral bending and torsion are likely to yield and then buckle inelastically, while stocky steel beams with high resistances will fail in some other mode, such as in-plane collapse when fully plastic.

The elastic buckling of simply supported beams is treated in this chapter, while Chapter 8 deals with restrained beams, Chapter 9 with cantilevers and Chapter 10 with braced and continuous beams. Some special topics on elastic buckling are treated in Chapter 16. The inelastic buckling of steel beams is discussed in Chapter 14, and the use of buckling predictions in the design of steel beams against flexural-torsional buckling is treated in Chapter 15.

The beams considered in this chapter are simply supported in-plane and out-of-plane. Beams which are simply supported in-plane are single span beams whose ends are fixed against in-plane transverse deflections ($v_0 = v_L = 0$) but are unrestrained against in-plane rotations v'_0, v'_L. The ends of beams which are simply supported out-of-plane are fixed against out-of-plane deflections and twist rotations ($u_0 = u_L = \phi_0 = \phi_L = 0$), but are unrestrained against minor axis rotations u'_0, u'_L (so that $u''_0 = u''_L = 0$) and against warping displacements proportional to ϕ'_0, ϕ'_L (so that $\phi''_0 = \phi''_L = 0$), as shown in Figure 7.5b.

The beams are assumed to be perfectly straight and untwisted before loading (crooked and twisted beams are treated in Chapter 15) and to be loaded by moments or loads which initially cause deflections only in the plane of loading. It is also assumed that the directions of the loads or of the planes of the applied moments remain unchanged during buckling (the effects of loads whose directions change are considered in Chapter 16).

Figure 7.1 Flexural-torsional buckling of a beam.

7.2 Uniform bending

7.2.1 DOUBLY SYMMETRIC SECTIONS

Uniform bending is induced in a simply supported beam of doubly symmetric cross-section by equal and opposite end moments M as shown in Figure 7.5a, so that $M_x = M$. If the beam buckles laterally u and twists ϕ into an adjacent position as shown in Figure 7.5c and d, then for this position to be one of equilibrium, the differential equilibrium equations (section 2.8.4.2)

$$(EI_y u'')'' + (M_x \phi)'' = 0 \tag{7.1}$$

and

$$(EI_w \phi'')'' - (GJ\phi')' + M_x u'' = 0, \tag{7.2}$$

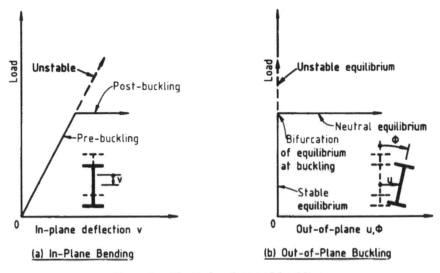

Figure 7.2 Elastic bending and buckling.

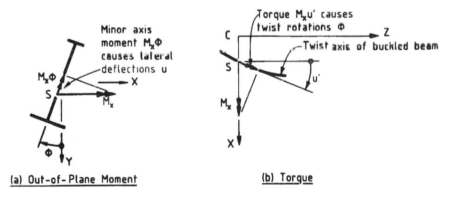

Figure 7.3 Interdependence of u and ϕ.

and the boundary conditions

$$u_{0,L} = u''_{0,L} = \phi_{0,L} = \phi''_{0,L} = 0 \tag{7.3}$$

must be satisfied. Equation 7.1 represents the equality at equilibrium between the out-of-plane bending action $-(M_x\phi)''$ and the flexural resistance $(EI_yu'')''$, and equation 7.2 represents the equality between the torsion action $-M_xu''$, and the warping and torsional resistances $(EI_w\phi'')''$ and $-(GJ\phi')'$.

It can be verified by substitution that these equations are satisfied by the buckled shapes

$$u/\delta = \phi/\theta = \sin \pi z/L \tag{7.4}$$

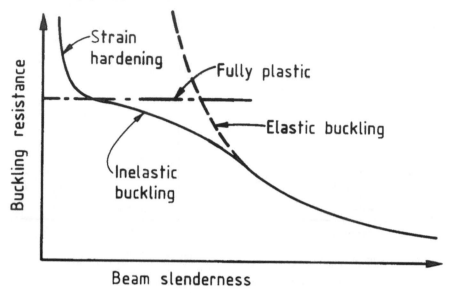

Figure 7.4 Effect of slenderness on buckling resistance.

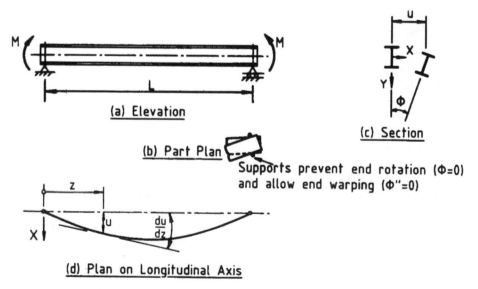

Figure 7.5 Doubly symmetric beam in uniform bending.

where δ and θ are the values of u and ϕ at mid-span, provided the value of the applied moments M is given by

$$M_{yz} = \sqrt{\{(\pi^2 EI_y/L^2)(GJ + \pi^2 EI_w/L^2)\}} \tag{7.5}$$

or

$$M_{yz} = r_0 \sqrt{(P_y P_z)}, \tag{7.6}$$

in which

$$r_0^2 = (I_x + I_y)/A, \tag{7.7}$$

$$P_y = \pi^2 EI_y/L^2, \tag{7.8}$$

and

$$P_z = (GJ + \pi^2 EI_w L^2)/r_0^2 \tag{7.9}$$

are the minor axis flexural and torsional buckling loads of a simply supported column (see section 5.2). It can also be verified that the magnitudes of the buckled shapes are related by

$$\delta/\theta = M_{yz}/P_y \tag{7.10}$$

which defines an axis of buckling rotation of the cross-section (see Figure 7.6) at a distance

$$y_c = u/\phi = M_{yz}/P_y \tag{7.11}$$

below the z axis.

These results may also be obtained by substituting the buckled shapes of

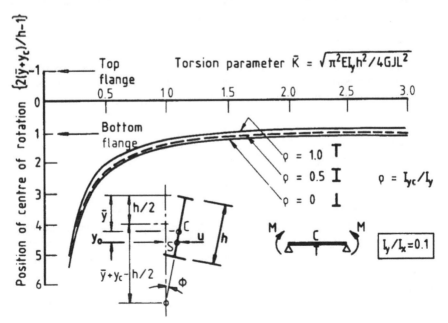

Figure 7.6 Approximate centres of rotation during buckling.

equation 7.4 into the energy equation (section 2.8.4.1)

$$\frac{1}{2}\int_0^L \{EI_y u''^2 + EI_w \phi''^2 + GJ\phi'^2\}\,dz = -\frac{1}{2}\int_0^L 2M_x \phi u''\,dz \qquad (7.12)$$

which represents the equality at buckling between the flexural, warping and torsional strain energy stored and the work done by the bending moment $M_x = M$.

At the buckling moment M_{yz} defined by equation 7.5, the buckling deformations are defined in shape by equations 7.4 and 7.10, but are indeterminate in magnitude, as indicated in Figure 7.2b. When the applied moments M are less than M_{yz}, then the only solution of equations 7.1 and 7.2 is

$$u = \phi = 0, \qquad (7.13)$$

indicating that the beam remains unbuckled until M_{yz} is reached, as shown in Figure 7.2b. Thus the state of equilibrium bifurcates at $M = M_{yz}$ from the stable position given by equation 7.13 to neutral equilibrium positions defined by equations 7.4 and 7.10.

The relationship given by equation 7.6 between the buckling resistance M_{yz} and the harmonic mean of the column buckling loads P_y, P_z demonstrates the interdependence of the flexural and torsional resistances in providing the beam buckling resistance.

The expression for the buckling resistance given by equation 7.5 demonstrates that the resistance decreases as the beam length L increases, and at a rate which depends on the value of the torsion parameter

$$K = \sqrt{(\pi^2 EI_w/GJL^2)}. \qquad (7.14)$$

For narrow rectangular section beams, $I_w = 0$, and so the buckling resistance M_{yz} is inversely proportional to L. In this case $I_y = dt^3/12$ and $J \approx dt^3/3$, while $E/G \approx 2.5$ for steel, so that equation 7.5 becomes

$$M_{yz} \approx 0.4\,P_y L. \qquad (7.15)$$

For I-section beams, $I_w = I_y h^2/4$, where h is the distance between flange centroids. For very thin-walled beams, $J \to 0$, and equation 7.5 approaches

$$M_{yz} \approx P_y h/2 \qquad (7.16)$$

so that M_{yz} is nearly inversely proportional to L^2. This is similar to the flexural buckling behaviour of columns, and has led to the modelling of beam buckling in terms of the column buckling resistance $P_y/2$ of the compression flange. Such a model is valid in this limiting case, since the axis of buckling rotation (at $y_c = M_{yz}/P_y \approx h/2$) is at the tension flange, which therefore does not make any contribution to the flexural and warping resistances.

However, this model is inaccurate for the usual range of hot-rolled steel I-section beams, for which both I_w and J are significant. In this case the buckling resistance varies in a less simple way with the beam length L, as indicated by equation 7.5, while the distance to the centre of rotation y_c varies between $0.4L$ approximately for narrow rectangular beams and $h/2$ for very thin-walled I-section beams. The variation of $2y_c/h$ with K ($K = \bar{K}$ for doubly symmetric I-beams) is shown by the dashed line in Figure 7.6.

The solution given by equation 7.5 for the buckling resistance of a doubly symmetric beam in uniform bending ignores the effect of the pre-buckling in-plane deflections, which transform the beam into a 'negative arch', and increase its buckling resistance. The resistance M is more accurately given by (see section 16.6)

$$\frac{M}{M_{yz}} = \frac{1}{\sqrt{\{(1 - EI_y/EI_x)(1 - [GJ + \pi^2 EI_w/L^2]/2EI_x)\}}}. \tag{7.17}$$

For many practical I-section beams, EI_x is much greater than EI_y, GJ, EI_w/L^2, and the more accurate value is only slightly larger than that given by equation 7.5, and so it is common to ignore this effect. However, this is not the case for hollow rectangular sections, where EI_y and GJ are often comparable with EI_x, and the buckling resistance may be greatly increased. Equation 7.17 indicates that the resistance of a circular hollow section ($EI_y = EI_x$) is infinitely large, and that beams bent about their minor axes ($EI_x < EI_y$) do not buckle. These conclusions for doubly symmetric beams in uniform bending are summarized in Figure 7.7a.

7.2.2 POINT SYMMETRIC SECTIONS

Sections which are point symmetric, such as zed-sections, have coinciding centroidal and shear centre axes ($x_0 = y_0 = 0$). When such a beam is bent in its

Section Type	Example	Buckling Equation	Comment	Section Type	Example	Buckling Equation	Comment
(a) Doubly Symmetric $x_o = y_o = 0$	 I	7.5	7.16 if very thin-walled	(c) Mono-symmetric $y_o = 0$	⊏	7.5	—
	ǀ	7.15	$I_w = 0$		<	7.5	$I_w = 0$
	▯	7.17	Rarely buckles		H	7.5	Rarely buckles
	O	(7.17)	Does not buckle	(d) Mono-symmetric $x_o = 0$	T	7.22	—
	▭	(7.17)	Does not buckle		⋃	(7.22)	Does not buckle (uniform bending)
	⊢⊣	(7.17)	Does not buckle		∧	(7.22)	Does not buckle (uniform bending)
(b) Point Symmetric	S	7.5	Principal plane bending is rare	(e) Asymmetric	ſ	7.22	Principal plane bending is rare

Figure 7.7 Effect of cross-section for uniform bending.

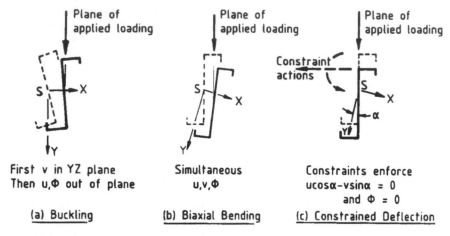

Figure 7.8 Behaviour of zed-beams.

stiffer YZ principal plane, it may buckle laterally u, ϕ out of this plane of bending as indicated in Figure 7.8a. The differential equilibrium equations (and the energy equation) for point symmetric beams are the same as those for doubly symmetric beams, and so their elastic buckling resistances under uniform bending are also given by equation 7.5, provided appropriate values of the section constants I_y, J, and I_w are substituted.

However, zed-beams bent in principal planes are rare in practice. More commonly, a beam is loaded initially in the plane of the web as shown in Figure 7.8b, in which case it bends biaxially, by simultaneously deflecting u, v, and twisting ϕ.

Alternatively, a zed-beam may be constrained to deflect in a particular plane such as that of the web, as shown in Figure 7.8c. In this case the constraints prevent any buckling effects.

7.2.3 MONOSYMMETRIC SECTIONS

7.2.3.1 Sections bent about an axis of symmetry

Sections which may be bent about an axis of symmetry ($y_0 = 0$) include channels, equal leg angles, and monosymmetric I-sections, as shown in Figure 7.7c. The differential equilibrium equations and the energy equation for these are the same as those for doubly symmetric beams and so the elastic buckling resistances under uniform bending are also given by Equation 7.5. For cases where EI_x is not large compared with EI_y, GJ, EI_w/L^2, then equation 7.17 may be used for a more accurate estimate of the buckling resistance.

7.2.3.2 Sections bent in a plane of symmetry

Sections which may be bent in a plane of symmetry include monosymmetric I-sections, trough sections, and equal leg angles, as shown in Figure 7.7d. For these sections, additional terms enter into the equilibrium equations

$$(EI_y u'')'' + (M_x \phi)'' = 0, \tag{7.18}$$

$$(EI_w \phi'')'' - (GJ\phi')' + M_x u'' - (M_x \beta_x \phi')' = 0, \tag{7.19}$$

and the energy equation

$$\frac{1}{2}\int_0^L (EI_y u''^2 + EI_w \phi''^2 + GJ\phi'^2)\,dz = -\frac{1}{2}\int_0^L (2M_x \phi u'' + M_x \beta_x \phi'^2)\,dz. \tag{7.20}$$

These additional terms are associated with a monosymmetry property of the cross-section

$$\beta_x = (1/I_x)\int_A y(x^2 + y^2)\,dA - 2y_0. \tag{7.21}$$

The buckled shapes which satisfy equations 7.18, 7.19, and the boundary conditions of equation 7.3 are again given by equation 7.4, but the uniform moment M at elastic buckling, which can be obtained by substituting equation 7.4 into equations 7.18 and 7.19 or into equation 7.20, is given by

$$\frac{M}{M_{yz}} = \pm \sqrt{\left\{1 + \left(\frac{\beta_x P_y}{2M_{yz}}\right)^2\right\}} + \left(\frac{\beta_x P_y}{2M_{yz}}\right) \tag{7.22}$$

in which M_{yz} and P_y are given by equations 7.5 and 7.8, while the centre of rotation during buckling lies at

$$y_c = M/P_y + y_0. \tag{7.23}$$

The positive solutions for the elastic buckling resistance M given by equation 7.22 are shown in Figure 7.9. These solutions correspond to positive moments M which cause compression in the top fibres of the section. The negative solution, which corresponds to a negative moment which causes tension in the top fibres, is identical with the positive solution when the beam is inverted so that β_x changes sign. For positive M, the buckling resistance increases with the dimensionless monosymmetry parameter $\beta_x P_y/2M_{yz}$. For negative values of $\beta_x P_y/2M_{yz}$, for which the shear centre S lies below the centroid C of the cross-section, the buckling resistance is low, while the reverse is true for positive values of $\beta_x P_y/2M_{yx}$ for which the shear centre is above the centroid.

A better understanding of this may be gained by noting that tensile longitudinal stresses reduce twisting, as indicated in Figure 7.10a by the action of the child's spinning toy, while compressive stresses increase twisting as demonstrated in Figure 7.10b by the torsional buckling of a water tower. This latter effect is the same as that which causes torsional buckling in cruciform-section columns (section 5.3). For doubly symmetric beams, the tensile and compressive bending

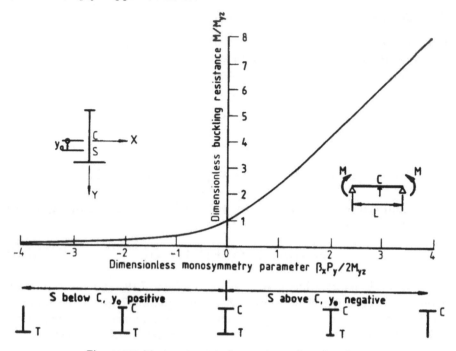

Figure 7.9 Monosymmetric beams in uniform bending.

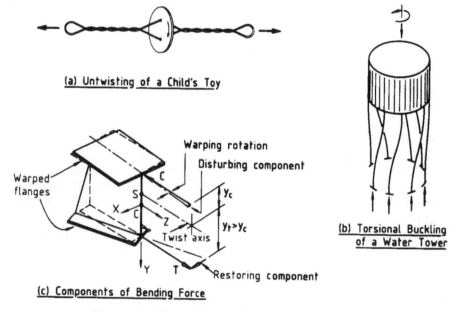

(a) Untwisting of a Child's Toy

(b) Torsional Buckling of a Water Tower

(c) Components of Bending Force

Figure 7.10 Effects of longitudinal stress on twisting.

stresses are equal, and so the increased buckling resistance caused by the tensile stresses is balanced by the increased buckling action caused by the compressive stresses. However, this balance is upset in monosymmetric beams, such as the I-section shown in Figure 7.10c. Here the compressive force C in the larger top flange is approximately equal to the tensile force T in the bottom flange, but because this top flange is closer to the axis of twist through the shear centre S, its warping rotations during twisting are less than those of the smaller bottom flange, and its disturbing transverse component is less than the restoring component of the tensile bottom flange force. The disturbing torque exerted by the top flange component about the close shear centre axis is even less than the restoring torque of the more distant bottom flange component, and so the effect of the force in the smaller flange dominates. In this case it is a tensile force, and so the buckling resistance is increased. Conversely, when the larger flange is in tension, the resistance to buckling is decreased, as shown in Figure 7.9.

The solution of equation 7.22 for the elastic buckling moment requires the use of the section constants I_y, J, I_w, and β_x. While these can be determined from the section dimensions, simple approximations which can be evaluated by hand are often sought. For monosymmetric I-sections, these can be expressed in terms of the value of

$$\rho = I_{yc}/I_y \qquad (7.24)$$

the ratio of the minor axis second moments of area of the compression flange and the complete cross-section. Thus the monosymmetry property β_x can be approximated by [1]

$$\beta_x \approx 0.9h(2\rho - 1)(1 - I_y^2/I_x^2) \qquad (7.25)$$

while

$$I_w = \rho(1 - \rho)I_y h^2 \qquad (7.26)$$

and

$$J \approx \Sigma b t^3/3 \qquad (7.27)$$

where b and t are the width and the thickness of each narrow rectangular element of the cross-section.

Approximate values of $ML/\sqrt{(EI_y GJ)}$ determined from these equations are plotted in Figure 7.11 against the torsion parameter

$$\bar{K} = \sqrt{(\pi^2 EI_y h^2/4GJL^2)} \qquad (7.28)$$

for selected values of ρ. For a tee-beam with the flange in compression ($\rho = 1.0$), the dimensionless buckling resistance is significantly higher than that for an equal flanged I-beam ($\rho = 0.5$) with the same value of $\bar{K}$, but the resistance is greatly reduced for a tee-beam with the flange in tension ($\rho = 0.0$).

The positions of the centre of rotation during buckling may also be approximated by using equation 7.25, and these are shown in Figure 7.6. It can be seen that these are close to the bottom flange for high values of $\bar{K}$, but move lower as $\bar{K}$ decreases.

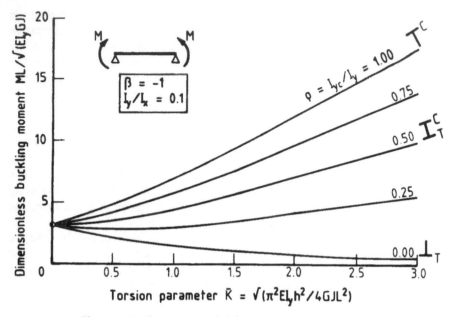

Figure 7.11 Monosymmetric I-beams in uniform bending.

7.2.4 ASYMMETRIC SECTIONS

The differential equilibrium equations and the energy equation for asymmetric sections (x_0, $y_0 \neq 0$) bent in their stiffer principal planes are the same as equations 7.18–7.20 for monosymmetric beams ($x_0 = 0$) bent in the plane of symmetry, and so the buckling resistances under uniform bending are also given by equation 7.22. The most common example of an asymmetric section is an unequal angle (Figure 7.7e).

However, angle beams bent in principal planes are rare in practice. More commonly, an angle beam is loaded in the plane of a leg, in which case it bends biaxially in the same way as does the zed-beam shown in Figure 7.8b. Alternatively, it may be constrained to deflect in the plane of a leg, in much the same way as the zed-beam of Figure 7.8c is constrained to the plane of the web, in which case the constraints prevent any buckling effects.

7.3 Moment gradient

A simply supported beam is under constant moment gradient when its bending moment distribution varies linearly, as in the case of the beam shown in Figure 7.12a, which is bent by end moments M and βM, so that

$$M_x = M - M(1 + \beta)z/L. \tag{7.29}$$

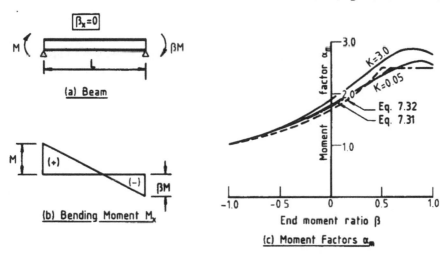

Figure 7.12 Doubly symmetric beams under moment gradient.

In this case, some of the terms in the differential equations of bending and torsion (equations 7.18 and 7.19) have variable coefficients, and it is much more difficult to obtain solutions than previously. Approximate solutions may be obtained by using the hand energy method discussed in Chapter 3, while more accurate solutions may be found by using finite element computer programs, such as that described in Chapter 4.

Some numerical solutions for doubly symmetric sections obtained by this latter method are shown in Figure 7.12c, in which the variations of a moment factor

$$\alpha_m = M/M_{yz} \tag{7.30}$$

are plotted against the end moment ratio β for selected values of the torsion parameter K (equation 7.14). It can be seen that while α_m is almost independent of K, there are substantial variations with the end moment ratio β, and that the buckling resistance of a beam in double curvature bending ($\beta = +1$) is approximately 2.5 times the resistance for uniform bending ($\beta = -1$). Approximate lower bound values for α_m are given by

$$\alpha_m = 1.75 + 1.05\beta + 0.3\beta^2 \leqslant 2.5 \tag{7.31}$$

or

$$\alpha_m = 1/(0.6 - 0.4\beta) \leqslant 2.5, \tag{7.32}$$

while approximate mean values are given by

$$\alpha_m = -1.81/\sqrt{(1 - 1.40\beta + 0.89\beta^2)} \leqslant 2.6. \tag{7.33}$$

The increases in buckling resistance that occur with increases in the end moment ratio β are associated principally with changes that occur in the buckled deflected shapes, which change from a symmetric half sine wave for uniform bending

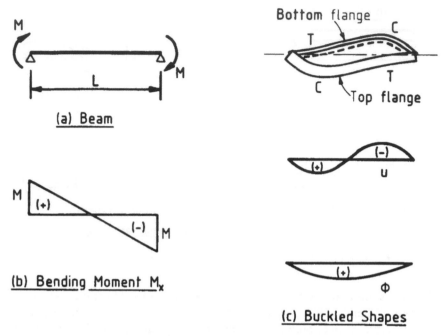

Figure 7.13 Buckled shapes for double curvature bending ($\beta = 1$).

($\beta = -1$), to the anti-symmetric double half wave for double curvature bending ($\beta = +1$) shown in Figure 7.13c. In this latter case, the anti-symmetry of the M_x distribution requires the deflection u to be anti-symmetric if the twist rotation ϕ is symmetric (see equations 7.1 and 7.2). Because of this anti-symmetric u, the flange deflections $u \pm h\phi/2$ are biased towards the compression regions of the beam (top left and bottom right), as shown in Figure 7.13c.

The buckling of monosymmetric I-beams under moment gradient has been studied in [2], and approximate solutions for double curvature bending ($\beta = +1$) are shown in Figure 7.14. In this case it does not matter which flange is the larger (i.e. the curves for ρ and $(1 - \rho)$ are identical), since each flange is similarly stressed. This is quite different from the behaviour shown in Figure 7.11 for beams in uniform bending ($\beta = -1$), where one flange acts only in compression, and the other only in tension. Approximate solutions for other moment gradients are given in [2], which suggests that these may be approximated by

$$\frac{ML}{\sqrt{(EI, GJ)}} = \alpha_m \pi \left\{ \sqrt{\left[1 + 4\rho(1 - \rho) f_1 \bar{K}^2 + \left(\frac{f_2 \bar{K} \beta_x}{h} \right)^2 \right]} + \frac{f_2 \bar{K} \beta_x}{h} \right\}, \quad (7.34)$$

where

$$f_1 = 1 + 3(\beta_x/h)^2 (1 + \beta)^n/2^n, \quad (7.35)$$

$$f_2 = \alpha_m(1 - \beta)/2 + (2\beta_x/h)(1 + \beta)^n/2^n, \quad (7.36)$$

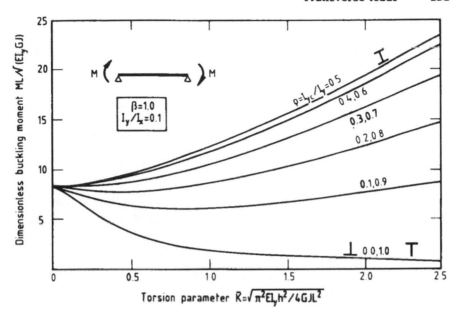

Figure 7.14 Monosymmetric I-beams in double curvature bending.

and

$$\eta = 9(1 + \rho)/(1 + \bar{K}).$$ (7.37)

7.4 Transverse loads

7.4.1 CONCENTRATED LOADS

The bending moment M_x in a beam with transverse load varies along the beam, and so the differential equilibrium equations again have some variable coefficients, and are difficult to solve. Numerical solutions obtained by the finite element method for doubly symmetric beams with central concentrated loads Q are shown in Figure 7.15. For the case where the load acts at the shear centre, these may be closely approximated by using

$$M_m = \alpha_m M_{yz},$$ (7.38)

with

$$\alpha_m = 1.35$$ (7.39)

and

$$M_m = QL/4.$$ (7.40)

When the shear centre load Q acts at a distance a away from mid-span, then equation 7.38 can again be used, but with

$$\alpha_m = 1.35 + 0.4(2a/L)^2$$ (7.41)

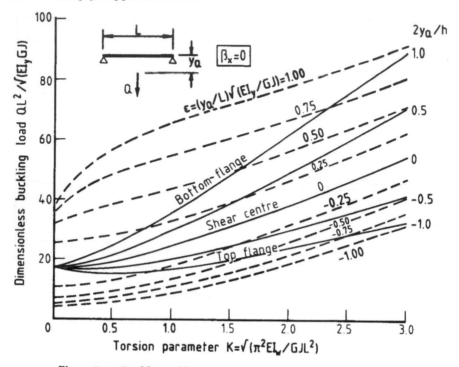

Figure 7.15 Buckling of beams with central concentrated loads.

and

$$M_m = (QL/4)(1 - 4a^2/L^2). \qquad (7.42)$$

Equation 7.41 is plotted in Figure 7.16, and it can be seen that the resistance is lowest when the load is at mid-span, so that the maximum bending moment is also at mid-span.

When there are two shear centre loads Q, each acting at a distance a from the centre, then

$$\alpha_m = 1.0 + 0.35(1 - 2a/L)^2 \qquad (7.43)$$

and

$$M_m = (QL/2)(1 - 2a/L). \qquad (7.44)$$

Equation 7.43 is also plotted in Figure 7.16, and it can be seen that the moment factor α_m is lowest when $a = L/2$ so that the beam is in uniform bending, and highest when $a = 0$ so that the length of the uniform bending region is zero. However, even though α_m decreases as $2a/L$ increases, the value of Q at buckling increases. The approximate solutions of equations 7.38–7.44 are collected together in Figure 7.17.

Solutions for the buckling of monosymmetric I-beams with transverse loads are reported in [3–5]. These solutions indicate that the interaction between the

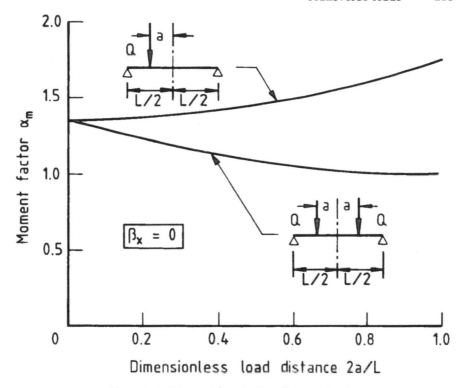

Figure 7.16 Moment factors for off-centre loads.

moment distribution and monosymmetry is complex, especially for T-beams, and that it is not easy to develop simple approximations for the buckling resistance. Nevertheless, a wide range of solutions is presented non-dimensionally in [3–5] in tabular or graphical form. Approximate equations are also given in [4], and for beams with central concentrated loads Q at the shear centre, these may be expressed as

$$\frac{M_m}{M_{yz}} = \alpha_m \left\{ \sqrt{\left[1 + \left(\frac{0.4\alpha_m f_3 \beta_x}{2M_{yz}/P_y} \right)^2 \right]} + \frac{0.4\alpha_m f_3 \beta_x}{2M_{yz}/P_y} \right\}$$ (7.45)

with $M_m = QL/4$, $\alpha_m = 1.35$, and

$$f_3 = \pi^2/8 - \tfrac{1}{2} \approx 0.73.$$ (7.46)

Approximations are also given for off-centre loads, and for two point loads. These approximate equations are reported to be of good accuracy generally, except for very monosymmetric sections for which ρ approaches 0 or 1.

Beam Segment	Moment Distribution	α_m	Range
 $M \,(\!\!\!-\!\!\!-\!\!\!-\!\!\!)\,\beta M$	 $M \,\cdots\, \beta M$	$1.75+1.05\beta+0.3\beta^2 \not> 2.5$	$-1<\beta<1$
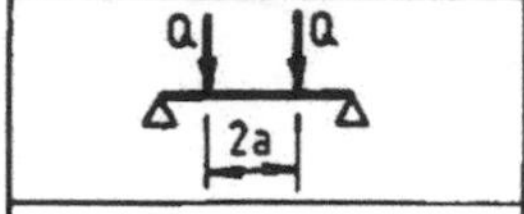 $Q\ Q$ with $2a$	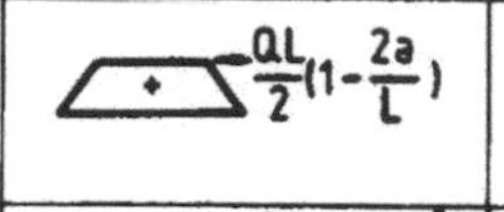 $\frac{QL}{2}(1-\frac{2a}{L})$	$1.0+0.35(1.0-2a/L)^2$	$0<\frac{2a}{L}<1$
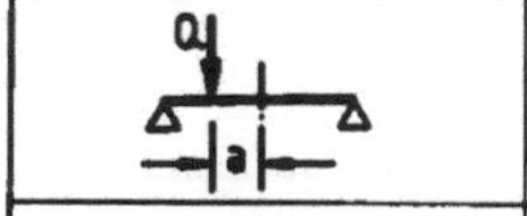 Q with a	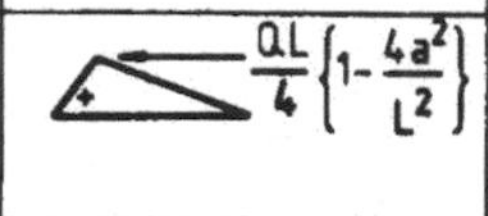 $\frac{QL}{4}\left\{1-\frac{4a^2}{L^2}\right\}$	$1.35+0.4(2a/L)^2$	$0<\frac{2a}{L}<1$
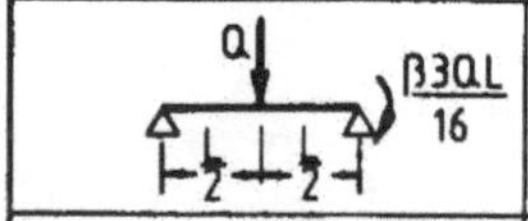 Q, $\beta\frac{3QL}{16}$	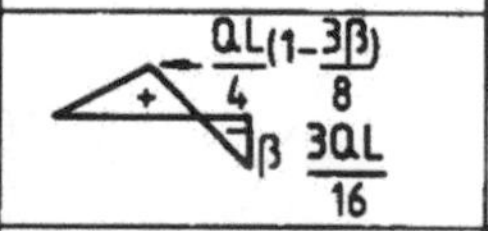 $\frac{QL}{4}\frac{(1-3\beta)}{8}$, $\beta\frac{3QL}{16}$	$1.35+0.15\beta$ $-1.2+3.0\beta$	$0<\beta<0.89$ $0.89<\beta<1$
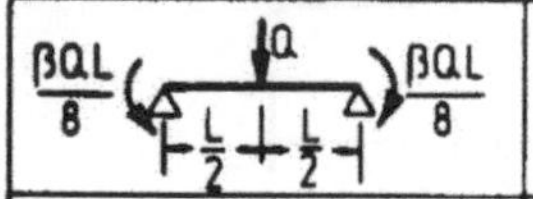 $\frac{\beta QL}{8}\,(\!-\!)\,\frac{\beta QL}{8}$	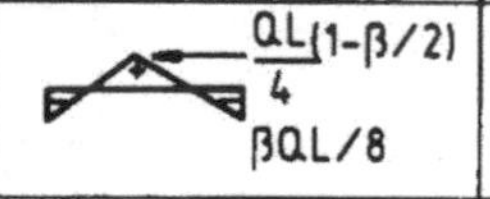 $\frac{QL(1-\beta/2)}{4}$, $\beta QL/8$	$1.35+0.36\beta$	$0<\beta<1$
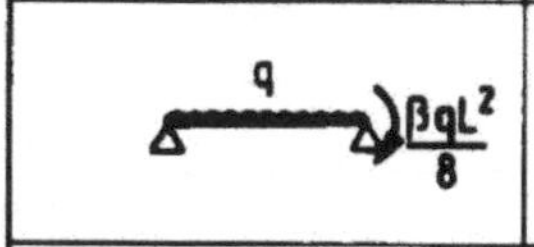 q, $\beta\frac{qL^2}{8}$	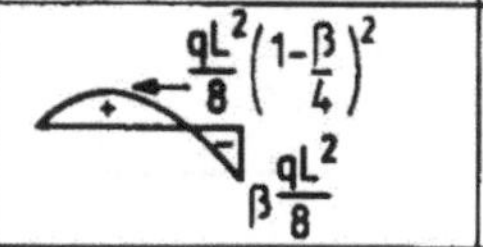 $\frac{qL^2}{8}\left(1-\frac{\beta}{4}\right)^2$, $\beta\frac{qL^2}{8}$	$1.13+0.10\beta$ $-1.25+3.5\beta$	$0<\beta<0.7$ $0.7<\beta<1$
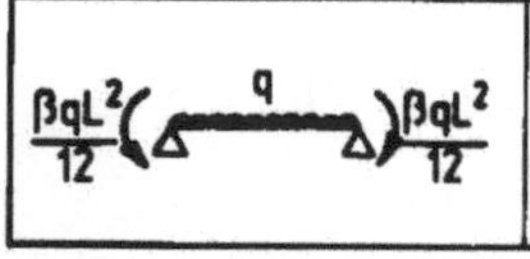 $\frac{\beta qL^2}{12}\,(\!-\!q\!-\!)\,\frac{\beta qL^2}{12}$	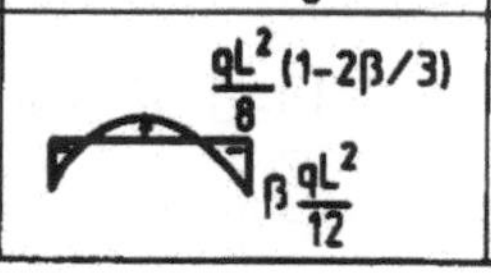 $\frac{qL^2}{8}(1-2\beta/3)$, $\beta\frac{qL^2}{12}$	$1.13+0.12\beta$ $-2.38+4.8\beta$	$0<\beta<0.75$ $0.75<\beta<1$

Figure 7.17 Moment factors for doubly symmetric beams (loads at shear centre).

7.4.2 UNIFORMLY DISTRIBUTED LOADS

The buckling resistances of doubly symmetric beams with uniformly distributed loads q are shown in Figure 7.18. For shear centre loading, these may be approximated by using equation 7.38 with

$$\alpha_m = 1.13 \tag{7.47}$$

and

$$M_m = qL^2/8. \tag{7.48}$$

This result is lower than the buckling resistance given by equation 7.39 for beams with central concentrated loads, because the bending moment distribution for distributed loading is more nearly uniform. This reinforces the conclusion reached in section 7.4.1 that the moment factor α_m decreases as the moment distribution approaches that of uniform bending. The approximate solution of equations 7.47 and 7.48 is included in Figure 7.17.

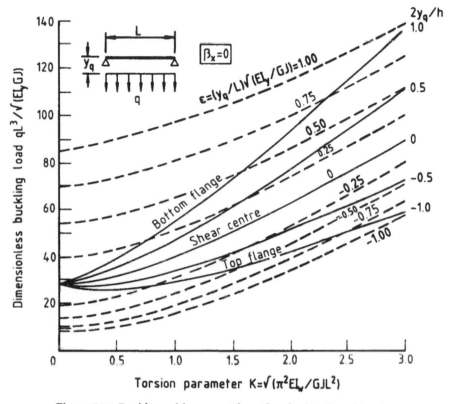

Figure 7.18 Buckling of beams with uniformly distributed loads.

Solutions for monosymmetric I-beams are reported in [3–5]. For beams with uniformly distributed shear centre loading, these can be approximated [4] by using equation 7.45 with $M_m = qL^2/8$, $\alpha_m = 1.13$, and

$$f_3 = \pi^2/6 - \tfrac{1}{2} \approx 1.14. \qquad (7.49)$$

Once again, these are reported to be of good accuracy, except for very monosymmetric sections.

7.5 Transverse loads and end moments

7.5.1 GENERAL

While isolated simply supported beams rarely have in-plane end moments, they may be used to represent the individual segments or spans of braced or continuous beams, as shown in Figure 7.19. The approximations for out-of-plane simple supports ($u = 0 = \phi$, $u'' = 0 = \phi''$) ignore any continuity actions which may

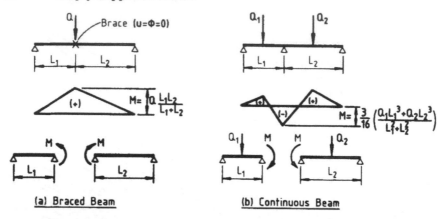

(a) Braced Beam   (b) Continuous Beam

Figure 7.19 Simple beam approximations for braced and continuous beams.

restrain buckling of the individual segment or span. The effects of buckling restraints are treated in Chapter 8, while approximate methods of estimating their effects in braced and continuous beams are given in Chapter 10. When these buckling restraint effects are ignored, then approximate estimates of the elastic buckling resistances can be obtained from section 7.5.2 or 7.5.3 following.

7.5.2 CENTRAL CONCENTRATED LOADS

The magnitudes of the in-plane end moments acting on a beam may be conveniently expressed in terms of the corresponding end moments of a built-in beam. Thus for a beam with equal moments at both ends, the corresponding fixed end moments for central concentrated load Q are $QL/8$, so that the actual end moment may be expressed as $\beta QL/8$, where $0 \leqslant \beta \leqslant 1$. The maximum moment M_m always occurs at mid-span, and is given by

$$M_m = (1 - \beta/2)QL/4. \tag{7.50}$$

For the case where the load Q acts at the shear centre of a doubly symmetric I-section beam, the value of M_m at elastic buckling may be approximated by using equation 7.38 with

$$\alpha_m = 1.35 + 0.36\beta. \tag{7.51}$$

For a beam with zero moment at one end, the fixed end moment is $3QL/16$, and so the actual end moment may be expressed as $3\beta QL/16$, where $0 \leqslant \beta \leqslant 1$. The maximum moment occurs at mid-span while $0 \leqslant \beta < 8/9$, and is given by

$$M_m = (1 - 3\beta/8)QL/4 \tag{7.52}$$

and occurs at the end while $8/9 \leqslant \beta \leqslant 1$, and is given by

$$M_m = 3\beta QL/16. \tag{7.53}$$

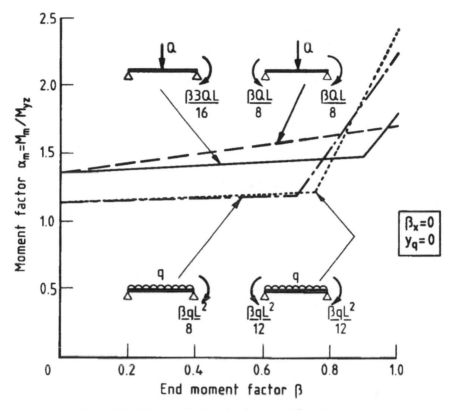

Figure 7.20 Moment factors for beams with end moments.

For the case where the load Q acts at the shear centre of a doubly symmetric I-section beam, the value of M_m at elastic buckling may be approximated by using equation 7.38 with

$$\alpha_m = 1.35 + 0.15\beta \tag{7.54}$$

while $0 \leqslant \beta < 0.89$, and by

$$\alpha_m = -1.2 + 3.0\beta \tag{7.55}$$

while $0.89 < \beta \leqslant 1$.

The approximations of equations 7.50–7.55 are collected together in Figure 7.17, and plotted in Figure 7.20.

7.5.3 UNIFORMLY DISTRIBUTED LOADS

For a beam with equal end moments and uniformly distributed load q, the fixed end moments are $qL^2/12$, and so the actual end moment may be expressed as

$\beta qL^2/12$, where $0 \leqslant \beta \leqslant 1$. The maximum moment occurs at mid-span while $0 \leqslant \beta \leqslant 0.75$ and is given by

$$M_m = (1 - 2\beta/3)qL^2/8 \tag{7.56}$$

and occurs at the end while $0.75 \leqslant \beta \leqslant 1$ and is given by

$$M_m = \beta qL^2/12. \tag{7.57}$$

For the case where the load q acts at the shear centre of a doubly symmetric I-section beam, the value of M_m at elastic buckling may be approximated by using equation 7.38 with

$$\alpha_m = 1.13 + 0.12\beta \tag{7.58}$$

while $0 \leqslant \beta < 0.75$, and by

$$\alpha_m = -2.38 + 4.8\beta \tag{7.59}$$

while $0.75 < \beta \leqslant 1$.

Similar approximations have been developed for beams with zero moment at one end, and these are collected together with those of equations 7.56–7.59 in Figure 7.17, and plotted in Figure 7.20.

7.6 Effects of load height

7.6.1 CONCENTRATED LOADS

The buckling resistance of a simply supported beam may be significantly affected by the distances of transverse loads from the shear centre axis. Most examples of off-shear centre loading involve freely swinging gravity loads, as for example in the case where a crane runway girder supports loads acting at its top flange, or when a monorail supports loads acting at its bottom flange.

When a transverse concentrated load Q acts at a distance $(y_Q - y_0)$ below the shear centre and moves with the beam during buckling, as shown in Figure 7.21a, it exerts an additional torque $-Q(y_Q - y_0)\phi$ about the shear centre axis. This additional torque opposes the twist rotations ϕ of the beam, and increases the resistance to buckling, as indicated in Figure 7.15. Conversely, when the load acts above the shear centre, then the additional torque amplifies the twist rotations, and reduces the buckling resistance of the beam.

This latter effect may be very dangerous, as for example in the case of a beam with $I_y > I_x$, which might not be expected to buckle. If the torsional stiffness of such a beam is low, then buckling may occur in a predominantly torsional mode. In the limiting case in which there is no lateral deflection and the work done by the bending moment M_x is negligible, then the buckling load may be evaluated by setting the work $-Q(y_Q - y_0)\phi^2/2$ done by the load equal to the strain energy $\frac{1}{2}\alpha\phi^2$ stored in torsion, where α is the torsional stiffness of the system, so

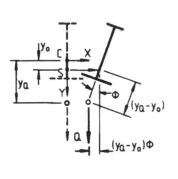

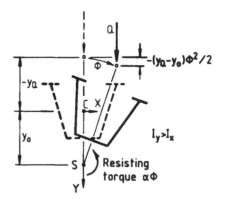

(a) Flexural-Torsional Buckling   (b) Torsional Buckling

Figure 7.21 Effects of load height.

that

$$Q = -\alpha/(y_Q - y_0). \qquad (7.60)$$

The negative sign in this equation indicates that downwards load causes buckling when it acts above the shear centre so that $(y_Q - y_0)$ is negative. For simply supported doubly symmetric beams with central concentrated load, the torsional stiffness is

$$\alpha = (4GJ/L)/\{1 - [\tan h(\pi/2K)]/(\pi/2K)\}. \qquad (7.61)$$

This stiffness may be reduced in monosymmetric beams with negative values of β_x, such as the trough girder shown in Figure 7.21b, by the effect of the monosymmetry term $-\frac{1}{2}\int_0^L M_x \beta_x \phi'^2 \, dz$ in the energy equation (equation 7.20), in which case the buckling resistance will be correspondingly reduced.

The torsional buckling resistance given by equation 7.60 is reduced when the minor axis flexural rigidity EI_y is low, in which case significant lateral deflections u occur during buckling. Finite element solutions for doubly symmetric I-beams with central concentrated loads Q are shown in Figure 7.15, in which the dimensionless buckling loads $QL^2/\sqrt{(EI_yGJ)}$ are plotted against the torsion parameter K. Two sets of curves are shown, one suitable for I-section beams in which the load height is represented by the parameter

$$\zeta = 2y_Q/h \qquad (7.62)$$

so that $\zeta = \pm 1$ for bottom or top flange loading, and the other of more general application which uses the parameter

$$\varepsilon = \frac{y_Q}{L} \sqrt{\left(\frac{EI_y}{GJ}\right)}. \qquad (7.63)$$

For doubly symmetric I-section beams,

$$\varepsilon = \xi K/\pi. \tag{7.64}$$

Figure 7.15 indicates that the effect of the load height parameter $2y_Q/h$ increases with the torsion parameter K, demonstrating that load height effects are more important in short beams with high EI_w and low GJ.

The solutions shown in Figure 7.15 for doubly symmetric beams may be closely approximated by using

$$\frac{M_m}{M_{yz}} = \alpha_m \left\{ \sqrt{\left[1 + \left(\frac{0.4\alpha_m y_Q}{M_{yz}/P_y} \right)^2 \right]} + \frac{0.4\alpha_m y_Q}{M_{yz}/P_y} \right\} \tag{7.65}$$

and $M_m = QL/4$ and $\alpha_m = 1.35$. The variations of $(M_m/\alpha_m M_{yz})$ with $(0.4\alpha_m P, y_Q/M_{yz})$ are shown in Figure 7.22, which again demonstrates the importance of load height and its effect on buckling resistance.

Numerical solutions for doubly symmetric beams with off-centre and two-point loads are given in [4]. These indicate that equation 7.65 can again be used to obtain approximate solutions, provided the appropriate expressions for M_m and α_m given in Section 7.4.1 are used. Additional numerical solutions for beams with end moments and central concentrated loads are given in [6–9].

The effects of load height on the buckling resistances of monosymmetric I-beams have been tabulated in [3, 5] and graphed in [4]. This latter study suggests that good approximations can be obtained for central concentrated

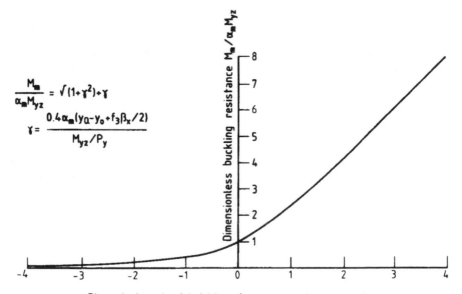

Figure 7.22 Approximation for load-height effects on monosymmetric I-beams.

loads on monosymmetrical beams (except for extreme monosymmetry) by using

$$\frac{M_m}{M_{yz}} = \alpha_m\{\sqrt{(1+\gamma^2)} + \gamma\} \tag{7.66}$$

where

$$\gamma = \frac{0.4\alpha_m(y_Q - y_0 + f_3\beta_x/2)}{M_{yz}/P_y} \tag{7.67}$$

and f_3 is given by equation 7.46. Equation 7.66 is plotted in Figure 7.22.

Approximations for the effects of load height on the buckling resistances of monosymmetric beams with off-centre or two point loading are also given in [4].

7.6.2 UNIFORMLY DISTRIBUTED LOADS

The effects of load height on the buckling resistances of simply supported beams with uniformly distributed loads q are similar to those of concentrated loads. Numerical solutions for beams of doubly symmetric cross-section are shown in Figure 7.18, and these may be closely approximated by using equation 7.65 with

$$\alpha_m = 1.13 \tag{7.68}$$

and

$$M_m = qL^2/8. \tag{7.69}$$

Numerical solutions for monosymmetric I-beams have been tabulated in [3, 5] and graphed in [4]. This latter study suggests that these may be approximated (except for very monosymmetric beams) by using equations 7.66 and 7.67 with equations 7.68, 7.69 and 7.49. Additional numerical solutions for beams with end moments and uniformly distributed loads are given in [6–9].

7.7 Problems

PROBLEM 7.1

Determine the elastic flexural-torsional buckling moment of a beam whose properties are given in Figure 7.23, and which is in uniform bending and simply supported over a span of 12.0 m, as shown in Figure 7.24a.

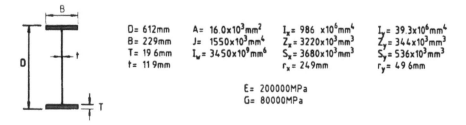

D= 612mm	A= 16.0x10³mm²	I_x= 986 x10⁶mm⁴	I_y= 39.3x10⁶mm⁴
B= 229mm	J= 1550x10³mm⁴	Z_x= 3220x10³mm³	Z_y= 344x10³mm³
T= 19.6mm	I_w= 3450x10⁹mm⁶	S_x= 3680x10³mm³	S_y= 536x10³mm³
t= 11.9mm		r_x= 249mm	r_y= 49.6mm

E= 200000MPa
G= 80000MPa

Figure 7.23 Example section and properties.

142   Simply supported beams

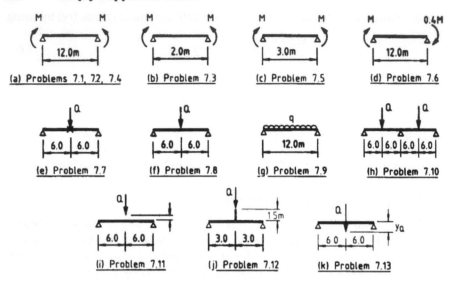

Figure 7.24 Problems 7.1–7.13.

PROBLEM 7.2

Evaluate the effect of the pre-buckling in-plane deflections on the elastic flexural-torsional buckling moment of Problem 7.1.

PROBLEM 7.3

Determine the elastic flexural-torsional buckling moment of a narrow rectangular (50 mm × 10 mm) steel beam ($E = 200\,000\,\text{MPa}$, $G = 80\,000\,\text{MPa}$) which is in uniform bending and simply supported over a span of 2.0 m, as shown in Figure 7.24b.

PROBLEM 7.4

A monosymmetric beam consists of an I-section member whose properties are given in Figure 7.23 with a 300 mm × 20 mm plate welded to one flange. Determine the elastic flexural-torsional buckling moments for the beam which is in uniform bending and simply supported over a span of 12.0 m, as shown in Figure 7.24a,

(a) when the plated flange is in compression;
(b) when the plated flange is in tension.

PROBLEM 7.5

A tee-section beam 305 mm deep is formed by cutting the I-section shown in Figure 7.23 along its centre line. Determine the elastic flexural-torsional buckling moments for the beam which is in uniform bending and simply supported over a span of 3.0 m, as shown in Figure 7.24c,

(a) when the flange is in compression;
(b) when the flange is in tension.

PROBLEM 7.6

Determine the elastic flexural-torsional buckling moment M of a beam whose properties are given in Figure 7.23, and which is simply supported over a span of 12.0 m with unequal end moments M and $0.4 M$ which cause double curvature bending, as shown in Figure 7.24d.

PROBLEM 7.7

Determine the elastic flexural-torsional buckling load of a beam whose properties are given in Figure 7.23, and which is simply supported over a span of 12.0 m. The beam has a concentrated load at mid-span, where bracing prevents lateral deflection and twist, as shown in Figure 7.24e.

PROBLEM 7.8

Determine the elastic flexural-torsional buckling load of a beam whose properties are given in Figure 7.23, and which is simply supported over a span of 12.0 m. The beam has a concentrated load at the shear centre at mid-span, which is unrestrained, as shown in Figure 7.24f.

PROBLEM 7.9

Determine the elastic flexural-torsional buckling load of a beam whose properties are given in Figure 7.23, and which is simply supported over a span of 12.0 m. The beam has uniformly distributed shear centre loading, as shown in Figure 7.24g.

PROBLEM 7.10

Determine the elastic flexural-torsional buckling load Q of a beam whose properties are given in Figure 7.23, and which is continuous over two equal spans of 12.0 m. Each span has a central concentrated load Q acting at the shear centre, which is unbraced, as shown in Figure 7.24h.

PROBLEM 7.11

Determine the elastic flexural-torsional buckling load of a beam whose properties are given in Figure 7.23, and which is simply supported over a span of 12.0 m. The beam has a concentrated load acting on the top flange at mid-span which is unbraced, as shown in Figure 7.24i.

PROBLEM 7.12

Determine the elastic flexural-torsional buckling of a steel beam ($E = 200\,000$ MPa, $G = 80\,000$ MPa) of circular hollow cross-section (outside diameter 610 mm, wall thickness 12.7 mm, $I_y = 1060 \times 10^6$ mm^4, $J = 2130 \times 10^6$ mm^4), and which is simply supported over a span of 6.0 m. The beam has a concentrated load acting at a height of 1500 mm above the beam axis through a rigid stub column welded to the beam at mid-span, as shown in Figure 7.24j.

PROBLEM 7.13

The monosymmetric beam of Problem 7.4 is simply supported over a span of 12.0 m, as shown in Figure 7.24k. The beam has a concentrated load acting

   (i) at the top flange; or
  (ii) at the bottom flange

at mid-span, which is unbraced. Determine the elastic flexural-torsional buckling loads

   (a) when the plated flange is in compression;
   (b) when the plated flange is in tension.

7.8 References

1. Kitipornchai, S. and Trahair, N.S. (1980) Buckling properties of monosymmetric I-beams. *Journal of the Structural Division, ASCE*, **106**(ST5), 941–57.
2. Kitipornchai, S. Wang, C.M. and Trahair, N.S. (1986) Buckling of monosymmetric I-beams under moment gradient. *Journal of Structural Engineering, ASCE*, **112**(4), 781–99.
3. Anderson, J.M. and Trahair, N.S. (1972) Stability of monosymmetric beams and cantilevers, *Journal of the Structural Division, ASCE*, **98**(ST1), 269–86.
4. Wang, C.M. and Kitipornchai, S. (1986) Buckling capacities of monosymmetric I-beams. *Journal of Structural Engineering, ASCE*, **112**(11), 2373–91.
5. Roberts, T.M. and Burt, C.A. (1985) Instability of monosymmetric I-beams and cantilevers. *International Journal of Mechanical Sciences*, **27**(5), 313–24.
6. Austin, W.J. Yegian, S. and Tung, T.P. (1955) Lateral buckling of elastically end-restrained beams. *Proceedings, ASCE*, **81** (Separate No. 673), 1–25.
7. Trahair, N.S. (1965) Stability of I-beams with elastic end restraints. *Journal of the Institution of Engineers, Australia*, **37**(6), 157–68.
8. Trahair, N.S. (1966) Elastic stability of I-beam elements in rigid-jointed structures. *Journal of the Institution of Engineers, Australia*, **38**(7–8), 171–80.
9. Trahair, N.S. (1968) Elastic stability of propped cantilevers. *Civil Engineering Transactions*, Institution of Engineers, Australia, **CE10**(1), 94–100.

8 Restrained beams

8.1 General

A beam is often connected to other elements which participate in the buckling action, and significantly influence its buckling resistance. Braces are provided specifically for the purpose of increasing the buckling resistance (Figure 8.1a), but many other elements, such as sheeting, which are primarily intended for other purposes, also have important restraining actions.

These elements may induce restraining end moments which act in the plane of loading (Figure 8.1b), as for example in built-in beams and propped cantilevers, and also in continuous beams. These restraining end moments change the in-plane bending moment distribution and modify the buckling resistance. The influence of in-plane end restraining moments on the elastic buckling of simply supported beams is discussed in section 7.5, and of continuous beams in Chapter 10.

Out-of-plane restraining actions (Figure 8.2) restrict the buckled shape of the beam, and increase its buckling resistance. Discrete restraints act at points where braces or other restraining elements are connected to the beam, and induce actions which resist the buckling deflections, rotations, and warping displacements. These restraints are usually assumed to be elastic, in which case they may be characterized by their elastic stiffnesses.

In some cases the discrete restraints may be assumed to be rigid, so that they prevent one or more of the buckling deformations. In the special case where rigid restraints prevent lateral deflections and twist rotations of the restrained cross-sections, then the beam may be described as a braced beam which consists of beam elements between the restraint points and the supports. The elastic buckling of braced beams is treated in Chapter 10.

Continuous restraints are usually considered to be uniform along the length of a beam, and are often used to approximate the actions of restraining elements which are connected to the beam at closely spaced intervals, as in the case of roof sheeting (Figure 8.3). Continuous restraints are similar to discrete restraints, in that they induce actions which restrain the buckling deformations (Figure 8.2). They also are usually assumed to be elastic, and so may be characterized by their elastic stiffnesses. When continuous restraints can be assumed to be rigid, then they enforce a longitudinal axis about which the beam cross-sections rotate during buckling.

This chapter is concerned with the influence of out-of-plane restraints on the elastic buckling of simply supported beams. The buckling of restrained columns

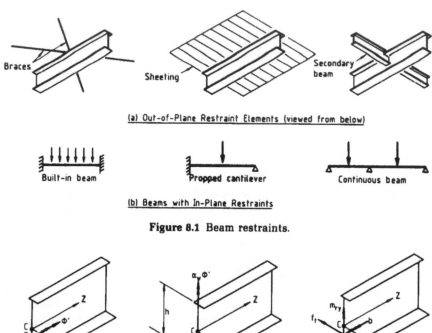

Figure 8.2 Continuous restraint actions.

is treated in Chapter 6, of restrained cantilevers in Chapter 9, of continuous beams in Chapter 10, and of restrained beam-columns in section 11.5.

8.2 Restraint stiffnesses

8.2.1 CONTINUOUS RESTRAINTS

A beam may be continuously restrained as shown in Figure 8.2b by a translational restraint of stiffness α_t which acts at a distance y_t below the centroid, by a minor axis rotational restraint of stiffness α_{ry} which acts at a distance y_r below the centroid, by a torsional restraint of stiffness α_{rz}, and by a warping restraint of stiffness α_w.

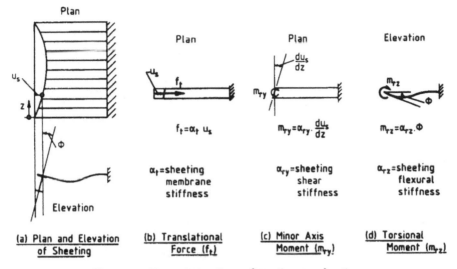

Figure 8.3 Restraint actions of continuous sheeting.

The actions exerted by these restraints can be replaced by the shear centre actions

$$\{r\} = \{f_t, m_{ry}, m_{rz}, b\}^t \qquad (8.1)$$

shown in Figure 8.2c, in which b is the bimoment per unit length. For this set of actions to be statically equivalent to those of the restraints, they must be related to the shear centre deformations (Figure 8.2a)

$$\{d\} = \{u, u', \phi, \phi'\}^T \qquad (8.2)$$

by

$$\{r\} = [\alpha_b] \{d\}, \qquad (8.3)$$

in which the continuous restraint stiffness matrix is

$$[\alpha_b] = \begin{bmatrix} \alpha_t & 0 & -\alpha_t(y_t - y_0) & 0 \\ 0 & \alpha_{ry} & 0 & -\alpha_{ry}(y_r - y_0) \\ -\alpha_t(y_t - y_0) & 0 & \{\alpha_t(y_t - y_0)^2 + \alpha_{rz}\} & 0 \\ 0 & -\alpha_{ry}(y_r - y_0) & 0 & \{\alpha_{ry}(y_r - y_0)^2 + \alpha_w\} \end{bmatrix}. \qquad (8.4)$$

8.2.2 DISCRETE RESTRAINTS

A beam may have discrete restraints which have translational stiffness α_T (at a distance y_T below the centroid), minor axis rotational stiffness α_{Ry} (at a distance y_R below the centroid), torsional stiffness α_{Rz} and warping stiffness α_W.

The actions exerted by these restraints can be replaced by the shear centre actions

$${R} = {F_T, M_{Ry}, M_{Rz}, B}^T \qquad (8.5)$$

in which B is the bimoment. For this set of actions to be statically equivalent to those of the restraints, they must be related to the shear centre deformations

$${D} = {u, u', \phi, \phi'}^T \qquad (8.6)$$

by

$${R} = [\alpha_B]{D}, \qquad (8.7)$$

in which the discrete restraints' stiffness matrix is

$$[\alpha_B] = \begin{bmatrix} \alpha_T & 0 & -\alpha_T(y_T - y_0) & 0 \\ 0 & \alpha_{Ry} & 0 & -\alpha_{Ry}(y_R - y_0) \\ -\alpha_T(y_T - y_0) & 0 & {\alpha_T(y_T - y_0)^2 + \alpha_{Rz}} & 0 \\ 0 & -\alpha_{Ry}(y_R - y_0) & 0 & {\alpha_{Ry}(y_R - y_0)^2 + \alpha_W} \end{bmatrix}.$$

$$(8.8)$$

8.3 Buckling analysis

In the general case of restrained beams, exact solutions for the buckling loads cannot be obtained, and a numerical method must be used to obtain approximate solutions. If the approximate energy method is used (Chapter 3), then the strain energy stored in an element during buckling should be increased to

$$\frac{1}{2}\delta^2 U = \frac{1}{2}\int_0^L {EI_y u''^2 + EI_w \phi''^2 + GJ\phi'^2} dz$$

$$+ \frac{1}{2}\sum {D}^T [\alpha_B]{D} + \frac{1}{2}\int_0^L {d}^T [\alpha_b]{d} dz \qquad (8.9)$$

to account for the strain energy stored in the discrete and continuous restraints acting on the element.

If the finite element method of computer analysis is used (Chapter 4), then each element stiffness matrix should be augmented by transforming the term $\frac{1}{2}\int_0^L {d}^T[\alpha_b]{d}dz$ which accounts for continuous restraints [1] and the beam stiffness matrix should be augmented by including a transformation of the $\frac{1}{2}{D}^T[\alpha_B]{D}$ term for each discrete restraint.

8.4 Continuous restraints

8.4.1 UNIFORM BENDING

A simply supported beam ($u_{0,L} = \phi_{0,L} = 0, u''_{0,L} = \phi''_{0,L} = 0$) in uniform bending with uniform continuous restraints buckles into n half sine waves [2] so that

$$u/\delta = \phi/\theta = \sin n\pi z/L. \tag{8.10}$$

The moment which causes elastic buckling can be obtained by substituting these buckled shapes into the energy equation

$$\tfrac{1}{2}(\delta^2 U + \delta^2 V) = 0 \tag{8.11}$$

in which the strain energy $\tfrac{1}{2}\delta^2 U$ stored during buckling is given by equation 8.9 and the work done $\tfrac{1}{2}\delta^2 V$ on the applied loads is given by

$$\frac{1}{2}\delta^2 V = \frac{1}{2}\int_0^L \{2M\phi u'' + M\beta_x \phi'^2\}\,dz. \tag{8.12}$$

This leads to

$$\begin{aligned}
(M &+ \alpha_{ry}(y_r - y_0) + \alpha_t(y_t - y_0)L^2/n^2\pi^2)^2 \\
&= (n^2 P_y + \alpha_{ry} + \alpha_t L^2/n^2\pi^2)(GJ + n^2\pi^2 EI_w/L^2 + M\beta_x \\
&+ \alpha_{ry}(y_r - y_0)^2 + \alpha_w + \{\alpha_t(y_t - y_0)^2 + \alpha_{rz}\}L^2/n^2\pi^2).
\end{aligned} \tag{8.13}$$

In general, a number of trials must be made before the integer value of n which leads to the lowest value of the buckling moment M can be determined.

Doubly symmetric beams ($\beta_x = 0 = y_0$) with continuous warping and torsional restraints only ($\alpha_t = 0 = \alpha_{ry}$) buckle in a single half wave ($n = 1$) at a moment

$$M = \sqrt{\{P_y(GJ + \pi^2 EI_w/L^2 + \alpha_w + \alpha_{rz} L^2/\pi^2)\}}. \tag{8.14}$$

It can be seen that these restraints contribute directly to the effective torsional stiffness of the beam.

Doubly symmetric beams with continuous minor axis rotational restraints also buckle in a single half wave ($n = 1$) at a moment given by

$$\left\{\frac{M}{M_{yz}} + \frac{\alpha_{ry}}{P_y}\frac{y_r P_y}{M_{yz}}\right\}^2 = \left\{1 + \frac{\alpha_{ry}}{P_y}\right\}\left\{1 + \frac{\alpha_{ry}}{P_y}\left(\frac{y_r P_y}{M_{yz}}\right)^2\right\} \tag{8.15}$$

in which

$$M_{yz} = \sqrt{\{P_y(GJ + \pi^2 EI_w/L^2)\}} \tag{8.16}$$

and

$$P_y = \pi^2 EI_y/L^2. \tag{8.17}$$

Solutions obtained from equation 8.15 for the dimensionless buckling resistance M/M_{yz} are plotted in Figure 8.4 against the dimensionless restraint position $y_r P_y/M_{yz}$ for specified values of the dimensionless restraint stiffness α_{ry}/P_y.

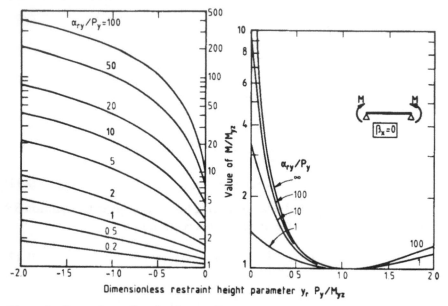

Figure 8.4 Beams in uniform bending with continuous minor axis rotational restraints.

When the restraint acts above the shear centre ($y_r P_y/M_{yz}$ is negative) the elastic buckling moment M increases indefinitely with the restraint stiffness. However, for beams with restraints which act below the shear centre ($y_r P_y/M_{yz}$ is positive), the buckling moment increases with increasing restraint stiffness towards a limiting value M_∞ which is given by

$$\frac{M_\infty}{M_{yz}} = \frac{1}{2}\left[\frac{y_r P_y}{M_{yz}} + \frac{M_{yz}}{y_r P_y}\right].$$ (8.18)

This limiting value, which corresponds to the case where the beam buckles with an enforced centre of rotation (at the restraint position y_r), is also shown graphically in Figure 8.4. It is of interest to note that M_∞/M_{yz} is a minimum ($= 1.0$) when $y_r P_y/M_{yz} = 1$. In this case, the restraint has no effect because it acts at the axis of cross-section rotation (see Figure 7.6) of an unrestrained beam. For beams with restraints which act below the unrestrained axis of rotation, the restraints are comparatively ineffective unless they act at some considerable distance below the unrestrained axis of rotation.

Doubly symmetric beams with continuous translational restraints buckle in n half waves at a moment given by

$$\left\{\frac{M}{M_{yzn}} + \frac{\alpha_t L^2}{n^4\pi^2 P_y}\frac{n^2 y_t P_y}{M_{yzn}}\right\}^2 = \left\{1 + \frac{\alpha_t L^2}{n^4\pi^2 P_y}\right\}\left\{1 + \frac{\alpha_t L^2}{n^4\pi^2 P_y}\left(\frac{n^2 y_t P_y}{M_{yzn}}\right)^2\right\}$$ (8.19)

where

$$M_{yzn}^2 = n^2 P_y(GJ + n^2\pi^2 EI_w/L^2).$$ (8.20)

Equation 8.19 has the same form as equation 8.15 for beams with minor axis rotational restraints, and so Figure 8.4 can also be used for beams with translational restraints, provided $M/M_{yz}, \alpha_{ry}/P_y$, and $y_r\,P_y/M_{yz}$ are replaced by M/M_{yzn}, $\alpha_t L^2/n^4\pi^2 P_y$, and $n^2 y_t P_y/M_{yzn}$ respectively. Figure 8.4 can therefore be interpreted for beams with translational restraints in the same way as for beams with rotational restraints.

However, equations 8.19 and 8.20 indicate that there are a number of different elastic buckling moments whose values vary with the number n of half waves into which the beam buckles. The lowest value of M for given values of $\alpha_t L^2/\pi^2 P_y$ and $y_t P_y/M_{yz}$ may be determined by calculating successive values of M/M_{yz} for $n = 1, 2, 3, \ldots$. In general, these will decrease at first until the minimum is reached and then increase. Thus the successive calculations may be terminated as soon as there is an increase in M/M_{yz}. Some solutions of equation 8.19 are plotted in Figure 8.5. It can be seen that in general, the number n of half waves at buckling increases with the dimensionless restraint stiffness $\alpha_t L^2/\pi^2 P_y$, but decreases as the dimensionless restraint position $y_t P_y/M_{yz}$ and the torsion parameter K increase. Thus the number of half waves at buckling will be high in beams with low warping rigidity EI_w which are highly restrained at points high above the shear centre.

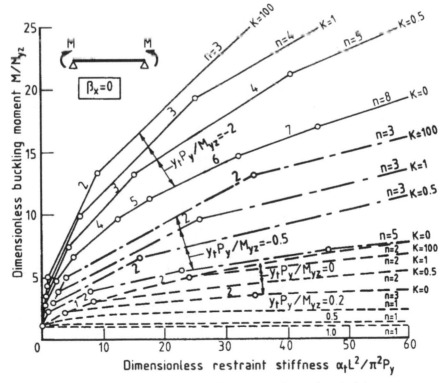

Figure 8.5 Beams with continuous translational restraints.

8.4.2 UNIFORMLY DISTRIBUTED LOADS

The elastic buckling of beams with uniformly distributed loads q acting through sheeting which also provides continuous restraints has been studied in [3, 4]. Solutions for simply supported beams ($u_{0,L} = \phi_{0,L} = 0$, $u''_{0,L} = \phi''_{0,L} = 0$) which are loaded and restrained against minor axis rotation at the same distance y_r from the shear centre are shown in Figure 8.6. The two different sets of curves shown for $K = 0.01, 5$ indicate that the effects of the torsion parameter K on these curves are very small, and may be neglected.

The similarity between the results shown in Figure 8.6 and those in Figure 8.4 for uniform bending has led to the approximation

$$\left\{\frac{qL^2}{8M_{yz}} + \frac{\gamma_a\alpha_{ry}}{P_y}\frac{\gamma_y y_r P_y}{M_{yz}}\right\}^2 = \left\{\frac{q_0L^2}{8M_{yz}} + \frac{\gamma_a\alpha_{ry}}{P_y}\right\}\left\{\frac{q_0L^2}{8M_{yz}} + \frac{\gamma_a\alpha_{ry}}{P_y}\left(\frac{\gamma_y y_r P_y}{M_{yz}}\right)^2\right\} \quad (8.21)$$

in which

$$\frac{q_0L^2}{8M_{yz}} = \alpha_m\left\{\sqrt{\left[1 + \left(\frac{0.4\alpha_m y_r}{M_{yz}/P_y}\right)^2\right]} + \left(\frac{0.4\alpha_m y_r}{M_{yz}/P_y}\right)\right\}, \quad (8.22)$$

$$\alpha_m = 1.13, \quad (8.23)$$

$$\gamma_a = \left\{1.09 - \left(\frac{0.011K^2}{0.08 + K^2}\right)\right\} - \left\{0.331 - \left(\frac{0.112K^2}{0.08 + K^2}\right)\right\}\left\{\frac{\alpha_{ry}/P_y}{10 + \alpha_{ry}/P_y}\right\}, \quad (8.24)$$

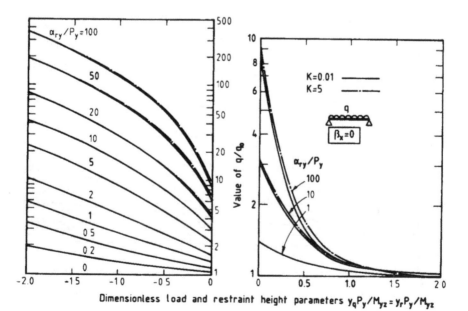

Figure 8.6 Uniformly loaded beams with continuous minor axis rotational restraints.

and

$$\gamma_y = 0.55 - 0.05\,K^2/(0.08 + K^2) \text{ for } y_r \leqslant 0,$$

or

$$\gamma_y = 0.45 + \left(2 - \frac{y_r P_y}{M_{yz}}\right)\left\{0.15 - \left(\frac{0.05 K^2}{0.08 + K^2}\right)\right\} \text{ for } y_r > 0,$$

(8.25)

which is reasonably accurate provided that $-2 \leqslant y_r P_y/M_{yz} \leqslant 2$ and $\alpha_{ry}/P_y \leqslant 100$.

Solutions for beams with uniformly distributed loads and major axis restraining moments acting at one or both ends are also given in [3]. These indicate that the high efficiency of minor axis rotational restraints acting above the shear centre falls off as the amount of negative bending caused by the end moments increases. It is also demonstrated that minor axis rotational restraints are most effective near the supports where $(u' - y_r\phi')$ is usually greatest, and least effective near mid-span.

The elastic buckling of very thin-walled beams $(K \to \infty)$ with rigid continuous restraints which enforce an axis of buckling rotation at the loaded top flange was studied in [4]. The predicted maximum moments M_m at the elastic buckling of beams which are prevented from deflecting and twisting at their ends $(u_{0,L} = 0 = \phi_{0,L})$ are shown non-dimensionally in Figure 8.7 against the load parameter $qL^2/8M$ and the end moment ratio β. The relationships between M_m and the maximum end moment M or the distributed load q are shown in Figure 8.8. The dimensionless buckling resistance $M_m L^2/\{\pi^2 \sqrt{(EI, EI_w)}\}$ shown in Figure 8.7 is usually higher for gravity loading (when the restrained top flange is generally in compression) than it is for uplift loading (when the unrestrained bottom flange is often in compression).

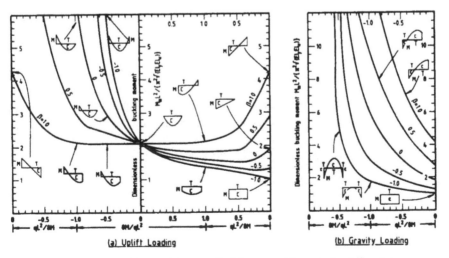

Figure 8.7 Buckling of thin-walled beams with restrained flanges.

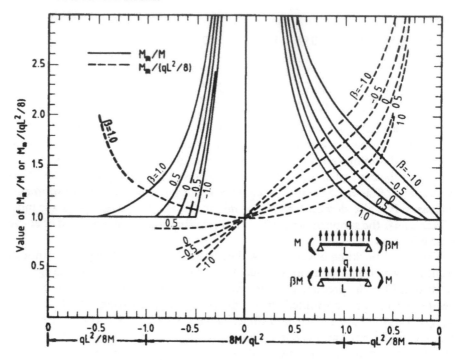

Figure 8.8 Maximum moments in purlin segments.

8.5 Discrete restraints

8.5.1 UNIFORM BENDING

8.5.1.1 *End restraints*

Restraints acting at the supported ends ($u_{0,L} = \phi_{0,L} = 0$) of a beam may restrain minor axis rotations u' and warping displacements proportional to ϕ'. The particular case of a beam in uniform bending whose four flange ends have equal minor axis rotational end restraints of stiffness α_R is shown in Figure 8.9a. These exert restraining moments $M_{T,B} = \alpha_R u'_{T,B}$ where the flange rotations at one end are $u'_{T,B} = u' \pm h\phi'/2$, and store strain energy during buckling

$$\tfrac{1}{2}\delta^2 U_R = \frac{1}{2}\alpha_R\{(u' + h\phi'/2)^2 + (u' - h\phi'/2)^2\}. \tag{8.26}$$

If this is compared with

$$\tfrac{1}{2}\delta^2 U_R = \tfrac{1}{2}\{D\}^T[\alpha_B]\{D\} \tag{8.27}$$

obtained using equation 8.8, then the restraint stiffnesses at one end can be expressed as

$$\left. \begin{array}{l} \alpha_T = \alpha_{R_r} = 0, \\ \alpha_{Ry} = 4\alpha_w/h^2 = 2\alpha_R, \end{array} \right\} \tag{8.28}$$

with $y_R = 0$.

In this case, the buckled shape is defined by [5]

$$\frac{u}{\delta} = \frac{\phi}{\theta} = \frac{\cos(\pi z/kL - \pi/2k) - \cos(\pi/2k)}{1 - \cos(\pi/2k)} \tag{8.29}$$

where k satisfies

$$\frac{\alpha_R L}{EI_y} = -\frac{\pi}{2k}\cot\frac{\pi}{2k}. \tag{8.30}$$

These equations are the same as the corresponding ones for braced columns with equal rotational end restraints (section 6.5), and define a beam effective length kL equal to the distance between the inflexion points of the buckled shape, as shown in Figure 8.9b. Note that this effective length (of the buckled shape) is not related to the length between any inflexion points of the in-plane deflected shape.

Substitution of equations 8.27–8.30 into the energy equation

$$\frac{1}{2}\int_0^L \{EI_y u''^2 + EI_w \phi''^2 + GJ\phi'^2\}\, dz + \frac{1}{2}\delta^2 U_R + \frac{1}{2}\int_0^L 2M\phi u''\, dz = 0 \tag{8.31}$$

leads to

$$M = \sqrt{\left\{ \frac{\pi^2 EI_y}{k^2 L^2}\left(GJ + \frac{\pi^2 EI_w}{k^2 L^2} \right) \right\}} \tag{8.32}$$

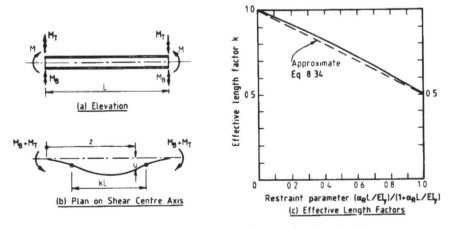

Figure 8.9 Buckling of beams with equal end restraints.

and

$$\frac{\delta}{\theta} = \frac{M}{\pi^2 EI_y/k^2 L^2} \tag{8.33}$$

which can also be obtained from equations 7.5 and 7.10 for an unrestrained beam by substituting the effective length kL for the actual length L.

The relationship obtained from equation 8.30 between the effective length factor k and the dimensionless end restraint stiffness $\alpha_R L/EI_y$, is shown in Figure 8.9c. It can be seen that k varies almost linearly from 1.0 to 0.5 as $(\alpha_R L/EI_y)/(1 + \alpha_R L/EI_y)$ increases from zero (for an unrestrained beam) to unity (for a rigidly restrained beam). This suggests that k may be closely approximated by

$$k \approx \frac{2 + \alpha_R L/EI_y}{2 + 2\alpha_R L/EI_y} \tag{8.34}$$

as shown in Figure 8.9c.

The correspondence of equations 8.29 and 8.30 with equations 6.31 and 6.32 for columns with equal end restraints has led to the suggestion that the effects of unequal flange end restraints α_{RA}, α_{RB}, at the ends A, B of a beam should be approximated by using the solutions shown in Figure 8.10b obtained from Figure 6.6b for braced columns with unequal end restraints. This suggestion has proved to be satisfactory for a wide range of braced and continuous beams (see Chapter 10).

Sometimes constraints may act at the ends of a beam [6, 7] which effectively prevent end warping ($\phi'_{0,L} = 0$). The increased buckling resistance may be

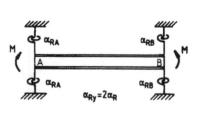

(a) Flange End Restraints

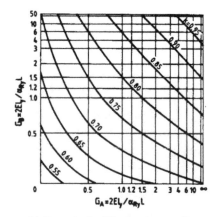

(b) Approximate Effective Length Factors

Figure 8.10 Buckling of beams with unequal end restraints.

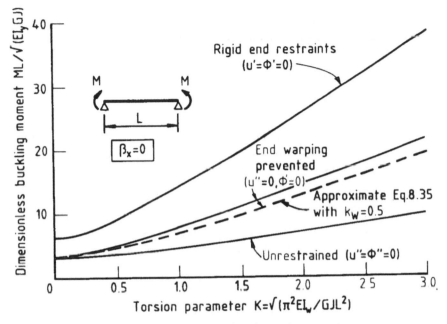

Figure 8.11 Effect of rigid end warping restraints.

approximated by

$$M = \sqrt{\left\{\frac{\pi^2 EI_y}{L^2}\left(GJ + \frac{\pi^2 EI_w}{k_w^2 L^2}\right)\right\}} \tag{8.35}$$

in which the warping restraint effective length factor is given by $k_W = 0.5$, as shown in Figure 8.11.

If elastic end restraints against minor axis rotation and warping act simultaneously, the buckling resistance may be estimated from

$$M = \sqrt{\left\{\frac{\pi^2 EI_y}{k_{R_y}^2 L^2}\left(GJ + \frac{\pi^2 EI_w}{k_w^2 L^2}\right)\right\}} \tag{8.36}$$

in the case where the two ends are identically restrained. The flexural and warping effective length factors k_{R_y}, k_W may be approximated by

$$k_{R_y} \approx \frac{4 + \alpha_{R_y} L/EI_y}{4 + 2\alpha_{R_y} L/EI_y} \tag{8.37}$$

and

$$k_W \approx \frac{4 + (\alpha_W + \alpha_{R_y} h^2/4) L/EI_w}{4 + 2(\alpha_W + \alpha_{R_y} h^2/4) L/EI_w}. \tag{8.38}$$

It is usually assumed that the twist rotations ϕ at the ends of a beam are prevented by rigid torsional end restraints. When these restraints are not rigid, then the buckling resistance is decreased. Numerical solutions for the reduced resistance may be approximated by using

$$\frac{M}{M_{yz}} \approx \sqrt{\left\{ \frac{\alpha_{Rz}L/GJ}{(4.9 + 4.5K^2)\alpha_{Rz}L/GJ} \right\}}. \tag{8.39}$$

These approximate solutions are shown in Figure 8.12. It can be seen that the resistance falls off rapidly for beams with high values of K, and that comparatively high values of torsional restraint stiffness are required to prevent significant reductions below M_{yz} for such beams.

8.5.1.2 Intermediate restraints

Intermediate restraints acting between the points of support may restrain lateral deflections u and twist rotations ϕ. The particular case of a beam in uniform

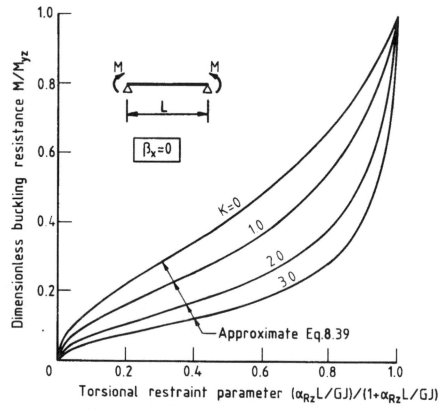

Figure 8.12 Effect of elastic torsional end restraints.

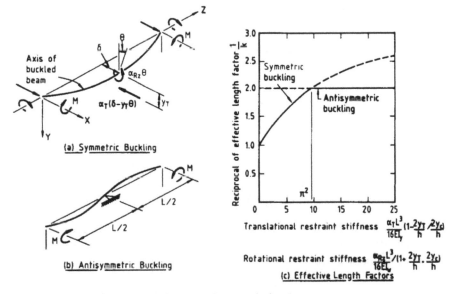

Figure 8.13 Beam with central elastic restraints.

bending with central restraints is shown in Figure 8.13a. When the restraints are rigid so that $u = \phi = 0$ at the restraint point, the beam buckles antisymmetrically into two half waves, each having an effective length of $L/2$, as shown in Figure 8.13b. Thus the elastic buckling resistance is given by

$$M = \sqrt{\left\{\frac{\pi^2 EI_y}{k^2 L^2}\left(GJ + \frac{\pi^2 EI_w}{k^2 L^2}\right)\right\}}
\qquad (8.40)$$

with $k = 0.5$.

When the restraints are elastic, the beam may buckle in a symmetrical mode, as shown in Figure 8.13a. An analytical solution can be obtained for the special case when the restraint stiffnesses are related by

$$\frac{\alpha_T L^3}{EI_y}\left(1 - \frac{2y_T/h}{2y_c/h}\right) = \frac{\alpha_{Rz} L^3}{EI_w}\bigg/\left(1 + \frac{2y_T}{h}\frac{2y_c}{h}\right) = \frac{\alpha_{Re} L^3}{EI_y}
\qquad (8.41)$$

in which

$$\frac{\alpha_{Re} L^3}{16 EI_y} = \frac{(\pi/2k)^3 \cot(\pi/2k)}{(\pi/2k)\cot(\pi/2k) - 1}
\qquad (8.42)$$

$$y_c = M/(\pi^2 EI_y/k^2 L^2)
\qquad (8.43)$$

and M is given by equation 8.40.

Equations 8.40–8.43 correspond to the symmetric buckled shape defined by [8]

$$\frac{u}{\delta} = \frac{\phi}{\theta} = \frac{z}{L} - \frac{\sin(\pi z/kL)}{(\pi/k)\cos(\pi/2k)} \tag{8.44}$$

while $0 \leqslant z \leqslant L/2$, and to an effective length kL. This buckled shape is the same as that for a column with a central elastic translational restraint. Equations 8.40–8.43 with $y_c = \delta/\theta$ satisfy the energy equation for the restrained beam.

The relationship obtained from equation 8.42 between the reciprocal of the effective length factor k and the dimensionless central restraint stiffness $\alpha_{Re}L^3/16EI_y$, is shown in Figure 8.13c. It can be seen that $1/k$ increases from 1.0 towards 2.86 approximately as the restraint stiffness increases from zero towards infinity.

Also shown in Figure 8.13c is the solution $1/k = 2$ for the antisymmetric buckling mode of a beam with rigid restraints (Figure 8.13b), which intersects the symmetric mode solution at a limiting restraint stiffness given by

$$\alpha_{ReL}L^3/16EI_y = \pi^2. \tag{8.45}$$

Since buckling always occurs in the mode which has the lowest buckling load, a beam with $\alpha_{Re} > \alpha_{ReL}$ always buckles antisymmetrically in two half waves, and at a moment obtained from equation 8.40 by using $k = 0.5$. On the other hand, a beam with $\alpha_{Re} < \alpha_{ReL}$ buckles symmetrically in a single half wave, and at a moment obtained from equation 8.40 by using the solution of equation 8.42. This solution may be closely approximated by

$$k = \frac{1}{1 + 1.4\alpha_{Re}/\alpha_{ReL} - 0.4(\alpha_{Re}/\alpha_{ReL})^2} \tag{8.46}$$

in the range $0 \leqslant \alpha_{Re} \leqslant \alpha_{ReL}$.

Equations 8.41 and 8.45 indicate that the values of α_{TL} and α_{RzL} required to cause antisymmetric buckling decrease as y_T decreases so that the restraint point moves higher. The value of α_{RzL} decreases to zero when

$$y_T = \frac{-h^2}{4y_c} = -\frac{h}{2}\frac{2K}{\sqrt{(1+4K^2)}} \geqslant -\frac{h}{2} \tag{8.47}$$

for which the value of α_{TL} is given by

$$\frac{\alpha_{TL}L^3}{16EI_y} = \frac{\pi^2(1+4K^2)}{(1+8K^2)} \leqslant \pi^2. \tag{8.48}$$

Thus it can be concluded that a top (compression) flange translational restraint $(y_T = -h/2)$ alone of stiffness $\alpha_T = 16\pi^2 EI_y/L^3$ will be sufficient to cause antisymmetric buckling with $k = 0.5$.

When a central translational restraint acts alone, then the limiting restraint stiffness α_{TL} required to cause antisymmetric buckling increases with the restraint position y_T. The solutions of [9] for beams with shear centre restraint ($y_T = 0$) suggest that in this case the limiting restraint stiffness may be approximated by

$$\frac{\alpha_{T_1} L^3}{EI_y} \approx \frac{153(1 + 5K^2)}{(1 + 0.74K^2)}. \tag{8.49}$$

For $\alpha_T > \alpha_{TL}$, then $k = 0.5$, while approximate solutions for M for $\alpha_T < \alpha_{TL}$ may be obtained from

$$\frac{M}{M_{yz}} \approx \sqrt{\left\{ \frac{2400 + 51\alpha_T L^3/EI_y}{2400 + \alpha_T L^3/EI_y} \right\}}. \tag{8.50}$$

The effect of a central torsional restraint acting alone has been studied in [9, 10]. The buckling mode was found to change from a symmetric half wave to two antisymmetric half waves at limiting restraint stiffnesses α_{RzL} which may be approximated by

$$\frac{\alpha_{RzL} L}{GJ} \approx 30 + 100K^2. \tag{8.51}$$

For $\alpha_{Rz} > \alpha_{RzL}$, then $k = 0.5$, while approximate solutions for $\alpha_{Rz} < \alpha_{RzL}$ may be obtained from

$$\left(\frac{M}{M_{yz}} \right)^2 = 1 + \frac{\alpha_{Rz}}{\alpha_{RzL}} \left\{ \left(\frac{M_{yzL}}{M_{yz}} \right)^2 - 1 \right\} \tag{8.52}$$

in which M_{yzL} is the value of M for $k = 0.5$.

The influence of n_r equal stiffness restraints spaced at equal intervals $s = L/(n_r + 1)$ has been studied in [11–13]. For high restraint values, the beam may buckle between restraints into $(n_r + 1)$ half waves at a moment given by equation 8.40 with

$$k = s/L. \tag{8.53}$$

The limiting stiffness α_{RL} of each restraint required to cause this mode of buckling may be determined approximately by treating the beam as continuously restrained (section 8.4) and determining the limiting stiffness α_{rL} of the continuous restraints required to cause buckling at the moment determined from equations 8.40 and 8.53. The limiting stiffness α_{RL} of each discrete restraint may then be approximated by 'concentrating' the continuous stiffness α_{rL} according to

$$\alpha_{RL} = L\alpha_{rL}/n_r. \tag{8.54}$$

For $\alpha_R < \alpha_{RL}$, the beam buckles in a complex mode which is characterized by twisting at some or all of the restraint points. The buckling resistance may be approximated by 'smearing' the discrete restraints into equivalent uniform continuous restraints of stiffness

$$\alpha_r = n_r \alpha_R / L \tag{8.55}$$

and using this in the continuous restraint solution of equation 8.13.

This method is generally conservative for low values of n_r, the number of restraint points. For example, the limiting stiffness calculated by this method for a beam with a central $(n_r = 1)$ translational restraint at the shear centre is given by

$$\frac{\alpha_{TL} L^3}{EI_y} \approx \frac{3\pi^4(1 + 5K^2)}{(1 + K^2)} \tag{8.56}$$

which varies between 1.41 and 1.91 times the value obtained from equation 8.49, which the buckling resistance for less than the limiting stiffness is given by

$$\left(\frac{M}{M_{yz}}\right)^2 = 1 + \frac{\alpha_T}{\alpha_{TL}}\left\{\left(\frac{M_{yzL}}{M_{yz}}\right)^2 - 1\right\}. \tag{8.57}$$

Similarly, the limiting stiffness calculated for a beam with a central rotational restraint is

$$\alpha_{RzL} L/GJ = 3\pi^2(1 + 5K^2) \tag{8.58}$$

which varies between 0.99 and 1.48 times the value obtained from equation 8.51, while the buckling resistance for less than the limiting stiffness is given by equation 8.52. It can be expected that the accuracy of the method will increase with the number of restraint points n_r.

8.5.2 NON-UNIFORM BENDING

8.5.2.1 End restraints

There have been a number of studies [14–19] of the effects of elastic end restraints against minor axis rotation (α_{Ry}), warping (α_W), and twist rotation (α_{Rz}) on the elastic buckling of beams in non-uniform bending. The scope of these studies is summarized in Figure 8.14.

Generally, the studies indicate that these effects are qualitatively similar to those on beams in uniform bending (section 8.5.1.1). This observation leads to the suggestion that the effects of elastic flange end restraints might be approximated by using

$$M_m = \alpha_m \sqrt{\left\{\frac{\pi^2 EI_y}{k^2 L^2}\left(GJ + \frac{\pi^2 EI_w}{k^2 L^2}\right)\right\}} \tag{8.59}$$

for the maximum moment at elastic buckling, in which the moment factor α_m is approximated as in sections 7.3, 7.4 or 7.5 for simply supported beams with centroidal loading $(y_e, y_Q = 0)$, and the effective length factor k is approximated as in Figure 8.10.

A similar suggestion may be made for estimating the effect of load height (see section 7.6) on the buckling of end-restrained beams from equations of the form of equation 7.65, but with M_{yz}, P_y calculated using the approximate effective

Loading	β	Ends Restrained	Restraint Types α_{Ry}	α_w	α_{Rz}	Reference
	-1.0	Equally	Variable	0	∞	[15]
	-1.0	" "	" "	∞	∞	[15]
	-1.0	" "	0	0	Variable	[15, 19]
	-1.0	" "	0	∞	" "	[15]
	Variable	" "	Variable	$h^2\alpha_{Ry}/4$	∞	[18]
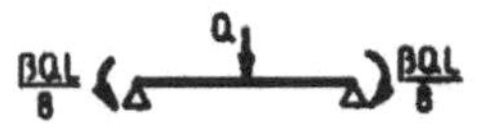	0	Equally	0	0	Variable	[15]
	0	" "	0	∞	" "	[15]
	Variable	" "	Variable	Variable	∞	[14, 16]
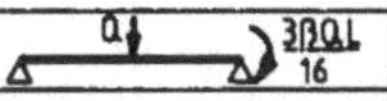	Variable	RH end	Variable	$h^2\alpha_{Ry}/4$	∞	[17]
	0	Equally	0	0	Variable	[15]
	0	" "	0	∞	" "	[15]
	Variable	" "	Variable	Variable	∞	[14, 16]
	Variable	" "	Variable	$h^2\alpha_{Ry}/4$	∞	[17]

Figure 8.14 Studies of beams with end restraints.

length kL determined from Figure 8.10 in place of the actual length. This leads to predictions of somewhat variable accuracy, but these may be tested by comparing them with the numerical solutions reported in the studies summarized in Figure 8.14.

8.5.2.2 Intermediate restraints

There have been a number of studies [8–11, 13, 20] of the effects of central translational or torsional restraints on the buckling of simply supported beams in non-uniform bending, including both the limiting stiffnesses required to cause antisymmetrical buckling, and also the buckling resistances when the restraint stiffnesses are less than the limiting values. The scope of these studies is summarized in Figure 8.15. While quantitative solutions for many specific cases can be obtained from these references, some more general but qualitative conclusions can be drawn.

First of all, when the central restraint exceeds the limiting stiffness, then the buckling resistance is determined for a half beam which generally has a less uniform moment distribution than the complete beam. Because of this, the benefits of central limiting restraints are usually greater for beams in non-uniform bending than they are for uniform bending. As a result of this, the limiting restraint stiffnesses for beams in non-uniform bending are generally higher than for uniform bending.

Loading	Load Height	Restraint Points	Restraint Types			Reference
			α_T	γ_T	α_{Rz}	
M ⟨△————△⟩ M	–	1	Varies	Varies	Varies	[8, 9]
	–	1	0	–	" "	[10]
	–	n_r	Varies	Varies	" "	[13]
	–	n_r	∞	" "	" "	[11]
M ⟨△————△⟩ βM	–	n_r	∞	Varies	Varies	[11]
 ↓Q △———————△	Varies	1	Varies	Varies	Varies	[8, 9]
	Centroid	1	0	–	" "	[10]
 q ↓↓↓↓↓↓ △———————△	Varies	1	Varies	Varies	Varies	[9]
	Centroid	1	0	–	" "	[10]
	Varies	n_r	Varies	Varies	" "	[13]

Figure 8.15 Studies of beams with intermediate restraints.

The limiting translational restraint stiffnesses α_{TL} required to cause antisymmetrical buckling increase as the restraint point moves lower, or as the load point moves higher. Antisymmetrical buckling often cannot be achieved when the restraint point is below the load point, but top flange restraints with $\alpha_T L^3/EI_y \geqslant 16\pi^2$ will cause antisymmetrical buckling when the loads act at or below the top flange. Approximate solutions for translational restraints with $\alpha_T < \alpha_{TL}$ may be obtained from

$$\left(\frac{M}{M_m}\right)^2 = 1 + \frac{\alpha_T}{\alpha_{TL}}\left\{\left(\frac{M_L}{M_m}\right)^2 - 1\right\} \tag{8.60}$$

in which M_m is the maximum moment in an unrestrained beam at elastic buckling, and M_L is the maximum moment for antisymmetrical buckling.

The limiting torsional restraint stiffnesses α_{RzL} required to cause antisymmetrical buckling appear to increase as the load point moves higher. Approximate solutions for torsional restraints with $\alpha_{Rz} < \alpha_{RzL}$ may be obtained from

$$\left(\frac{M}{M_m}\right)^2 = 1 + \frac{\alpha_{Rz}}{\alpha_{RzL}}\left\{\left(\frac{M_L}{M_m}\right)^2 - 1\right\} \tag{8.61}$$

where M_m, M_L are defined similarly to the values used in equation 8.60 above.

The effects of equal stiffness restraints spaced at equal intervals have been studied in [13]. For centroidal loading, these may be approximated by assuming that the beam is in uniform bending and using the approximate method proposed in section 8.5.1.2. The accuracy of this will increase with the number of restraints, since the bending moment distributions between the restraints will become more uniform.

8.6 Problems

PROBLEM 8.1

A beam whose properties are given in Figure 7.23 is simply supported over a span of 12.0 m, as shown in Figure 8.16a. It has a uniformly distributed uplift load acting at the top flange, which is continuously restrained by elastic minor axis rotational restraints of stiffness $\alpha_{ry}/P_y = 10$. Determine the elastic flexural-torsional buckling load.

PROBLEM 8.2

Determine the effects of rigid end warping restraints on the elastic flexural-torsional buckling moment of a beam whose properties are given in Figure 7.23, and which is in uniform bending and simply supported over a span of 12.0 m, as shown in Figure 8.16b.

PROBLEM 8.3

A beam whose properties are given in Figure 7.23 is in uniform bending over a span of 12.0 m, as shown in Figure 8.16b. One end of the beam is simply supported out-of-plane, and the other has elastic minor axis rotational and warping end restraints whose stiffnesses are defined by $\alpha_{R_y} = 4\alpha_w/h^2 = 4EI_y/L$. Determine the elastic flexural-torsional buckling moment.

PROBLEM 8.4

A beam whose properties are given in Figure 7.23 is in uniform bending over a span of 12.0 m, as shown in Figure 8.16b. Both ends are prevented from deflecting laterally, free to rotate laterally and to warp, but are elastically restrained against twist rotation by restraints whose stiffnesses are defined by $\alpha_{R_z} L/GJ = 0.8$.

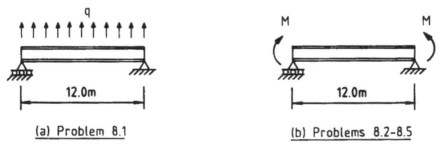

(a) Problem 8.1 (b) Problems 8.2–8.5

Figure 8.16 Problems 8.1–8.5.

Determine the reduction in the elastic flexural-torsional buckling moment from the value for full restraint against twist rotation.

PROBLEM 8.5

A beam whose properties are given in Figure 7.23 is in uniform bending over a span of 12.0 m, as shown in Figure 8.16b. Both ends are simply supported, and the mid-span is elastically restrained against lateral deflection and twist rotation by restraints whose stiffnesses are defined by

$$\frac{\alpha_T L^3}{16EI_y}\left(1 - \frac{2y_T}{h} \Big/ \frac{2y_c}{h}\right) = \frac{\alpha_{Rz} L^3}{16EI_w}\Big/\left(1 + \frac{2y_T}{h}\frac{2y_c}{h}\right) = 3.$$

Determine the increase in the elastic flexural-torsional moment caused by the restraints.

8.7 References

1. Hancock, G.J. and Trahair, N.S. (1978) Finite element analysis of the lateral buckling of continuously restrained beam-columns. *Civil Engineering Transactions*, Institution of Engineers, Australia, CE **20**, (2), 120–27.
2. Trahair, N.S. (1979) Elastic lateral buckling of continuously restrained beam-columns, *The Profession of a Civil Engineer* (eds. D. Campbell-Allen and E.H. Davis), Sydney University Press, Sydney, pp. 61–73.
3. Hancock, G.J. and Trahair, N.S. (1979) Lateral buckling of roof purlins with diaphragm restraints. *Civil Engineering Transactions*, Institution of Engineers, Australia, CE **21**, 10–15.
4. Ings, N.L. and Trahair, N.S. (1984) Lateral buckling of restrained roof purlins. *Thin-Walled Structures*, **2**, 285–306.
5. Trahair, N.S. and Bradford, M.A. (1991) *The Behaviour and Design of Steel Structures*, revised 2nd edition, Chapman and Hall, London.
6. Vacharajittiphan, P. and Trahair, N.S. (1974) Warping and distortion at I-section joints. *Journal of the Structural Division, ASCE*, **100** (ST3), 547–64.
7. Ojalvo, M. and Chambers, R.S. (1977) Effect of warping restraints on I-beam buckling. *Journal of the Structural Division, ASCE*, **103** (ST12), 2351–60.
8. Mutton, B.R. and Trahair, N.S. (1973) Stiffness requirements for lateral bracing. *Journal of the Structural Division, ASCE*, **99** (ST10), October, 2167–82.
9. Nethercot, D.A. (1973) Buckling of laterally or torsionally restrained beams. *Journal of the Engineering Mechanics Division, ASCE*, **99** (EM4), 773–91.
10. Taylor, A.C. and Ojalvo, M. (1966) Torsional restraint of lateral buckling. *Journal of the Structural Division, ASCE*, **92** (ST2), 115–29.
11. Horne, M.R. and Ajmani, J.L. (1969) Stability of columns supported laterally by side-rails. *International Journal of Mechanical Sciences*, **11**, 159–74.
12. Milner, H.R. (1975) The buckling of equal flanged beams under uniform moment restrained torsionally by stiff braces, in *Proceedings, 5th Australasian Conference on the Mechanics of Structures and Materials*, Melbourne, pp. 405–20.

13. Medland, I.C. (1980) Buckling of interbraced beam systems. *Engineering Structures*, 2 (2), 90–6.
14. Austin, W.J. Yegian, S. and Tung, T.P. (1955) Lateral buckling of elastically end restrained I-beams. *Proceedings, ASCE*, 81 (Separate No. 673), 1–25.
15. Trahair, N.S. (1965) Stability of I-beams with elastic end restraints. *Journal of the Institution of Engineers, Australia*, 37 (6), 157–68.
16. Trahair, N.S. (1966) Elastic stability of I-beam elements in rigid-jointed structures. *Journal of the Institution of Engineers, Australia*, 38 (7–8), 171–80.
17. Trahair, N.S. (1968) Elastic stability of propped cantilevers, *Civil Engineering Transactions*, Institution of Engineers, Australia, CE10 (1), 94–100.
18. Nethercot, D.A. and Trahair, N.S. (1976) Lateral buckling approximations for elastic beams. *The Structural Engineer*, 54 (6), 197–204.
19. Trahair, N.S. (1983) Lateral buckling of overhanging beams. *Instability and Plastic Collapse of Steel Structures* (ed. L.J. Morris), Granada, London, pp. 503–18.
20. Trahair, N.S. and Nethercot, D.A. (1984) Bracing requirements in thin-walled structures, in *Developments in Thin-Walled Structures–2* (eds J. Rhodes and A.C. Walker), Elsevier Applied Science Publishers, pp. 93–130.

9 Cantilevers

9.1 General

Cantilevers are usually considered to be flexural members which in the plane of loading are built-in at the support ($v_0 = v_0' = 0$) and free at the other end (v_L, v_L' unrestrained), as shown in Figure 9.1a. However, in this book, the word cantilever is used to describe the out-of-plane end conditions of a flexural member. Thus a cantilever is restrained against out-of-plane deformations at the support ($u_0 = u_0' = \phi_0 = \phi_0' = 0$) and free at the other end ($u_L, u_L', \phi_L, \phi_L'$ unrestrained), as shown in Figure 9.1b. Such a member is often also free in-plane at this unrestrained end (v_L, v_L' unrestrained), since any in-plane support usually provides restraints against the out-of-plane deformations u_L, ϕ_L.

On the other hand, if there are out-of-plane restraints which prevent end deflection and twist ($u_L = \phi_L = 0$) as shown in Figure 9.1c, then the behaviour is similar to that of a simply supported beam which has restraints against minor axis rotation and warping at one end ($u_0' = \phi_0' = 0$). This case is treated in Chapter 8.

The absence of out-of-plane restraints at the free end of a cantilever substantially changes the buckling mode and buckling resistance, as shown in Figure 1.1. In an end loaded cantilever it is the tension flange which buckles the further, rather than the compression flange of a simply supported beam (Figure 7.1).

This chapter is concerned with the elastic buckling of cantilevers of doubly symmetric or monosymmetric cross-section under end moments, end loads, or uniformly distributed loads, and the effects of load height from the shear centre axis. The effects of concentrated and continuous restraints are also discussed.

The buckling of a cantilever which is continuous with the rest of the structure at its support ($u_0' \neq 0$, $\phi_0' \neq 0$) is treated in Chapter 10, while the inelastic buckling of cantilevers is referred to in section 14.5.3, and the buckling of cantilevers under directed loading in section 16.11.

9.2 Uniform bending

9.2.1 DOUBLY SYMMETRIC SECTIONS

Cantilevers with end moments which cause uniform bending are rare in practice, since most loading actions are by loads rather than moments. One possible source of an end moment is an axial load which acts eccentrically, so that it is

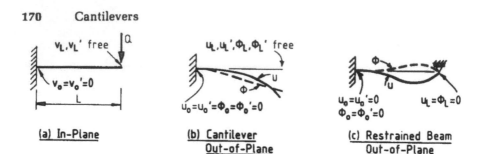

(a) In-Plane

(b) Cantilever
Out-of-Plane

(c) Restrained Beam
Out-of-Plane

Figure 9.1 Cantilever boundary conditions.

statically equivalent to a concentric axial load and an end moment. A member with such a loading is a beam-column, and so the cantilever represents the limiting case where the load tends to zero and its eccentricity to infinity. Despite its rare practical occurrence, end moment loading is at least of theoretical interest, since the closed form solution for simply supported beams with equal and opposite end moments (section 7.2.1) is so widely used as a basis for design.

Some care must be taken to define the plane of action of a moment acting at the free end of a cantilever (Figure 9.2a). The general expression $-\frac{1}{2}\int_0^L 2M_x\phi u''\,dz$ derived in section 17.4 for the work $-\frac{1}{2}\delta^2 V$ done during buckling is consistent with an end moment M which does not rotate ϕ_L with the end of the cantilever, as shown in Figure 9.2c, but which remains in a plane parallel to the original undeformed YZ plane. While it is difficult to imagine a practical loading situation where the end moment does not rotate ϕ_L with the end of the cantilever, this case

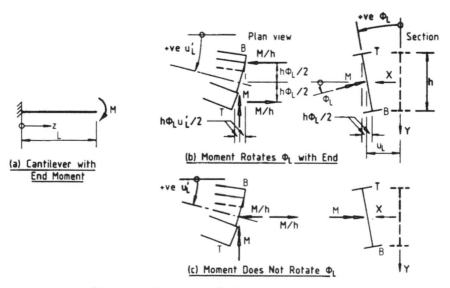

(a) Cantilever with
End Moment

(b) Moment Rotates ϕ_L with End

(c) Moment Does Not Rotate ϕ_L

Figure 9.2 Moments at the free end of a cantilever.

is still one of interest, since it is closely related to those of cantilevers with transverse loads only, which have no end moments to do work.

It is much more likely that an end moment would rotate ϕ_L with the end of a cantilever, as shown in Figure 9.2b. In this case the work done during buckling must be increased by the work $(M_L \phi_L u'_L)$ done by the end moment to

$$-\frac{1}{2}\delta^2 V = -\frac{1}{2}\int_0^L 2M_x \phi u'' \, dz + M_L \phi_L u'_L. \qquad (9.1)$$

In the case of the cantilever of Figure 9.2, $M_L = -M$.

The additional work $-M\phi_L u'_L$ done by the end moment M can also be evaluated by representing the end moment M by statically equivalent end forces $\pm M/h$ acting at the top and bottom flange end centroids, as shown in Figure 9.2b. These flange end centroids displace relatively in the transverse direction by $\pm h\phi_L/2$, and longitudinally $\mp (h\phi_L/2)u'_L$, so that the work done by the end forces is $2(\pm M/h)(\mp h\phi_L u'_L/2)$, or $-M\phi_L u'_L$ as in equation 9.1. (It may be noted that Figure 9.2b is drawn for positive ϕ_L, although in this case the actual ϕ_L is negative, as shown in Figure 9.3a.)

When the end moment rotates with the cantilever, then the relative positions of the buckled cantilever and the end moment are the same as those of half of a simply supported beam in uniform bending (see Figure 7.5) (the actual positions correspond to a rigid body deflection u_L and rotation ϕ_L of the half beam and moment). Thus the buckled shapes of the cantilever can be derived from those given by equation 7.4 for a simply supported beam as

$$u/\delta = \phi/\theta = 1 - \cos(\pi z/2L). \qquad (9.2)$$

Substitution of these into the energy equation (section 2.8.4.1 or 17.4.7)

$$\frac{1}{2}\int_0^L \{EI_y u''^2 + EI_w \phi''^2 + GJ\phi'^2\} \, dz + \frac{1}{2}\int_0^L 2M_x \phi u'' \, dz + M\phi_L u'_L = 0 \qquad (9.3)$$

leads to

$$\frac{ML}{\sqrt{(EI_y GJ)}} = \frac{\pi}{2}\sqrt{(1 + K^2/4)} \qquad (9.4)$$

where

$$K = \sqrt{(\pi^2 EI_w/GJL^2)}, \qquad (9.5)$$

and to

$$\delta/\theta = -4M/P_y \qquad (9.6)$$

where

$$P_y = \pi^2 EI_y/L^2. \qquad (9.7)$$

Solutions of equation 9.4 are compared in Figure 9.3b with those of

$$M_{yz}L/\sqrt{(EI_y GJ)} = \pi\sqrt{(1 + K^2)} \qquad (9.8)$$

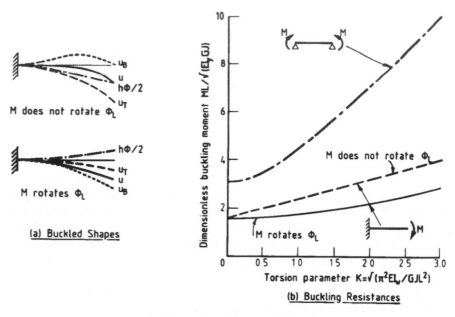

Figure 9.3 Buckling of cantilevers with end moments.

derived from equation 7.5 for simply supported beams. It can be seen that the cantilever buckling resistance M is half of M_{yz} for $K = 0$, and that it increases only slowly with the torsion parameter K.

Numerical solutions for the elastic buckling resistances and buckled shapes of cantilevers with end moments which do not rotate ϕ_L are also shown in Figure 9.3. The buckling resistance varies almost linearly with the torsion parameter K, and may be approximated by

$$ML/\sqrt{(EI_y GJ)} = 1.6 + 0.8\,K. \qquad (9.9)$$

These resistances are somewhat greater than those of equation 9.4 for cantilevers whose end moments rotate ϕ_L, but still less than those of equation 9.8 for simply supported beams.

The buckled shapes of cantilevers with $K = 1$ are compared in Figure 9.3a. These shapes have been normalized so that the value of $u_L = \delta$ is unity. It can be seen that for end moments which do not rotate ϕ_L, the lateral deflections u are somewhat less than for moments which rotate ϕ_L, but the twists ϕ are generally greater and of opposite sign. For equal flange I-beams, the buckled shapes may be shown as the top and bottom flange deflections $u_{T,B} = u \pm h\phi/2$. It can be seen that while both flanges of the cantilever with rotating end moment deflect in the same sense, the reverse is the case for the cantilever whose end moment does not rotate. When the moment rotates, it is the bottom (compression) flange which buckles the further, but when the moment does not rotate, then it is the top (tension) flange which buckles the further.

9.2.2 MONOSYMMETRIC SECTIONS

Numerical solutions for the dimensionless buckling resistances $ML/\sqrt{(EI_yGJ)}$ of monosymmetric cantilevers with end moments M which do not rotate ϕ_L are reported in [1], and these are plotted in Figure 9.4 against the beam parameter

$$\bar{K} = \sqrt{(\pi^2 EI_y h^2/4GJL^2)} \tag{9.10}$$

for selected values of the monosymmetry parameter

$$\rho = I_{yc}/I_y. \tag{9.11}$$

These curves are somewhat similar to those shown in Figure 7.11 for simply supported beams in uniform bending, in that the buckling resistance generally increases with the monosymmetry parameter ρ (that is, as the relative stiffness of the compression flange increases). However, there is a reversal near $\rho = 0.9$, and the buckling resistances of tee-sections with their flanges in compression ($\rho = 1.0$) are less than those of some slightly less monosymmetric sections. This is because the tension flange is more important in a cantilever than it is in a simply supported beam (in an equal flanged cantilever with an end moment that does not rotate, it is the tension flange which buckles the further, as shown in Figure 9.3a, whereas it is the compression flange which buckles the further in a beam, as shown in Figure 7.1). Thus the monosymmetry benefit which results from having a very large compression flange is somewhat reduced by the disadvantage of having no tension flange in a tee-section cantilever.

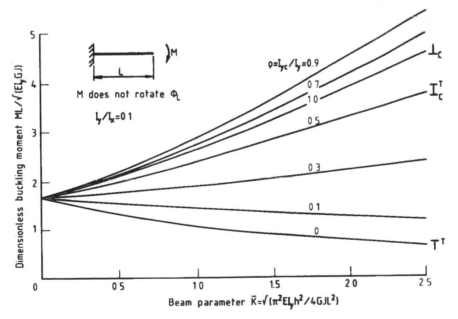

Figure 9.4 Monosymmetric cantilevers in uniform bending.

9.3 End loads

9.3.1 DOUBLY SYMMETRIC SECTIONS

Numerical solutions for the buckling resistances of doubly symmetric cantilevers
with end loads Q acting at distances y_Q below their centroids are reported in [2],
and are shown in Figure 9.5. These solutions may be approximated by

$$\frac{QL^2}{\sqrt{(EI_yGJ)}} = 11\left\{1 + \frac{1.2\varepsilon}{\sqrt{(1 + 1.2^2\varepsilon^2)}}\right\} + 4(K - 2)\left\{1 + \frac{1.2(\varepsilon - 0.1)}{\sqrt{(1 + 1.2^2(\varepsilon - 0.1)^2)}}\right\}$$

(9.12)

in which

$$\varepsilon = \frac{y_Q}{L}\sqrt{\left(\frac{EI_y}{GJ}\right)} = \frac{2y_Q}{h}\frac{K}{\pi}.$$

(9.13)

For centroidal loading ($\varepsilon = 0$), the variation of the buckling resistance with the
torsion parameter K is approximately linear, as it is for end moments which do
not rotate ϕ_L. A comparison of Figures 9.3b and 9.5 indicates that the buckling
resistance under end load is substantially greater than under end moment. This is
so because the bending moment M_x caused by end load is high only near the
restrained support.

The effect of load height y_Q is demonstrated in Figure 9.5, and it can be seen
that while bottom flange loading significantly increases the buckling resistance,
top flange loading may reduce it substantially, especially for cantilevers with high
values of K. The non-linear effect of load height is suggested by the approximate
formulation of equation 9.12. This effect is a little different from that shown in
Figure 7.22 for simply supported beams, where the buckling resistance increases
indefinitely with y_Q. For cantilevers, the approximation of equation 9.12 suggests
that the resistance may approach an upper bound.

9.3.2 MONOSYMMETRIC SECTIONS

Numerical solutions for the buckling resistances of monosymmetric cantilevers
with end loads Q acting at distances $(y_Q - y_0)$ from the shear centre are reported
in [1, 3, 4], and approximate equations in [1].

For shear centre loading, the buckling resistances vary with the beam parameter $\bar{K}$ and the monosymmetry parameter ρ in a somewhat similar manner to
that shown in Figure 9.4 for uniform bending, although the resistances are
generally higher for end load because the moment M_x is high only near the
support.

For bottom flange loading, the buckling resistance is generally increased over
that for shear centre loading, with much the same pattern of variation with $\bar{K}$

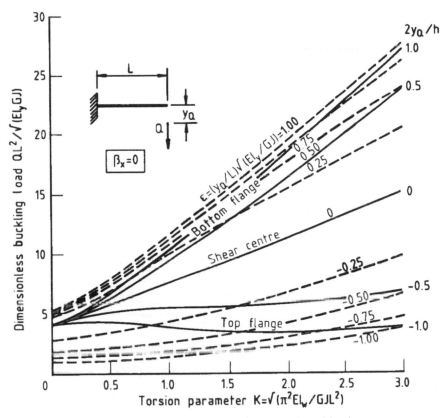

Figure 9.5 Buckling of cantilevers with end loads.

and ρ. However, this pattern changes considerably for top flange loading, with very significant reductions in stiffness for high values of ρ (i.e., for cantilevers with small tension flanges). It appears that the maximum resistance is achieved for sections with nearly equal flanges ($\rho \approx 0.5$), and that the low resistances of tee-sections are almost independent of their attitude ($\rho = 1$, flange in compression, or $\rho = 0$, flange in tension).

9.4 Uniformly distributed loads

9.4.1 DOUBLY SYMMETRIC SECTIONS

Numerical solutions for the buckling resistances of doubly symmetric cantilevers with uniformly distributed loads q acting at distances y_q from their centroids are reported in [2], and are shown in Figure 9.6. These solutions may be

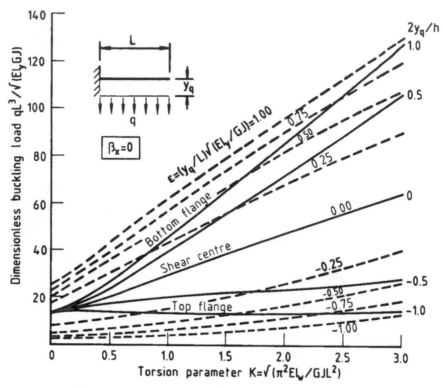

Figure 9.6 Buckling of cantilevers with distributed loads.

approximated by

$$\frac{qL^3}{2\sqrt{(EI_y GJ)}} = 27\left\{1 + \frac{1.4(\varepsilon - 0.1)}{\sqrt{(1 + 1.4^2(\varepsilon - 0.1)^2)}}\right\} + 10(K - 2)\left\{1 + \frac{1.3(\varepsilon - 0.1)}{\sqrt{(1 + 1.3^2(\varepsilon - 0.1)^2)}}\right\}$$

(9.14)

in which $\varepsilon = (2y_q/h)K/\pi$ is a dimensionless load height parameter. For centroidal loading ($\varepsilon = 0$), the buckling resistance again varies almost linearly with the torsion parameter K, and is higher than the resistance for end load (Figure 9.5), because the bending moment is generally lower. For loading away from the centroid, the resistance changes non-linearly with the dimensionless load height ε, as suggested by the approximate formulation of equation 9.14.

9.4.2 MONOSYMMETRIC SECTIONS

Numerical solutions for the buckling resistances of monosymmetric cantilevers with uniformly distributed loads q are reported in [1, 3], and approximate equations in [1].

Again, the buckling resistance for shear centre loading varies with the beam parameter $\bar{K}$ and the monosymmetry parameter ρ in a somewhat similar manner to that shown in Figure 9.4 for uniform bending. For bottom flange loading, the resistance is generally increased over that for shear centre loading, with much the same pattern of variation with $\bar{K}$ and ρ. For top flange loading, the pattern changes considerably, and it again appears that the resistance of equal flanged sections is close to optimal, and that the resistances of tee-sections are almost independent of their attitude (flange in compression or tension).

9.5 Continuous restraints

The actions of continuous restraints on cantilevers may be modelled as in section 8.2.1, and their effects on elastic buckling may be analysed numerically by using the adaptation of the finite element method described in section 8.3.

The effects of continuous elastic restraints on the elastic buckling of doubly symmetric cantilevers were studied in [5]. The results reported for end loads acting at the centroid indicate that in many cases a cantilever acts as if rigidly restrained if the translational restraint stiffness per unit length exceeds $\alpha_{tL} = 2500EI_y/L^4$ approximately, provided the torsion parameter $K = \sqrt{(\pi^2 EI_w/GJL^2)}$ is not small (i.e. $K > 0.5$).

Rigid continuous translational restraints enforce an axis of buckling rotation at the height y_t at which the restraints act. The results of [5] indicate that for cantilevers with high values of K (i.e. $K > 1.5$), tension flange restraints are markedly more effective than compression flange restraints, and that the buckling resistances are significantly higher than those for unrestrained cantilevers, especially when the end load acts at or below the centroid.

9.6 Discrete restraints

9.6.1 SUPPORT RESTRAINTS

9.6.1.1 Minor axis rotation restraint

The support restraints for cantilevers are generally assumed to be rigid, so that minor axis end rotations u_0' and end warping displacements proportional to ϕ_0' are prevented, in addition to the usual support conditions that lateral deflections u_0 and twist rotations ϕ_0 are prevented.

However, a rigid restraint is not required to prevent the minor axis end rotation u_0' when end twisting ϕ_0 is prevented, since under these circumstances the minor axis end moment $-(M_x\phi)_0$ is automatically zero. Because of this, any non-zero elastic restraint stiffness will be sufficient to ensure $u_0' = 0$. Indeed, even the limiting case of zero rotation restraint, for which the cantilever theoretically

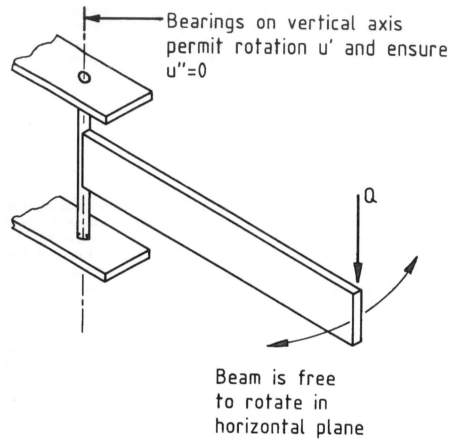

Figure 9.7 Rigid body buckling mode $(Q = 0)$.

buckles at zero applied load in a rigid body mode by rotating around the support as shown in Figure 9.7, does not lead to structural failure, since the cantilever remains unstrained. Instead, it is possible for the cantilever to support non-zero loads without structural failure, even though it may rotate u'_0 as a rigid body about the support. Thus it may be assumed that $u'_0 = 0$ so that rigid body rotation is prevented, even where there is no minor axis restraint, provided end twisting ϕ_0 is prevented.

9.6.1.2 Warping restraint

End warping is usually assumed to be prevented at the supports of cantilevers by rigid restraints to the flanges. Warping restraints make very significant contributions to the buckling resistance of cantilevers, and when these are removed so that $\phi''_0 = 0$, then the buckling resistances are greatly reduced, as is demonstrated in Figures 9.8, 9.9 and 9.10. These figures show that for centroidal loading, the buckling resistance is almost independent of the torsion parameter K, indicating that end warping restraint is required to ensure that the warping rigidity EI_w contributes to the buckling resistance. Cantilevers which are free to warp at the support (or 'overhanging segments') buckle with an almost linear

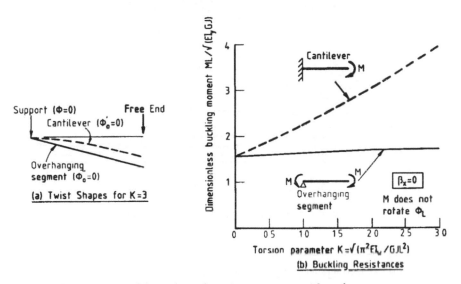

Figure 9.8 Buckling of overhanging segments with end moments.

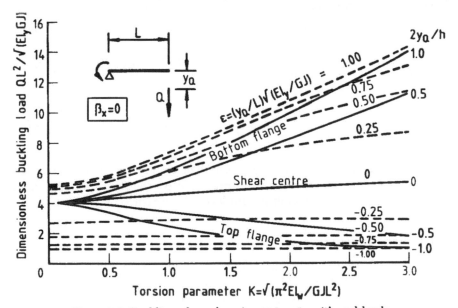

Figure 9.9 Buckling of overhanging segments with end loads.

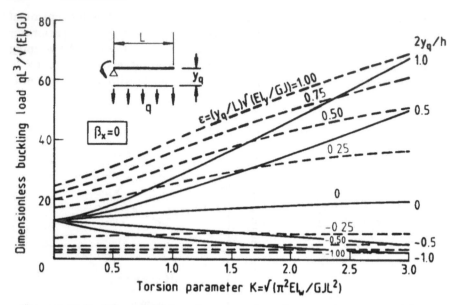

Figure 9.10 Buckling of overhanging segments with uniformly distributed loads.

variation in the twist rotation ϕ, as shown in Figure 9.8a, so that the warping strain energy stored $\frac{1}{2}\int_0^L EI_w(\phi'')^2\,dz$ is very small.

The elastic buckling resistances of overhanging segments with $\phi_0'' = 0$ shown in Figures 9.8, 9.9 and 9.10 have been approximated by [2]

$$ML/\sqrt{(EI_yGJ)} = 1.6 + 0.05K, \tag{9.15}$$

$$\frac{QL^2}{\sqrt{(EI_yGJ)}} = 6\left\{1 + \frac{1.5(\varepsilon - 0.1)}{\sqrt{[1 + 1.5^2(\varepsilon - 0.1)^2]}}\right\} + 1.5(K - 2)\left\{1 + \frac{3(\varepsilon - 0.3)}{\sqrt{[1 + 3^2(\varepsilon - 0.3)^2]}}\right\}, \tag{9.16}$$

$$\frac{qL^3}{2\sqrt{(EI_yGJ)}} = 15\left\{1 + \frac{1.8(\varepsilon - 0.3)}{\sqrt{[1 + 1.8^2(\varepsilon - 0.3)^2]}}\right\} + 4(K - 2)\left\{1 + \frac{2.8(\varepsilon - 0.4)}{\sqrt{[1 + 2.8^2(\varepsilon - 0.4)^2]}}\right\}, \tag{9.17}$$

in which $\varepsilon = (2y_q/h)K/\pi$ is a dimensionless load height parameter. These resistances are significantly lower than those predicted by the corresponding equations 9.9, 9.12 and 9.14 for built-in cantilevers for which $\phi_0' = 0$.

9.6.1.3 Torsional restraint

Supports which are capable of providing full minor axis rotation and warping restraints will usually be capable of providing full torsional restraint also, so that $\phi_0 = 0$. However, when a support is not capable of providing any warping restraint, then it is possible that full torsional restraint may not be available. Any

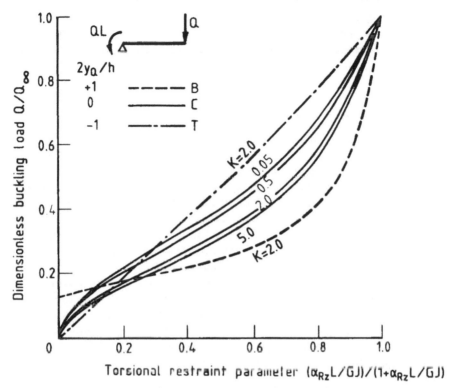

Figure 9.11 Overhanging segments with torsional restraints.

reduction in the stiffness of the torsional restraint at the support will cause a corresponding reduction in the buckling resistance.

The reduced buckling end loads Q of overhanging segments with elastic torsional support restraints are shown by the non-dimensional ratios Q/Q_∞ in Figure 9.11, in which Q_∞ is the buckling load for an overhanging segment with rigid torsional support restraint (see equation 9.16). These may be approximated by

$$\frac{Q}{Q_\infty} = \left\{1 - \frac{2y_q}{h}K\frac{(\alpha_{R_z}L/GJ)^2}{(1+\alpha_{R_z}L/GJ)^3}\right\} \bigg/ \sqrt{\left\{\frac{\alpha_{R_z}L/GJ}{[5+4K^2/(1+K^2)+\alpha_{R_z}L/GJ]}\right\}} \quad (9.18)$$

which is accurate for bottom flange and centroidal loading while $K \leqslant 2$, and conservative for top flange loading while $K \leqslant 2$ and $\alpha_{R_z}L/GJ \geqslant 2.5$.

9.6.2 END RESTRAINTS

The effects of rigid restraints acting at the free ends of cantilevers have been studied in [6, 7]. Cantilevers with free end restraints which prevent lateral displacement u_L and twist rotation ϕ_L are similar to simply supported beams with end restraints against minor axis rotation u'_0 and warping deflections proportional to ϕ'_0. They may be treated as in section 8.5.2.1, as may cantilevers with full end restraints ($u_L = \phi_L = u'_L = \phi'_L = 0$).

9.6.3 INTERMEDIATE RESTRAINTS

The elastic buckling of doubly symmetric and monosymmetric cantilevers with rigid intermediate restraints was studied in [6, 8]. The cantilevers had either concentrated end loads Q, or uniformly distributed loads q, acting at the top flange, centroid, or bottom flange ($2y_q/h = -1, 0, 1$). For doubly symmetric cantilevers, the optimum distance from the support of a restraint which prevented both lateral deflection and twist rotation ($u = \phi = 0$) varied between $0.35L$ and $0.85L$ approximately, depending on the loading, load height, and torsion parameter K, and the benefits from optimum bracing were substantial, especially for cantilevers with high values of K and top flange loading.

The effectiveness of rigid translational or torsional intermediate restraints was also studied in [6]. In general, translational restraints ($u = 0$) were found to be comparatively ineffective, except when acting at the end of the top flange. Torsional restraints ($\phi = 0$) were generally more effective, especially when they acted at a distance from the support between $0.45L$ and $0.75L$ approximately, although not as effective as optimum restraints against both translation and twist rotation.

9.7 Problems

PROBLEM 9.1

Determine the elastic flexural-torsional buckling moment of a cantilever whose properties are given in Figure 7.23, and which is in uniform bending over a cantilevered length of 6.0 m, as shown in Figure 9.12a,

(a) when the moment rotates with the twist rotation of the free end of the cantilever;
(b) when the moment does not rotate.

PROBLEM 9.2

Determine the elastic flexural-torsional buckling moments of a cantilever with the monosymmetric section of Problem 7.4, and which has a cantilevered length of 6.0 m as shown in Figure 9.12b, and which has end moments which do not rotate with the twist rotation of the free end.

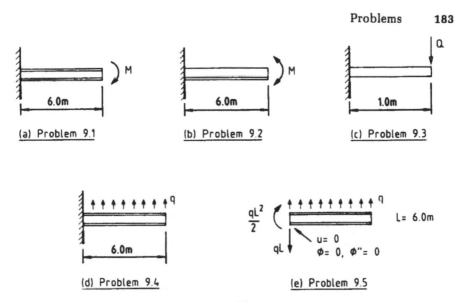

Figure 9.12 Problems 9.1–9.5.

PROBLEM 9.3

Determine the elastic flexural-torsional buckling load of a narrow rectangular
(50 mm × 10 mm) steel cantilever ($E = 200\,000$ MPa, $G = 80\,000$ MPa) which has
a cantilevered length of 1.0 m and a concentrated end load acting on the top
surface of the cantilever, as shown in Figure 9.12c.

PROBLEM 9.4

Determine the elastic flexural-torsional buckling load of a cantilever whose
properties are given in Figure 7.23, which has a cantilever length of 6.0 m as
shown in Figure 9.12d, and which has uniformly distributed uplift loading acting
along the top flange.

PROBLEM 9.5

Determine the elastic flexural-torsional buckling load for an overhanging segment having the same properties, length, and loading as the cantilever of
Problem 9.4, but which is free to warp at the support, as indicated in Figure 9.12e.

PROBLEM 9.6

Determine the stiffness of the torsional restraint required at the support of the
overhanging segment of Problem 9.5 if the reduction in the elastic flexural-
torsional buckling load below that for full torsional restraint is not to exceed 10%.

9.8 References

1. Wang, C.M. and Kitipornchai, S. (1986) On stability of monosymmetric cantilevers. *Engineering Structures*, 8(3), 169–180.
2. Trahair, N.S. (1983) Lateral buckling of overhanging beams, in *Instability and Plastic Collapse of Steel Structures*, (ed. L.J. Morris), Granada, London, pp. 503–18.
3. Anderson, J.M. and Trahair, N.S. (1972) Stability of monosymmetric beams and cantilevers. *Journal of the Structural Division, ASCE*, 98(ST1), 269–86.
4. Roberts, T.M. and Burt, C.A. (1985) Instability of monosymmetric I-beams and cantilevers. *International Journal of Mechanical Sciences*, 27(5), 313–24.
5. Assadi, M. and Roeder, C.W. (1985) Stability of continuously restrained cantilevers. *Journal of Engineering Mechanics, ASCE*, 111(12), 1440–56.
6. Kitipornchai, S., Dux, P.F. and Richter, N.J. (1984) Buckling and bracing of cantilevers, *Journal of Structural Engineering, ASCE*, 110(ST9), 2250–62.
7. Nethercot, D.A. (1973) The effective lengths of cantilevers as governed by lateral buckling. *The Structural Engineer*, 51(5), 161–8.
8. Wang, C.M., Kitipornchai, S. and Thevendran, V. (1987) Buckling of braced mono-symmetric cantilevers. *International Journal of Mechanical Sciences*, 29(5), 321–37.

10 Braced and continuous beams

10.1 General

The restraining actions between adjacent segments in continuous beams are not easily predicted, because there are interactions between the segments at buckling, as indicated in Figure 10.1. These interactions affect the stiffnesses of the restraining segments, which depend not only on their geometric and material properties, but also on the relative importance of their loading. Because of this, the information given in Chapters 8 and 9 for elastically restrained beams and cantilevers cannot be used directly to predict the elastic buckling loads of continuous beams.

This chapter is concerned with these buckling restraining actions, and presents approximate methods for predicting their effects. The methods may also be applied to elastic beams which are prevented from deflecting laterally and twisting ($u = \phi = 0$) at supports and brace points, as shown in Figure 10.2b. An extension of these methods to beams with overhanging segments (Figure 10.2c) is

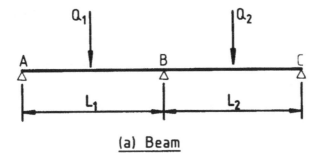

(a) Beam

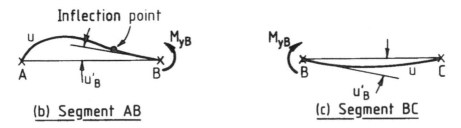

(b) Segment AB　　　　(c) Segment BC

Figure 10.1 Interaction buckling.

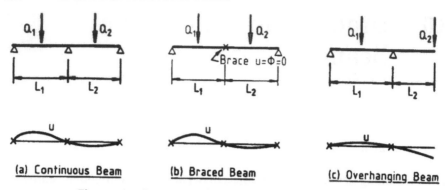

Figure 10.2 Continuous, braced, and overhanging beams.

also presented. The extension of the methods to inelastic beams is referred to in sections 14.5.4 and 14.5.5.

10.2 Interaction buckling

The continuity between the adjacent segments of a continuous beam during buckling induces interactions between them. Several different buckling modes are possible for the two-span beam shown in Figure 10.3a, depending on the loading

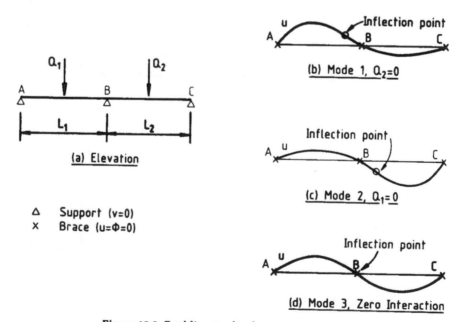

Figure 10.3 Buckling modes for a two-span beam.

arrangement. For the mode shown in Figure 10.3b, the left-hand segment is heavily loaded, and dominates in the buckling action. During buckling, this segment is elastically restrained by the right-hand segment, and there is a point of inflection in the deflected shape of the left-hand segment. For the mode shown in Figure 10.3c, the reverse is the case, and there is a point of inflection in the deflected shape of the heavily loaded right-hand segment.

When the segments are loaded so that they are equally important during buckling, then the inflection point is at the interior support, as shown in Figure 10.3d. In this special case there is no buckling interaction between the segments.

More generally, however, one segment is elastically restrained by the other, as shown in Figure 10.1b. In this case, the left-hand segment AB has the inflection point, and is positively restrained, in the sense that the restraining minor axis end moment M_{yB} provided by the right-hand segment BC opposes the end rotation u'_B of the left-hand segment. Thus the buckling resistance of the left-hand restrained segment is increased above the value for zero restraint, as indicated in Figure 10.4. This figure [1, 2] shows the variations of the dimensionless buckling load $QL^2/\sqrt{(EI_yGJ)}$ of one segment of a narrow rectangular section beam with the major and minor axis end restraint parameters

$$\beta_1 = (\alpha_{Rx}L/EI_x)/(3 + \alpha_{Rx}L/EI_x), \tag{10.1}$$

$$\beta_2 = (\alpha_{Ry}L/EI_y)/(3 + \alpha_{Ry}L/EI_y), \tag{10.2}$$

in which

$$\alpha_{Rx} = -M_{xB}/v'_B, \tag{10.3}$$

$$\alpha_{Ry} = -M_{yB}/u'_B, \tag{10.4}$$

are the equivalent in-plane and out-of-plane stiffnesses of the adjacent segment.

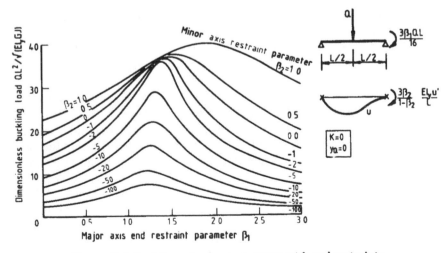

Figure 10.4 Buckling of a beam segment with end restraints.

Note that the values of α_R equal to 0 and ∞ corresponding to zero and rigid restraints lead to corresponding values of β_1 and β_2 equal to 0 and 1. Thus the buckling resistance increases from the zero minor axis restraint value (for $\beta_2 = 0$) to the rigidly restrained value (for $\beta_2 = 1$) as the restraint stiffness α_{Ry} increases from 0 to ∞.

When the left-hand segment of Figure 10.1a is positively restrained (so that the end moment M_{yB} opposes the minor axis end rotation u'_B), then the restraining right-hand segment is negatively restrained, in that the minor axis end moment M_{yB} acts in the same sense as the end rotation u'_B, as shown in Figure 10.1c. This negative restraining action decreases the buckling resistance of the restraining right-hand segment below the value for zero restraint ($\beta_2 = 0$), as indicated by the lower curves in Figure 10.4 for negative values of β_2.

The particular case where $Q_1 = 2Q_2$ and $L_1 = L_2$ may be examined in detail. Analysis of the in-plane bending of the beam leads to

$$(\beta_1)_1 = \frac{1 + Q_2 L_2^2 / Q_1 L_1^2}{1 + L_2/L_1} = 0.75$$

and

$$(\beta_1)_2 = \frac{1 + Q_1 L_1^2 / Q_2 L_2^2}{1 + L_1/L_2} = 1.50,$$

while minor axis continuity and equilibrium at the interior support require that $(\alpha_{Ry})_1 = -(\alpha_{Ry})_2$, which leads to

$$(\beta_2)_1 = \frac{-(\beta_2)_2 (L_1/L_2)}{1 - (\beta_2)_2 (1 + L_1/L_2)} = \frac{-(\beta_2)_2}{1 - 2(\beta_2)_2}.$$

Assuming $(\beta_2)_2 = 0$ for zero interaction leads to $(\beta_2)_1 = 0$ and $Q_2/Q_1 \approx 37.0/25.4 \approx 1.46$ (see Figure 10.4) instead of 0.5. A better guess is $(\beta_2)_2 = -20$ so that $(\beta_2)_1 \approx 0.49$ and $Q_2/Q_1 \approx 13.6/27.3 \approx 0.498$, which is very close to 0.5. Thus the effects of interaction between the two segments are to increase the dimensionless buckling load of the left-hand segment from 25.4 for zero interaction to 27.3 approximately and to decrease that of the right-hand segment from 37.0 to 13.6 approximately.

This example demonstrates the general characteristic of interaction buckling, that sufficient interaction takes place to increase the buckling resistance of the restrained segment and decrease that of the restraining segment until their ratio is equal to the ratio of the applied loads. Thus all segments participate in the buckling action, and buckle simultaneously, even though one segment is restrained by the other.

The buckling load combinations for the two-span beam of Figure 10.1 with $L_2 = L_1$ are shown as an interaction diagram in Figure 10.5. It can be seen that the buckling load Q_1 increases slowly with Q_2 until a maximum value is reached near the zero interaction point $Q_2 = Q_1$. Soon after this both Q_1 and Q_2 reduce, Q_1 rapidly and Q_2 slowly. The slow increase in Q_1 as Q_2 increases from zero is

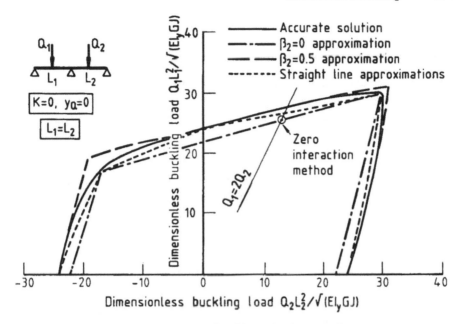

Figure 10.5 Interaction buckling of a two-span beam.

primarily due to the change in the major axis end restraint parameter $(\beta_1)_1$ which increases from 0.5 to 1.0 when Q_2/Q_1 increases from 0 to 1.0. Figure 10.4 indicates that the increase in the buckling resistance due to this cause is greater than the decrease due to the decrease in the minor axis restraint parameter $(\beta_1)_2$ from 0.5 to 0.

Also shown in Figure 10.5 are dash–dot curves for the buckling load predictions obtained by assuming zero interaction $(\beta_2 = 0)$, and using the lower of the buckling load factors for each span. These predictions are generally conservative, except at $Q_2 = Q_1$, where they are exact. In this case the beam buckles with zero interaction between the segments.

Another set of predictions is shown by the dashed curves in Figure 10.5 for which it is assumed that the restraint provided by the restraining segment is unaffected by its loading. Thus $\alpha_{Ry} = 3EI_y/L$ is used so that $\beta_2 = 0.5$. These predictions are generally unconservative, although they are close when either Q_1 or Q_2 is zero.

It can be seen from Figure 10.5 that in this case both of these assumptions produce approximations which are reasonably close to the accurate solutions. This is because in this case of a narrow rectangular beam, the minor axis restraints have their smallest effects. More generally, the minor axis restraint effects will be greater, and these approximations may not be sufficiently accurate. More accurate approximate methods are discussed in section 10.3.

10.3 Approximate methods

10.3.1 STRAIGHT LINE METHOD

The buckling load interaction diagram of Figure 10.5 suggests that close approximations can be obtained by drawing the straight dotted lines between the zero interaction load combination and the buckling loads obtained for the cases $Q_1 = 0$ and $Q_2 = 0$ by assuming that the restraint provided by the restraining segment is unaffected by its loading. Unfortunately, this method has proved to be too complex for use in routine design, possibly because the available tabulations of elastic buckling loads are not only insufficient to cover all the required loading and restraint conditions, but also because they are too detailed to enable them to be easily used.

10.3.2 ZERO INTERACTION METHOD

10.3.2.1 General

In the zero interaction method [3], it is assumed that there are no minor axis bending or warping interactions between adjacent segments at the brace and support points. Thus each segment buckles as if simply supported and independent of its adjacent segments. This is equivalent to assuming that each segment between adjacent brace or support points has an effective length factor of $k = 1$.

The buckling load of each segment may then be estimated by using the available data for simply supported beams (Chapter 7). It is usually convenient to express this as the buckling load factor given by

$$\lambda_n = \frac{M_{mn}}{M_{mni}} \tag{10.5}$$

in which M_{mn} is the maximum moment in the segment at elastic buckling, and M_{mni} is the corresponding moment caused by an initial load set.

This procedure leads to a number of different estimates λ_n for the buckling load factor, one for each segment. A lower bound estimate of the buckling load factor λ at which buckling actually takes place may then be obtained from the lowest of these estimates, so that

$$\lambda = (\lambda_n)_{min}. \tag{10.6}$$

The other values of λ_n generally overestimate the true buckling load factor.

Thus for the two-span beam of Figure 10.1 with $Q_1 = 2Q_2$, $L_1 = L_2$, $(\beta_1)_1 = 0.75$, $(\beta_1)_2 = 1.5$, and an initial load set corresponding to $Q_1 L_1^2/\sqrt{(EI_yGJ)} = 1.0$, $Q_2 L_2^2/\sqrt{(EI_yGJ)} = 0.5$, the zero interaction assumption of $(\beta_2)_1 = (\beta_2)_2 = 0$ leads to buckling load factors of $\lambda_1 = 25.4/1$ and $\lambda_2 = 37/0.5$ (see Figure 10.4). Thus the lower bound estimate is $\lambda = 25.4$, which is a little lower than the actual buckling load factor of 27.4. The value $\lambda_2 = 74$ is much higher.

The zero interaction method may be summarized as follows:

(a) Determine the properties EI_y, GJ, EI_w, L of each of the n segments.

(b) Analyse the in-plane bending moment distribution throughout the beam for an initial load set, and determine the initial maximum moment M_{mi} for each segment.

(c) Assume that the effective length factor k of each segment is unity.

(d) Estimate the maximum moment M_m at elastic buckling for each segment, and the corresponding buckling load factor $\lambda_n = (M_m/M_{mi})_n$.

(e) Determine a lower bound estimate of the beam buckling load λ as the lowest value of λ_n.

10.3.2.2 Example

The properties and loading of a braced beam are shown in Figure 10.6. The calculation steps (all in N and mm units) in the zero interaction method of approximating the elastic buckling moment are as follows:

(a) $EI_y = 2.0 \times 10^{12}$, $GJ = 49.33 \times 10^9$, $EI_w = 250.0 \times 10^{15}$, $L_1 = L_2 = 4000$.

(b) For an initial load set of $M = 1.0$, $M_{mi1} = 1.0$, $M_{mi2} = 0.5$.

(c) $k_1 = 1.0$, $k_2 = 1.0$.

(d) Using

$$M_m = \alpha_m \frac{\pi\sqrt{(EI_yGJ)}}{kL}\sqrt{\left(1 + \frac{\pi^2 EI_w}{GJk^2L^2}\right)}$$

and

$$= \alpha_m (348.9 \times 10^6)\sqrt{(1 + 1.768^2)}$$

$$\alpha_m = 1.75 + 1.05\beta + 0.3\beta^2 \leqslant 2.5$$

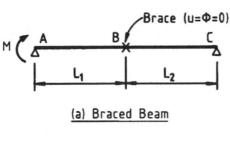

(a) Braced Beam

E=200,000 N/mm^2
G=80,000 N/mm^2
I_y= 10.0×10^{6}mm^4
J= 0.617×10^{6}mm^4
I_w=1.25×10^{12}mm^6
L_1=L_2=4000 mm

(c) Properties

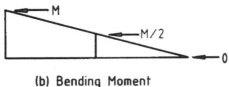

(b) Bending Moment

Figure 10.6 Braced beam example.

as in section 7.3, then

$$(\beta)_1 = -0.5, \alpha_{m1} = 1.30, M_{m1} = 921 \times 10^6, \lambda_1 = 921 \times 10^6,$$
$$(\beta)_2 = 0, \alpha_{m2} = 1.75, M_{m2} = 1240 \times 10^6, \lambda_2 = 2480 \times 10^6$$

(e) $\lambda = 921 \times 10^6$.

This result is 20% less than the accurate solution of $\lambda = 1156 \times 10^6$ determined by a finite element computer program of the type described in section 4.4.

10.3.3 IMPROVED METHOD

10.3.3.1 General

A more accurate estimate of the buckling load factor may be obtained by estimating the effects of the interactions which take place between adjacent segments during buckling [4–6]. Because all segments buckle simultaneously, these effects need only be estimated for one segment, and it is usually most convenient if this is the segment which has the lowest zero interaction buckling load factor. This segment is identified as the critical segment, and its adjacent segments as the restraining segments. The restraining segments generally provide positive restraints to the critical segment, and increase its buckling resistance.

The increased buckling resistance of a critical segment AB is estimated by determining its effective length factor k. It is assumed that this is related to the relative end stiffness ratios G_A, G_B as shown in Figure 10.7 in the same way as for

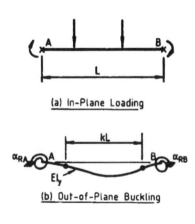

(a) In-Plane Loading

(b) Out-of-Plane Buckling

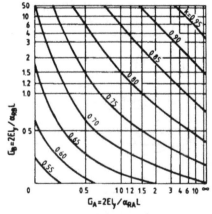

(c) Approximate Effective Length Factors

Figure 10.7 Buckling of restrainted segments.

compression members with elastic end restraints [5]. Thus

$$G_{A,B} = \frac{\alpha_c}{\alpha_{RA,RB}} \qquad (10.7)$$

in which

$$\alpha_c = (2EI_y/L)_c \qquad (10.8)$$

is the minor axis flexural stiffness of the critical segment, and α_{RA}, α_{RB} are the effective stiffnesses of the restraining segments.

The effective stiffness α_R of a restraining segment may be approximated by

$$\alpha_R = \gamma_F(EI_y/L)_R \gamma_M \qquad (10.9)$$

in which γ_M depends on the bending moment magnitude and γ_F depends on the conditions at the far end of the restraining segment. The factor γ_F is equal to 2 when the segment has to provide an equal restraining moment at its far end, is equal to 3 when there is zero moment at the far end, and is equal to 4 when the far end is built in, as shown in Figure 10.8.

The factor γ_M varies from 1 to 0 as the bending moment magnitude for the restraining segment varies from zero to a level sufficient to cause it to buckle with zero interaction. For restraining compression members [5], the corresponding factor is closely approximated by $(1 - P/P_y)$ in which P is the actual compression force in the restraining segment, and P_y is its (zero interaction) flexural buckling load. The accuracy of this approximation reflects the linear relationship between the buckling resistance and EI_y, and indicates that the stiffness available for

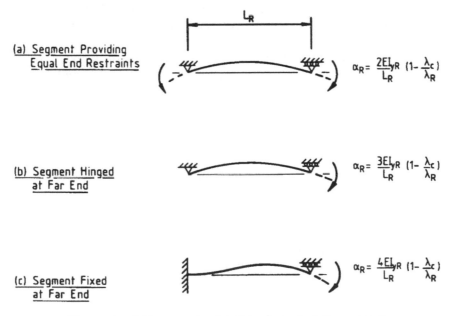

Figure 10.8 Stiffness approximations for restraining segments.

restraining the critical segment decreases almost linearly as the axial force increases towards P_y.

However, in flexural members, the relationship

$$(\alpha_m M_{yz})_R = \alpha_m \frac{\pi \sqrt{(EI_y GJ)_R}}{L_R} \sqrt{\left(1 + \frac{\pi^2 EI_y h^2}{4GJL^2}\right)_R} \tag{10.10}$$

for the buckling resistance of an I-section member varies between proportional to $\sqrt{(EI_y)_R}$ and proportional to $(EI_y)_R$ as $\pi^2 EI_y h^2/4GJL^2$ varies between 0 and ∞. It might therefore be expected that the factor γ_M for the stiffness available for restraining the critical segment should vary between $(1 - M_m^2/\alpha_m^2 M_{yz}^2)_R$ and $(1 - M_m/\alpha_m M_{yz})_R$.

Such expressions are not easy to use, since $(M_m)_R$ is unknown, and an iterative procedure must be used, starting with an assumed value of $(M_m)_R$. A simpler computational procedure is to assume

$$\gamma_M = (1 - \lambda_c/\lambda_R) \tag{10.11}$$

in which λ_c, λ_R are the zero interaction buckling load factors for the critical and restraining segments. This approximation is accurate when λ_c is equal to 0 or λ_R, and provides estimates which often lie between those of $(1 - M_m^2/\alpha_m^2 M_{yz}^2)_R$ and $(1 - M_m/\alpha_m M_{yz})_R$. It allows a direct calculation for γ_M to be made and avoids the iterative procedure.

The complete improved method may be summarized as follows:

(a–e) As for the zero interaction method (section 10.3.2.1).
(f)   Identify the segment with the lowest buckling load factor (λ_c) as the critical segment AB, and its adjacent segments as the restraining segments.
(g)   Use λ_c, λ_{RA}, λ_{RB}, and the stiffness approximations of Figure 10.8 in equation 10.9 to calculate α_{RA}, α_{RB} for the restraining segments.
(h)   Use equation 10.8 to calculate α_c for the critical segment.
(i)   Use equation 10.7 to calculate the stiffness ratios G_A, G_B.
(j)   Use Figure 10.7c to determine the effective length factor k for the critical segment.
(k)   Use k to estimate the maximum moment M_{mc} in the critical segment at elastic buckling, and the corresponding buckling load factor $\lambda = M_{mc}/M_{mei}$.

10.3.3.2 Example

The calculation steps (all in N and mm units) in the improved method of approximating the elastic buckling moment of the braced beam shown in Figure 10.6 are as follows.

(a–e) As in section 10.3.2.2.
(f)   AB is the critical segment, with $\lambda_c = 921 \times 10^6$, and $(\lambda_R)_{BC} = 2480 \times 10^6$.

(g) $\alpha_{RA} = 0$ by inspection

$\alpha_{RB} = (3EI/L)(1 - \lambda_c/\lambda_R) = 1.886\ EI/L$

(h) $\alpha_c = 2EI/L$

(i) $G_A = (2EI/L)/0 = \infty$

$G_B = (2EI/L)/(1.886\ EI/L) = 1.06$

(j) $k = 0.875$ from Figure 10.7

(k)

$$M_{mc} = \alpha_m \frac{\pi\sqrt{(EI_yGJ)}}{kL}\sqrt{\left(1 + \frac{\pi^2 EI_w}{GJk^2L^2}\right)}$$

$$= (1.30 \times 348.9 \times 10^6/0.875)\sqrt{(1 + 1.768^2/0.875^2)}$$

$$= 1169 \times 10^6$$

$$\lambda = 1169 \times 10^6$$

which is 27% higher than the zero interaction approximation, and within 1.2% of the accurate solution of $\lambda = 1156 \times 10^6$.

10.3.4 FURTHER IMPROVEMENTS AND EXTENSIONS

Further improvements in the approximate methods for calculating the elastic buckling loads of braced and continuous beams are given in [7]. A number of effective length charts are provided for segments under constant moment gradient, which are generally more accurate than the single chart of Figure 10.7c, and the bending moment distribution factor γ_F is approximated more accurately by

$$\gamma_F = (1 - M_m^2/\alpha_m^2 M_{yz}^2)_R. \tag{10.12}$$

Very accurate solutions are obtained after two or three iteration cycles.

The method described in section 10.3.3 has been extended [8,9] to allow the approximate analysis of the inelastic buckling of braced and continuous beams (see sections 14.5.4, 14.5.5).

10.4 Elastic buckling of braced and continuous beams

10.4.1 BRACED BEAMS

10.4.1.1 General

There have been a number of studies [4, 7, 10, 11] of the elastic buckling of braced beams, and the scope of these is summarized in Figure 10.9. Generally, these studies demonstrate the conservative nature of the zero interaction method of approximation (section 10.3.2), the increased accuracy of the improved method (section 10.3.3), and the high accuracy of the method given in [7].

Braced Beam	Loading	Load Height $\frac{2y_a}{h}$	Brace Position $\frac{a}{L}$	K	Reference
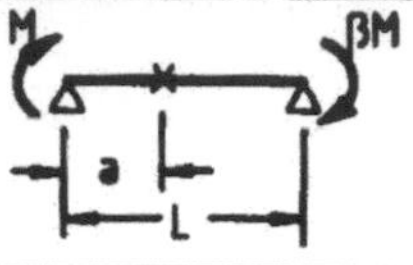 (diagram: M ... βM, a, L)	β=0.5,-1	—	0→1.0	0.14,2.22	[4]
	β=-1(0.5)+1	—	0→1.0	0.1,3.0	[10]
	β=-1→+1	—	Optimum	0.1,0.3,1,3	[10]
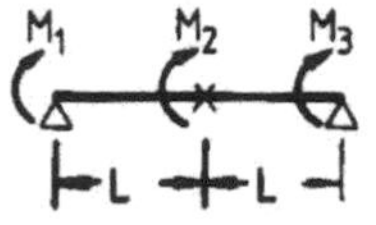 (diagram: M_1 M_2 M_3, L, L)	βM,2M,βM	—	Midspan	1.0	[11]
	M,2βM,M	—	Midspan	1.0	[11]
	M,(1+β)M,βM	—	Midspan	1.0	[11]
(diagram: Q_1 Q_2, a, L, a)	$Q_2=Q_1$	—	0→0.5	0.14,2.22	[4]
	$Q_2=Q_1$	—	0→0.5	0.1,1.0,3.0	[10]
	$Q_2=-Q_1$	—	0→0.5	0.14,2.22	[4]
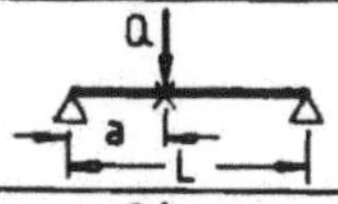 (diagram: Q, a, L)	—	—	0→0.5	0.1,1.0,3.0	[10]
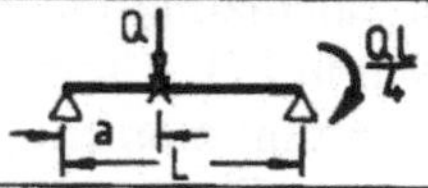 (diagram: Q, QL/4, a, L)	—	—	0→0.5	$0.1, 1 \times (1-\frac{a}{L})$	[7]
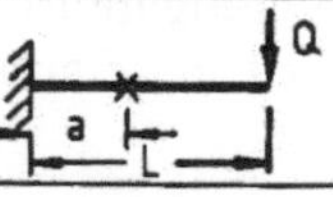 (diagram: cantilever Q, a, L)	—	—	0→1.0	1.0	[7]
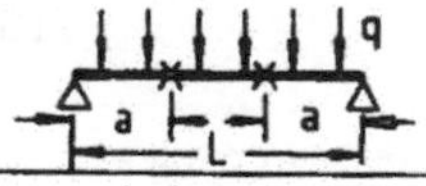 (diagram: q distributed, a, L, a)	—	0	0→0.5	0.1,1.0,3.0	[10]
	—	-1,0,+1	0→0.5	1.0	[10]
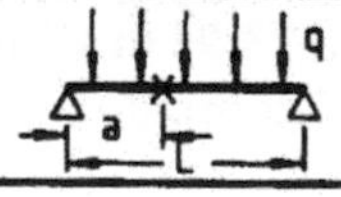 (diagram: q distributed, a, L)	—	0	0→0.5	0.1,1.0,3.0	[10]

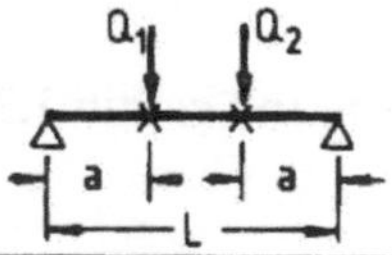

Figure 10.9 Studies of braced beams.

10.4.1.2 Optimum bracing

A study [10] has been made of the effects of the brace position on the buckling of a braced beam under moment gradient, and these are shown in Figure 10.10a as values of the dimensionless buckling resistance M/M_u, in which M_u is the buckling moment of the unbraced beam. It can be seen that the provision of a brace generally leads to a substantial increase in the buckling resistance, especially for beams with high values of K.

Optimum brace positions which maximize the buckling resistance are shown in Figure 10.10b. These lie in the half of the beam which has the higher end moment M, and appear to be close to the brace positions for zero buckling

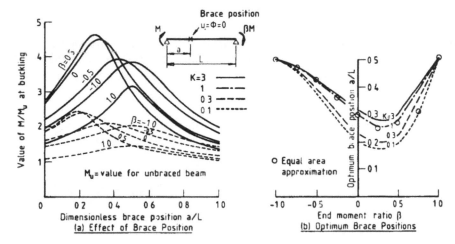

Figure 10.10 Braced beams under moment gradient.

interaction between the segments. A simple method of approximating the optimum brace position is suggested in [10] in which the brace is so placed that it divides the bending moment diagram into two equal areas (without regard for the sign of the moment).

10.4.1.3 Concentrated moments

The elastic buckling of braced beams with concentrated moments at the brace point (Figure 10.11) has been studied [11]. These beams develop internal restraints at the brace against either minor axis rotation or warping. These restraints are associated with the antisymmetry about the brace point of the bending moment diagram, and with the jump in the bending moment caused by the concentrated moment.

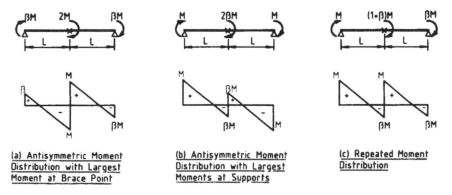

Figure 10.11 Braced beams with concentrated moments.

This behaviour is quite different from that of beams with symmetrical bending moment diagrams, whose buckled shapes u, ϕ tend to be antisymmetrical about the brace point. When the adjacent segments of these beams are identical and have identical symmetrical bending moment diagrams, the segments buckle with zero interaction, and both segments act as if unrestrained, as shown in Figure 10.12.

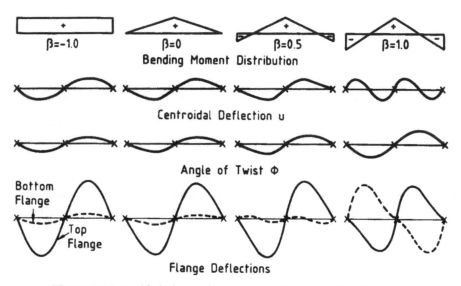

Figure 10.12 Buckled shapes for symmetrical moment distributions.

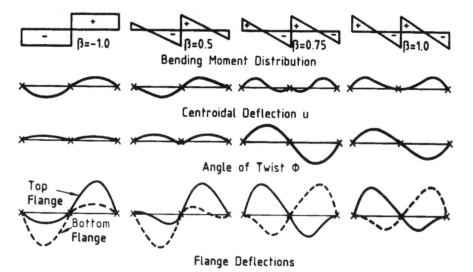

Figure 10.13 Buckled shapes for antisymmetrical moment distributions (Figure 10.11a).

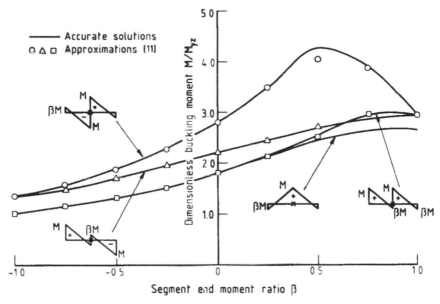

Figure 10.14 Buckling of braced beams with concentrated moments.

The buckled shapes of beams with antisymmetrical bending moment diagrams are shown in Figure 10.13. For these, either the lateral deflection u or the twist ϕ is symmetrical about the brace point, so that each segment acts as if fully restrained against either minor axis rotation ($u' = 0$) or warping ($\phi' = 0$).

A method of approximating the buckling resistances of braced beams with concentrated moments has been developed in [11], which has been shown to give accurate solutions for the beams shown in Figure 10.14, and conservative solutions otherwise.

10.4.2 CONTINUOUS BEAMS

10.4.2.1 General

There have been a number of studies [1, 2, 4, 12–14] of the elastic buckling of continuous beams, and the scope of these is summarized in Figure 10.15. Again, these studies demonstrate the conservative nature of the zero interaction method of approximation, and the increased accuracy of the improved method, as was shown for the two-span beam of Figure 10.5.

10.4.2.2 Continuous roof purlins

The elastic buckling of continuous thin-walled roof purlins which are continuously restrained against minor axis rotation at the loaded top flange has been

Continuous Beam	L_1/L_2	$\frac{Q_1}{Q_2}$ or $\frac{q_1}{q_2}$	Load Height $\frac{2y_a}{h}$	K_2	Reference
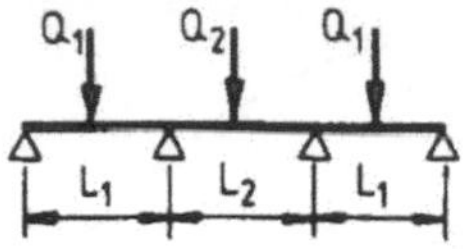 Q_1, Q_2, Q_1 point loads, spans L_1 L_2 L_1	Varies	0	0,±1	0,0.1,0.3,1,3	[14]
	0.2,1,5	Varies	0	0	[2]
	0.5	Varies	-1.18	0.168	[12]
	0.667,1	Varies	-1.18	0.126	[12]
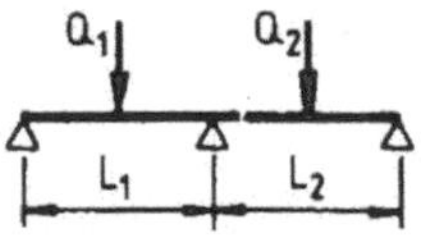 Q_1, Q_2 point loads, spans L_1 L_2	Varies	0	0,±1	0,0.1,0.3,1,3	[1]
	0.2,1	Varies	0	0	[2]
	0.2-5	1.0	0	0.1,3	[4]
	0.5	Varies	-1.18	0.168	[12]
	0.667,1	Varies	-1.18	0.126	[12]
	1	Varies	-1.04	0.512	[12]
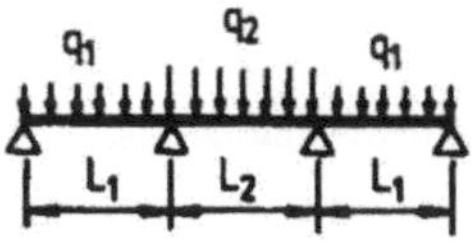 q_1, q_2, q_1 distributed loads, spans L_1 L_2 L_1	Varies	0	0,±1	0,0.1,0.3,1,3	[14]
	1	1	≈±1*	5	[13]
	0.2,1,5	Varies	0	0	[2]
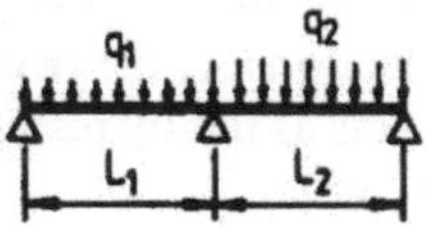 q_1, q_2 distributed loads, spans L_1 L_2	Varies	0	0,±1	0,0.1,0.3,1,3	[1]
	1	1	≈±1*	5	[13]
	0.2,1	Varies	0	0	[2]

* Continuous elastic restraints at load height

Figure 10.15 Studies of continuous beams.

studied [13]. The buckling resistances under gravity and uplift loading of single-, two-, and three-span purlins with $K = 5$ are compared in Figure 10.16 with those of single-span beams in uniform bending.

The resistance is almost independent of the restraint stiffness when uplift loading acts near the top flange, since this is close to the centre of buckling rotation of an unrestrained beam (see Figure 7.6). However, the resistances of purlins under gravity loading increase significantly with the restraint stiffness, especially for single span purlins. The effect is less pronounced for two- and three-span purlins, whose unrestrained bottom flanges are in compression near the interior supports.

It may be noted that the resistances of these continuous purlins are always higher for uplift than for gravity loading, despite the fact that uplift loading induces compression in the unrestrained bottom flange near mid-span. These higher resistances result from the uplift loading being applied at the restrained

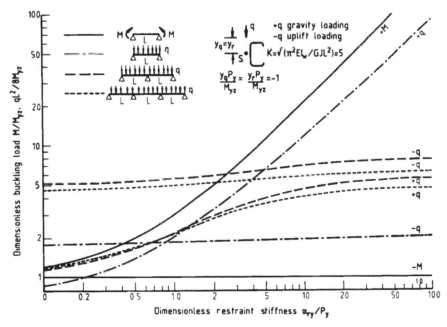

Figure 10.16 Buckling loads for thin-walled roof purlins.

top flange, so that the load height effect (section 7.6) is to resist twisting of the beam during buckling. Conversely, the load height effect of gravity loading at the top flange is to increase the twisting during buckling. The load height effect is evidently more important in these continuous purlins than that of the continuous restraints, as is demonstrated in Figure 10.16.

10.4.3 OVERHANGING BEAMS

The elastic buckling of overhanging beams has been studied in [15,16]. An overhanging beam consists of a supported segment and an overhanging segment (section 9.6.1.2) which is continuous over a support with the adjacent supported segment, as indicated in Figure 10.2c. The continuity between the segments generally leads to a restraining action against warping at the common support. The limiting cases are those where the supported segment is so stiff that it effectively prevents end warping ($\phi' = 0$) of the overhanging segment at the common support, and where the overhanging segment is so stiff that it effectively prevents end warping ($\phi' = 0$) of the supported segment. Between these two extreme cases lies the special zero interaction case for which there is no buckling restraint action at the support. In this case, both segments act as if free to warp ($\phi'' = 0$) at the support. The buckling resistances of overhanging segments free to warp are discussed in section 9.6.1.2, and of supported segments free to warp in Chapter 7.

More generally, the interaction between segments during buckling causes the less critically loaded segment to restrain the more critically loaded segment. In these cases, a lower bound estimate of the buckling resistance can be obtained by using the lower load factor calculated by assuming that each segment is free to warp ($\phi'' = 0$) at the support. This method is similar to the zero interaction method presented in section 10.3.2 for braced and continuous beams.

The results of the application of this lower bound method to symmetrical overhanging beams with equal end loads Q acting at the shear centre are shown by the dashed lines in Figure 10.17. For beams with $K = 2$ and short overhanging segments which restrain warping of the supported segments, the lower bound solutions are reasonably close to the accurate solutions shown by the solid line. (For beams with $K = 0$, the lower bounds are exact because these beams have no warping stiffness, and always act as if free to warp.) However, the lower bounds are increasingly conservative for beams with short supported segments which restrain the overhanging segments.

An upper bound solution may be obtained by assuming that both segments are prevented from warping ($\phi' = 0$). These solutions are shown by the dash-dot lines in Figure 10.17, and overestimate the buckling resistance, except in the limiting case where the supported segment is so short that it provides an effectively rigid restraint against warping.

A more accurate approximate method is suggested in [16], in which a warping effective length factor k_w for the overhanging segment is approximated by

$$k_w = 0.5 + 0.5\sqrt{(Q_0 L/M_{yz})} \tag{10.13}$$

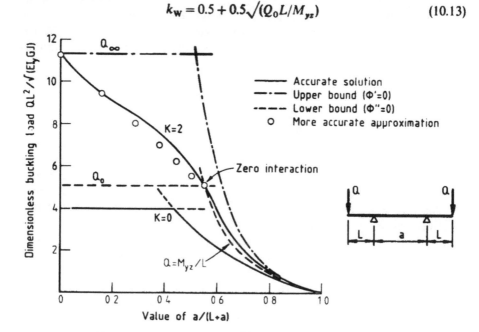

Figure 10.17 Interaction buckling of overhanging beams.

and the buckling load Q is approximated by

$$Q = Q_0(2 - 1/k_W) + Q_\infty(1/k_W - 1) \qquad (10.14)$$

in which Q_0 is the buckling load of an overhanging segment which is free to warp (section 9.6.1.2), Q_∞ is the buckling load of a cantilever prevented from warping at the support (section 9.3.1), and M_{yz} is the buckling moment of a supported segment which is free to warp (section 7.2.1). Thus Q_0 and M_{yz} are as calculated in the lower bound method by assuming $\phi'' = 0$, and Q_∞ is as calculated in the upper bound method by assuming $\phi' = 0$. The predictions of this approximate method are quite close to the accurate solutions, as is demonstrated in Figure 10.17.

10.5 Problems

PROBLEM 10.1

Determine the elastic flexural-torsional buckling load of a continuous beam whose properties are given in Figure 7.23. The beam has two equal spans of 12.0 m, and uniformly distributed shear centre loading, as shown in Figure 10.18a.

PROBLEM 10.2

Determine the value of Q at elastic flexural-torsional buckling for an overhanging beam whose properties are given in Figure 7.23. The supported span has a shear centre load of $4Q$ at mid-span, which is unbraced, and a load of Q at the end of the cantilever, which is braced, as shown in Figure 10.18b.

PROBLEM 10.3

Determine the value of Q at elastic flexural-torsional buckling for a braced beam whose properties are given in Figure 7.23. The beam span is 18.0 m, and is braced at each load point, as shown in Figure 10.18c.

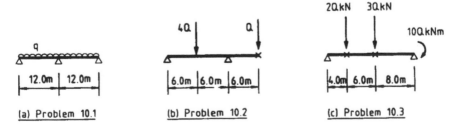

(a) Problem 10.1   (b) Problem 10.2   (c) Problem 10.3

Figure 10.18 Problems 10.1–10.3.

10.6 References

1. Trahair, N.S. (1968) Elastic stability of propped cantilevers, *Civil Engineering Transactions*, Institution of Engineers, Australia, **CE10**(1), 94–100.
2. Trahair, N.S. (1968) Interaction buckling of narrow rectangular continuous beams. *Civil Engineering Transactions*, Institution of Engineers, Australia, **CE10**(2), 167–172.
3. Salvadori, M.G. (1951) Lateral buckling of beams of rectangular cross-section under bending and shear, in *Proceedings*, First US National Congress of Applied Mechanics, pp. 403–5.
4. Nethercot, D.A. and Trahair, N.S. (1976) Lateral buckling approximations for elastic beams. *The Structural Engineer*, **54**(6), 197–204.
5. Trahair, N.S. and Bradford, M.A. (1991) *The Behaviour and Design of Steel Structures*, revised 2nd edition, Chapman and Hall, London.
6. Nethercot, D.A. and Trahair, N.S. (1977) Lateral buckling calculations for braced beams. *Civil Engineering Transactions*, Institution of Engineers, Australia, **CE19**(2), 211–4.
7. Dux, P.F. and Kitipornchai, S. (1982) Elastic buckling of laterally continuous I-beams. *Journal of the Structural Division, ASCE*, **108**(ST9), 2099–116.
8. Nethercot, D.A. and Trahair, N.S. (1976) Inelastic lateral buckling of determinate beams. *Journal of the Structural Division, ASCE*, **102**(ST4), 701–17.
9. Dux, P.F. and Kitipornchai, S. (1984) Buckling approximations for inelastic beams. *Journal of Structural Engineering, ASCE*, **110**(3), 559–74.
10. Kitipornchai, S. and Richter, N.J. (1978) Elastic lateral buckling of I-beams with discrete intermediate restraints. *Civil Engineering Transactions*, Institution of Engineers, Australia, **CE20**(2), 105–11.
11. Cuk, P.E. and Trahair, N.S. (1983) Buckling of beams with concentrated moments. *Journal of Structural Engineering, ASCE*, **109**(6), 1387–401.
12. Trahair, N.S. (1969) Elastic stability of continuous beams. *Journal of the Structural Division, ASCE*, **95**(ST6), 1295–312.
13. Hancock, G.J. and Trahair, N.S. (1979) Lateral buckling of roof purlins with diaphragm restraints. *Civil Engineering Transactions*, Institution of Engineers, Australia, **CE21**, 10–5.
14. Trahair, N.S. (1966) Elastic stability of I-beams elements in rigid-jointed structures. *Journal of the Institution of Engineers, Australia*, **38**(7–8), 171–80.
15. Nethercot, D.A. (1973) The effective lengths of cantilevers as governed by lateral buckling. *The Structural Engineer*, **51**(5), 161–8.
16. Trahair, N.S. (1983) Lateral buckling of beams, in *Instability and Plastic Collapse of Steel Structures* (ed. L.J. Morris), Granada, London, pp. 503–18.

11 Beam-columns

11.1 General

Beam-columns are members with compressive axial forces and transverse loads or moments. Beam-columns which are bent in a plane of symmetry may fail by excessive bending in that plane, or may buckle out of the plane by deflecting laterally u and twisting ϕ. This flexural-torsional buckling behaviour is an interaction between the buckling behaviour of columns discussed in Chapters 5 and 6 and that of beams discussed in Chapters 7 and 8 and extended to cantilevers in Chapter 9 and braced and continuous beams in Chapter 10.

The beam-columns considered in this chapter are initially straight and untwisted, the axial forces act concentrically through the centroid, and the transverse loads and moments act in the plane of section symmetry.

11.2 Uniform bending

11.2.1 DOUBLY SYMMETRIC SECTIONS

The beam-column of doubly symmetric cross-section shown in Figure 11.1 is simply supported and has compressive end forces P and equal and opposite end moments M which induce approximately uniform bending in the YZ plane. For in-plane equilibrium, the in-plane deflection v of the beam-column must satisfy the differential equilibrium equation (section 2.7.3)

$$(EI_x v'')'' + (Pv')' = 0 \tag{11.1}$$

and the boundary conditions

$$0 = v_{0,L} = v''_{0,L}. \tag{11.2}$$

The deflected shape which satisfies these equations is given by

$$v = (M/P)\{\cos \mu z + \sin \mu z \tan(\mu L/2) - 1\} \tag{11.3}$$

in which

$$\mu^2 = P/EI_x \tag{11.4}$$

and the bending moment $M_x = -EI_x v''$ is given by

$$M_x/M = \cos \mu z + \sin \mu z \tan (\mu L/2). \tag{11.5}$$

In many cases, beam-columns have compressive loads P which are substantially less than the in-plane column buckling loads

$$P_x = \pi^2 EI_x/L^2 \tag{11.6}$$

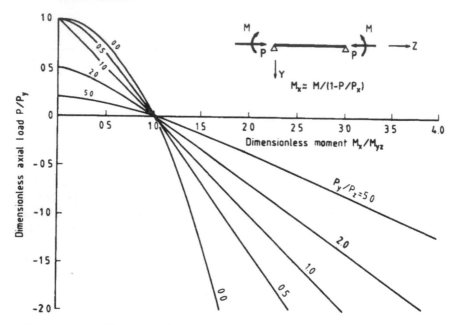

Figure 11.1 Buckling of doubly symmetric beam-columns in uniform bending.

in which case the bending moment M_x is reasonably well approximated by using the constant value

$$M_x \approx \frac{M}{1 - P/P_x}.$$ (11.7)

When the beam-column compressive load P and moments M are high enough, then it may buckle out-of-plane by deflecting laterally u and twisting ϕ. For the buckled position to be one of equilibrium, then the differential equilibrium equations (section 2.8.5.2)

$$(EI_y u'')'' + (Pu')' + (M_x \phi)'' = 0$$ (11.8)

and

$$(EI_w \phi'')'' - (GJ\phi')' + \{P(I_p/A)\phi'\}' + M_x u'' = 0$$ (11.9)

and the boundary conditions

$$0 = u_{0,L} = u''_{0,L} = \phi_{0,L} = \phi''_{0,L}$$ (11.10)

must be satisfied. Equation 11.8 represents the equality at equilibrium between the internal resistance $(EI_y u'')''$ to distributed transverse load and the sum of the transverse load components of the axial force P and the moment M_x, while equation 11.9 represents the equality between the sum of the internal resistances

$(EI_w\phi'')''$ and $-(GJ\phi')'$ to distributed torque and the sum of the distributed torque components of the axial force P and the moment M_x.

It can be verified by substitution that when M_x in these equations is taken as constant, as in the approximation of equation 11.7, then they are satisfied by buckled shapes given by

$$u/\delta = \phi/\theta = \sin \pi z/L \tag{11.11}$$

and

$$\delta/\theta = M_x/(P_y - P), \tag{11.12}$$

in which δ and θ are the values of u and ϕ at mid-height, provided the load P and moment M_x satisfy

$$\left(\frac{M_x}{M_{yz}}\right)^2 = \left(1 - \frac{P}{P_y}\right)\left(1 - \frac{P}{P_z}\right). \tag{11.13}$$

In this equation, P_y and P_z are the column flexural and torsional buckling resistances (section 5.2) given by

$$P_y = \pi^2 EI_y/L^2 \tag{11.14}$$

and

$$P_z = (GJ + \pi^2 EI_w/L^2)/r_1^2, \tag{11.15}$$

in which

$$r_1^2 = (I_x + I_y)/A + y_0^2 \tag{11.16}$$

in general (but with $y_0 = 0$ in this case of a doubly symmetric section), and M_{yz} is the beam uniform bending buckling resistance (section 7.2.1) given by

$$M_{yz} = r_1\sqrt{(P_y P_z)}. \tag{11.17}$$

This result may also be obtained by substituting the buckled shapes of equations 11.11 and 11.12 into the energy equation (section 2.8.5.1)

$$\frac{1}{2}\int_0^L \{EI_y u''^2 + EI_w \phi''^2 + GJ\phi'^2\}\,dz - \frac{1}{2}\int_0^L P\{u'^2 + (I_p/A)\phi'^2\}\,dz$$

$$+ \frac{1}{2}\int_0^L M_x\{2\phi u''\}\,dz = 0 \tag{11.18}$$

which represents the principle of conservation of energy during buckling through the sum of the increases in the strain energies of bending, warping, and uniform torsion and the increase in the work done by the compressive load P and the moment M_x.

The solutions of equation 11.13 for the moment M_x at buckling increase from zero as the axial compressive force P decreases from the lower of the column buckling loads P_y and P_z to the beam buckling moment M_{yz} when $P = 0$, as shown in Figure 11.1. Further increases take place after the axial load P becomes negative (tensile).

Although the derivation of equation 11.13 can include the approximation of equation 11.7 for the amplification by the axial force P of the in-plane bending moment from M to M_x, it neglects the effect of the pre-buckling deflections v on the curvature, which transform the beam-column into a 'negative arch', and increase its out-of-plane buckling resistance (see section 13.5.3). It has been shown [1] that the conditions at buckling are more accurately approximated by

$$\left(\frac{M}{M_{yzc}}\right)^2 = \left(1 - \frac{P}{P_x}\right)\left(1 - \frac{P}{P_y}\right)\left(1 - \frac{P}{P_z}\right) \tag{11.19}$$

in which

$$M_{yzc} = \frac{M_{yz}}{\sqrt{\{(1 - EI_y/EI_x)[1 - (GJ + \pi^2 EI_w/L^2)/2EI_x]\}}} \tag{11.20}$$

is a more accurate expression for the beam buckling resistance (sections 7.2.1 and 16.6) which includes the effect of the pre-buckling curvature.

Equation 11.19 is often simplified by replacing M_{yzc} by M_{yz}, and by replacing $(1 - P/P_z)$ by $(1 - P/P_x)(1 - P/P_y)$, which is conservative when $P_x > P_z > P_y$, as is often the case. When this is done, equation 11.19 can be replaced by the simpler

$$\frac{P}{P_y} + \frac{1}{(1 - P/P_x)}\frac{M}{M_{yz}} = 1. \tag{11.21}$$

This approximation may also be used for beam-ties with tensile axial forces (negative P), although it may become increasingly unconservative as P becomes more negative. In this case, a safer approximation (for negative P) is given by

$$\frac{P}{P_y} + \frac{M}{M_{yz}} = 1. \tag{11.22}$$

11.2.2 MONOSYMMETRIC SECTIONS

11.2.2.1 Sections bent about an axis of symmetry

Monosymmetric beam-columns which are bent about the x axis of symmetry ($y_0 = 0$), such as channels and equal leg angles, do not buckle suddenly out-of-plane, but bend biaxially and twist as soon as loading is commenced. This biaxial bending and twisting occurs because of the presence of additional terms $(Px_0\phi')'$ in the in-plane bending equilibrium equation, which cause coupling between the in-plane deflection v and the out-of-plane twist rotation ϕ at any non-zero value of P. These additional terms also appear in the corresponding equations for monosymmetric columns ($y_0 = 0$), where they lead to coupling between the buckling deformations v and ϕ (see sections 5.2 and 5.4).

11.2.2.2 Sections bent in a plane of symmetry

For a monosymmetric beam-column bent in a plane of symmetry ($x_0 = 0$), the additional terms described above in section 11.2.2.1 disappear, and there is no coupling between v and ϕ. However, other additional terms arise and lead to additional coupling between the buckling deformations u and ϕ, so that the differential equilibrium equations (section 2.8.5.2) become

$$(EI_y u'')'' + \{P(u' + y_0\phi')\}' + (M_x\phi)'' = 0 \tag{11.23}$$

$$(EI_w \phi'')'' - (GJ\phi')' + \{Py_0 u'\}' + \{P(I_P/A + y_0^2)\phi'\}'$$
$$+ M_x u'' - \{M_x(I_{Px}/I_x - 2y_0)\phi'\}' = 0 \tag{11.24}$$

and the energy equation (section 2.8.5.1) becomes

$$\frac{1}{2}\int_0^L \{EI_y u''^2 + EI_w \phi''^2 + GJ\phi'^2\}\,dz$$

$$-\frac{1}{2}\int_0^L P\{u'^2 + (I_P/A + y_0^2)\phi'^2 + 2y_0 u'\phi'\}\,dz$$

$$+\frac{1}{2}\int_0^L M_x\{2\phi u'' + (I_{Px}/I_x - 2y_0)\phi'^2\}\,dz = 0. \tag{11.25}$$

These additional terms also appear in the corresponding equations for mono-symmetric columns (sections 5.2 and 5.4) and in those for monosymmetric beams (section 7.2.3.2).

The buckled shapes which satisfy equations 11.23 and 11.24 and the boundary conditions of equation 11.10 are given by equation 11.11 and

$$\delta/\theta = (M_x + Py_0)/(P_y - P) \tag{11.26}$$

when M_x is taken as constant, and the conditions at buckling obtained from equations 11.23 and 11.24 or from equation 11.25 become

$$\left\{\frac{M_x}{M_{yz}}\left(1 + \frac{Py_0}{M_x}\right)\right\}^2 = \left(1 - \frac{P}{P_y}\right)\left\{1 - \frac{P}{P_z}\left(1 - \frac{M_x\beta_x}{Pr_1^2}\right)\right\} \tag{11.27}$$

in which

$$\beta_x = I_{Px}/I_x - 2y_0 \tag{11.28}$$

and $r_1^2 = r_0^2 + y_0^2$. This reduces to equation 11.13 for doubly symmetric sections for which $y_0 = 0$ and $\beta_x = 0$.

The solutions of equation 11.27 for the combinations of (M_x/M_{yz}) $(1 + Py_0/M_x)$ and P/P_y at buckling are shown in Figure 11.2 by curves for given values of $(P_y/P_z)(1 - M_x\beta_x/Pr_1^2)$. These curves are similar to those of Figure 11.1 for doubly symmetric beam-columns ($y_0 = 0, \beta_x = 0$), except for the additional curves shown for negative values of $(P_y/P_z)(1 - M_x\beta_x/Pr_1^2)$ which occur when $M_x\beta_x > Pr_1^2$. In these cases, the dimensionless moment $(M_x/M_{yz})(1 + Py_0/M_x)$ is not unbounded, but increases as P decreases from P_y until a maximum value is reached, and then decreases towards zero as P/P_z decreases towards $1/\{(P_y/P_z)$ $(1 - M_x\beta_x/Pr_1^2)\}$.

The buckling curves shown in Figure 11.2 are not easy to interpret physically because the value of $(P_y/P_z)(1 - M_x\beta_x/Pr_1^2)$ does not remain constant for a

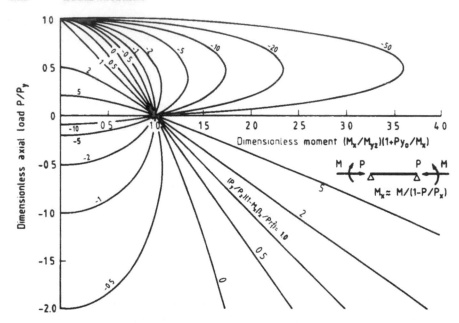

Figure 11.2 Buckling of monosymmetric beam-columns in uniform bending.

particular monosymmetric beam-column, but changes with the value of M_x/P. Simpler interpretations of the buckling of monosymmetric beam-columns may be made from interaction diagrams showing the variations of P/P_y and M_x/M_{yz} at buckling for particular beam-columns, such as those shown in Figures 11.3 and 11.4.

For the tee-section beam-columns of Figure 11.3, the maximum axial load at buckling is $P = P_y$ which occurs when $M_x = -P_y y_0$. If the moment is considered to correspond to that created by an axial force P acting at a distance y_P below the beam-column centroid so that $M_x = -P y_P$, then $P = P_y$ when $y_P = y_0$. Thus the buckling load P is equal to the maximum value P_y when the load acts through the shear centre y_0.

When the axial force is zero, the beam-column has two different buckling moments. The higher one of these is positive, and occurs when the moment causes compression in the flange of the tee-section, while the lower one is negative, corresponding to a reversal in the sense of the moment which causes tension in the flange. This behaviour is similar to the buckling of monosymmetric beams discussed in section 7.2.3.2. When the axial force is negative (tensile), it tends to stabilize the beam-column, and the moments at buckling are increased.

Similar interpretations may be made from the interaction diagrams shown in Figure 11.4 for lipped channel beam-columns. However, in this particular case $P_x < P_y$, and so large in-plane deflections will occur as P_x is approached, and P_y cannot be reached.

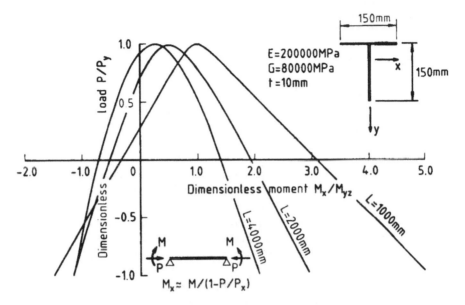

Figure 11.3 Buckling of a tee-section beam column in uniform bending.

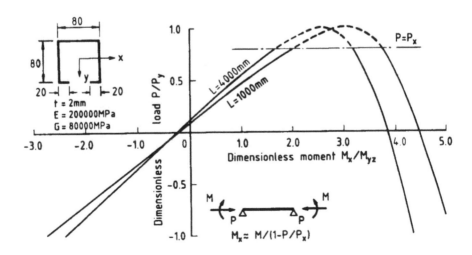

Figure 11.4 Buckling of a lipped channel beam-column in uniform bending.

The flexural-torsional buckling of beam-columns of monosymmetric hat section has been studied in [2], and of monosymmetric I-section in [3–5].

11.3 Moment gradient

The simply supported doubly symmetric beam-column shown in Figure 11.5a has unequal end moments M and βM, so that the variation of the bending moment M_x is approximately linear, and is given by

$$M_x = M - M(1 + \beta)z/L \tag{11.29}$$

when the components Pv due to the axial force P and the in-plane deflections v are neglected. In this case some of the terms in the differential equations of bending and torsion (equations 11.8 and 11.9) have variable coefficients, and it is much more difficult to obtain solutions than previously. Approximate solutions may be obtained by using the hand energy method discussed in Chapter 3, and more accurate solutions by using finite element computer programs such as that described in Chapter 4.

Numerical solutions for the buckling of beam-columns under moment gradient are given in [3–6]. These may be approximated by using the interaction equation

$$\left(\frac{M}{C_{bc}M_{yz}}\right)^2 = \left(1 - \frac{P}{P_y}\right)\left(1 - \frac{P}{P_z}\right) \tag{11.30}$$

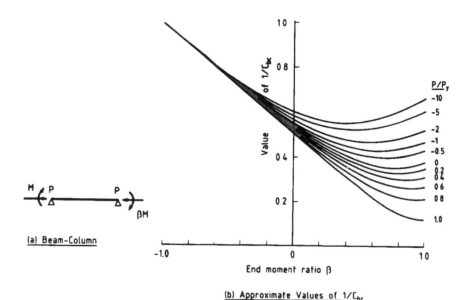

(a) Beam-Column

(b) Approximate Values of $1/C_{bc}$

Figure 11.5 Buckling of doubly symmetric beam-columns under moment gradient.

with values of the beam-column moment factor C_{bc} given by

$$\frac{1}{C_{bc}^2} = \left(\frac{1-\beta}{2}\right)^2 + \frac{(1 - 0.92 P/P_y)}{(1 - 0.23 P/P_y)}\left\{0.38\left(\frac{1+\beta}{2}\right)\right\}^2. \tag{11.31}$$

Equation 11.31 is plotted in Figure 11.5b, which shows that C_{bc} generally increases with the end moment ratio β (except at very high moment gradients) and with the axial load ratio P/P_y. The more accurate solutions of [4, 5] indicate that the value of C_{bc} also increases slightly with the value of $K = \sqrt{(\pi^2 EI_w/GJL^2)}$, but that the approximations of equations 11.30 and 11.31 are generally quite accurate, except for some conservatism at high values of K, β, and P/P_y.

11.4 Transverse loads

The elastic flexural-torsional buckling of simply supported beam-columns of doubly symmetric cross-section under central concentrated load Q or uniformly distributed load q has been studied [4]. When the transverse load acts at the shear centre, the elastic buckling conditions may be approximated with reasonable accuracy by using

$$\left(\frac{M_m}{C_{bc}M_{yz}}\right)^2 = \left(1 - \frac{P}{P_y}\right)\left(1 - \frac{P}{P_z}\right) \tag{11.32}$$

For central concentrated loads Q, the approximate maximum bending moment is

$$M_m \approx QL/4 \tag{11.33}$$

and the moment factor C_{bc}, which varies slightly with P/P_y, is approximated by

$$C_{bc} \approx 1.35. \tag{11.34}$$

For uniformly distributed load q,

$$M_m \approx qL^2/8 \tag{11.35}$$

and

$$C_{bc} \approx 1.13. \tag{11.36}$$

When the transverse load acts away from the shear centre at a distance y_Q below the centroid, then the elastic buckling solutions of [4] are reasonably well approximated by

$$\frac{M_m}{C_{bc}M_{yz}}\left\{\frac{M_m}{C_{bc}M_{yz}} - \frac{0.8 C_{bc} y_Q}{(M_{yz}/P_y)}\left(1 - \frac{P}{P_y}\right)\right\} = \left(1 - \frac{P}{P_y}\right)\left(1 - \frac{P}{P_z}\right). \tag{11.37}$$

This equation reduces to those given in section 7.6 for simply supported beams $(P = 0)$.

11.5 Restraints

11.5.1 GENERAL

A beam-column may have continuous or discrete restraints which restrict its buckled shape and increase its resistance to flexural-torsional buckling. These restraints may be translational (α_t), minor axis rotational (α_{ry}), torsional (α_{rz}), or warping (α_w), as discussed in section 8.2.

The effects of restraints on the buckling resistance of a beam-column may be determined after including additional terms $\frac{1}{2}\Sigma\{D\}^T[\alpha_B]\{D\}$ and $\frac{1}{2}\int_0^L\{d\}^T[\alpha_b]\{d\}\,dz$ in the energy equation of equation 11.18, in which $[\alpha_B],[\alpha_b]$ are the stiffness matrices for discrete and continuous restraints and $\{D\},\{d\}$ are the vectors of buckling deformations given in section 8.2.

11.5.2 CONTINUOUS RESTRAINTS

A simply supported beam-column in uniform bending with uniform continuous restraints buckles into n half sine waves [7], so that

$$u/\delta = \phi/\theta = \sin n\pi z/L. \tag{11.38}$$

The axial force P and moment M_x which cause elastic buckling can be obtained by substituting these buckled shapes into an energy equation obtained by augmenting equation 11.25 by restraint terms $\frac{1}{2}\int_0^L\{d\}^T[\alpha_b]\{d\}\,dz$ discussed in section 11.5.1. This leads to

$$\{M_x + Py_0 + \alpha_{ry}(y_r - y_0) + \alpha_t(y_t - y_0)L^2/n^2\pi^2\}^2$$
$$= \{n^2 P_y - P + \alpha_{ry} + \alpha_t L^2/n^2\pi^2\}[GJ + n^2\pi^2 EI_w/L^2 - Pr_1^2 + M_x\beta_x$$
$$+ \alpha_{ry}(y_r - y_0)^2 + \alpha_w + \{\alpha_t(y_t - y_0)^2 + \alpha_{rz}\}L^2/n^2\pi^2] \tag{11.39}$$

which is a quadratic in M_x and P, and can be solved directly for the elastic buckling values. In general, a number of trials must be made before the integer value of n which leads to the lowest values can be determined.

Doubly symmetric beam-columns ($y_0 = \beta_x = 0$) with continuous warping (α_w) and torsional restraints (α_{rz}) only ($\alpha_t = \alpha_{ry} = 0$) buckle in a single half wave ($n = 1$) when

$$M_x^2 = (P_y - P)(GJ + \pi^2 EI_w/L^2 - Pr_0^2 + \alpha_w + \alpha_{rz}L^2/\pi^2). \tag{11.40}$$

It can be seen that these restraints contribute directly to the effective torsional stiffness of the beam-column, by increasing its torsional column buckling load P_z to

$$P_z^* = P_z + (\alpha_w + \alpha_{rz}L^2/\pi^2)/r_0^2 \tag{11.41}$$

and its beam buckling moment M_{yz} to

$$M_{yz}^* = r_0\sqrt{\{P_y P_z^*\}}. \tag{11.42}$$

Thus the solutions shown in Figure 11.1 for unrestrained beam-columns can also be applied to restrained beam-columns by substituting P_z^* for P_z and M_{yz}^* for M_{yz}.

Doubly symmetric beam-columns with continuous minor axis rotational restraints (α_{ry}) only also buckle in a single half wave $(n = 1)$ when

$$(M_x + \alpha_{ry} y_r)^2 = (P_y - P + \alpha_{ry})(GJ + \pi^2 EI_w/L^2 - Pr_0^2 + \alpha_{ry} y_r^2). \quad (11.43)$$

This equation can be expressed as

$$\left\{ \frac{M_x}{M_{yzp}} + \frac{\alpha_{ry}}{P_{yp}} \frac{y_r P_{yp}}{M_{yzp}} \right\}^2 = \left\{ 1 + \frac{\alpha_{ry}}{P_{yp}} \right\} \left\{ 1 + \frac{\alpha_{ry}}{P_{yp}} \left(\frac{y_r P_{yp}}{M_{yzp}} \right)^2 \right\} \quad (11.44)$$

in which

$$P_{yp} = P_y - P, \quad (11.45)$$

$$M_{yzp} = \sqrt{\{P_{yp}(GJ + \pi^2 EI_w/L^2 - Pr_0^2)\}}, \quad (11.46)$$

which are of the same form as equations 8.15–8.17 for beams $(P = 0)$, but with P_{yp} substituted for P_y, and M_{yzp} for M_{yz}. Thus the solutions shown in Figure 8.4 for beams may also be applied to beam-columns by making these substitutions.

In the case when the continuous restraints are rigid, so that the beam-column buckles with an enforced centre of rotation at the restraint position y_r, the moment M_∞ at buckling can be obtained from equation 8.18 as

$$\frac{M_\infty}{M_{yzp}} = \frac{1}{2} \left\{ \frac{y_r P_{yp}}{M_{yzp}} + \frac{M_{yzp}}{y_r P_{yp}} \right\} \quad (11.47)$$

which has a minimum value of $M_\infty/M_{yzp} = 1$ when $y_r P_{yp}/M_{yzp} = 1$. In this case the restraints have no effect because they act at the axis of cross-section rotation of an unrestrained beam-column, so that no restraining actions are generated.

11.5.3 DISCRETE RESTRAINTS

11.5.3.1 End restraints

Restraints at the supported ends $(u = \phi = 0)$ of a beam-column may restrain minor axis rotations u' and warping displacements proportional to ϕ'. For a doubly symmetric beam-column in uniform bending whose four flange ends have equal minor axis rotational end restraints of stiffness α_R, the buckled shape is defined by (see also sections 6.5.1 and 8.5.1.1)

$$\frac{u}{\delta} = \frac{\phi}{\theta} = \frac{\cos(\pi z/kL - \pi/2k) - \cos(\pi/2k)}{1 - \cos(\pi/2k)} \quad (11.48)$$

where k satisfies

$$\frac{\alpha_R L}{EI_y} = -\left(\frac{\pi}{2k}\right) \cot\left(\frac{\pi}{2k}\right). \quad (11.49)$$

The axial force P and moment M_x which cause elastic buckling can be obtained by substituting these buckled shapes into an augmented equation 11.18 (see section 11.5.1), which leads to

$$\left(\frac{M_x}{M_{yzR}}\right)^2 = \left(1 - \frac{P}{P_{yR}}\right)\left(1 - \frac{P}{P_{zR}}\right) \tag{11.50}$$

in which

$$P_{yR} = \pi^2 EI_y/k^2L^2 \tag{11.51}$$

$$P_{zR} = (GJ + \pi^2 EI_w/k^2L^2)/r_0^2 \tag{11.52}$$

and

$$M_{yzR} = r_0\sqrt{(P_{yR} P_{zR})} \tag{11.53}$$

which can also be obtained from equations 11.13–11.17 for an unrestrained beam-column by substituting the effective length kL for the actual length L.

Solutions of equation 11.49 are shown in Figure 6.5b and Figure 8.9c, and are closely approximated by

$$k = \frac{2 + \alpha_R L/EI_y}{2 + 2\alpha_R L/EI_y}. \tag{11.54}$$

The elastic buckling of beam-columns with unequal rotational end restraints may be approximated using the same method discussed in section 8.5.1.1 for beams. Thus end restraint flexibility ratios $G = 2EI_y/\alpha_{R_y}L$ are calculated for each end of a beam-column, and used in Figure 8.10b to determine the effective length factor k to be used in equations 11.50–11.53.

11.5.3.2 Intermediate restraints

Intermediate restraints acting on a beam-column may restrain its lateral deflection u and twist rotation ϕ. When n_r equally spaced restraints are rigid so that $u = \phi = 0$ at the restraint points, the beam-column buckles between the restraint points at an axial load P and moment M_x that satisfy equations 11.13–11.17 with the member length L replaced by the distance between restraints $L/(n_r + 1)$.

The buckling resistances of beam-columns with n_r equally spaced elastic restraints of equal stiffness may be taken as the lower of the value determined for buckling between the restraint points, and that approximated by 'smearing' the restraints into equivalent uniform continuous restraints of stiffness

$$\alpha_r = n_r \alpha_R/L \tag{11.55}$$

and using these in equation 11.39. The effects of intermediate restraints away from the shear centre which prevent lateral deflection and restrain twist rotation have been studied in [8].

11.6 Problems

PROBLEM 11.1

A simply supported I-section beam-column whose properties are given in Figure 7.23 has a span of 6.0 m, axial compressions Q kN, and equal and opposite end moments $1.5 Q$ kN m, as shown in Figure 11.6a. Determine the value of Q at elastic flexural-torsional buckling,

  (a) using equation 11.13;
  (b) using equation 11.19;
  (c) using equation 11.21.

PROBLEM 11.2

A simply supported I-section beam-tie whose properties are given in Figure 7.23 has a span of 6.0 m, axial tensions Q kN, and equal and opposite end moments 1.5 Q kN m, as shown in Figure 11.6b. Determine the values of Q at elastic flexural-torsional buckling,

  (a) using equation 11.13;
  (b) using equation 11.19;
  (c) using equation 11.21;
  (d) using equation 11.22.

PROBLEM 11.3

A simply supported beam-column has the lipped channel section given in Figure 5.8a ($x_0 = -78.7$ mm, $A = 560$ mm^2, $I_x = 672.0 \times 10^3$ mm^4, $I_y = 536.4 \times$

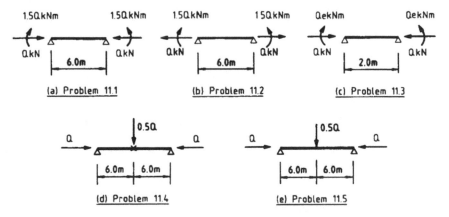

(a) Problem 11.1   (b) Problem 11.2   (c) Problem 11.3

(d) Problem 11.4   (e) Problem 11.5

Figure 11.6 Problems 11.1–11.5.

10^3 mm^4, $J = 0.7467 \times 10^3$ mm^4, $I_w = 1.0619 \times 10^6$ mm^6, $\beta_y = 171.6$ mm). The beam-column has a span of 2.0 m, axial compressions Q kN m, and equal and opposite end moments Qe kNm causing bending about the y axis, as shown in Figure 11.6c. Determine the variations of Q at elastic flexural-torsional buckling with e,

(a) when Qe causes compression in the lips;
(b) when Qe causes tension in the lips.

PROBLEM 11.4

A simply supported beam-column whose properties are given in Figure 7.23 has a span of 12.0 m, axial compressions Q kN, and a concentrated load $0.5 Q$ kN at mid-span, where the beam-column is braced, as shown in Figure 11.6d. Determine the value of Q at elastic flexural-torsional buckling.

PROBLEM 11.5

A simply supported beam-column whose properties are given in Figure 7.23 has a span of 12.0 m, axial compressions Q kN, and a concentrated shear centre load $0.5 Q$ kN at mid-span, where the beam-column is unbraced, as shown in Figure 11.6e. Determine the value of Q at elastic flexural-torsional buckling.

11.7 References

1. Vacharajittiphan, P., Woolcock, S.T. and Trahair, N.S. (1974) Effect of in-plane deformation on lateral buckling. *Journal of Structural Mechanics*, **3**(1), 29–60.
2. Pekoz, T.B. and Winter, G. (1969) Torsional-flexural buckling of thin-walled sections under eccentric load. *Journal of the Structural Division, ASCE*, **95**(ST5), 941–63.
3. Kitipornchai, S. and Wang, C.M. (1988) Out-of-plane buckling formulae for beam-columns/tie-beams. *Journal of Structural Engineering, ASCE*, **114**(12), 2773–89.
4. Kitipornchai, S. and Wang, C.M. (1988) Flexural-torsional buckling of monosymmetric beam-columns/tie-beams. *The Structural Engineer*, **66**(23), 393–9.
5. Wang, C.M. and Kitipornchai, S. (1989) New set of buckling parameters for monosymmetric beam-columns/tie beams. *Journal of Structural Engineering, ASCE*, **115**(6), 1497–513.
6. Cuk, P.E. and Trahair, N.S. (1981) Elastic buckling of beam-columns with unequal end moments. *Civil Engineering Transactions*, Institution of Engineers, Australia, **CE23**(3), 166–71.
7. Trahair, N.S. (1979) Elastic lateral buckling of continuously restrained beam-columns, in *The Profession of a Civil Engineer* (eds D. Campbell-Allen and E.H. Davis), Sydney University Press, Sydney, pp. 61–73.
8. Horne, M.R. and Ajmani, J.L. (1969) Stability of columns supported laterally by side-rails. *International Journal of Mechanical Sciences*, **11**, 159–74.

12 Plane frames

12.1 General

Plane frames are planar structures consisting of a number of individual members connected together at joints. The members are usually subjected to both axial force and bending actions, and are often described as beam-columns, with beams (no axial forces), and columns (no bending actions) being the limiting cases. The flexural-torsional buckling behaviour of individual isolated members was discussed in Chapters 5 (columns), 7 and 9 (beams and cantilevers), and 11 (beam-columns).

The flexural-torsional buckling of a plane frame loaded in its plane will generally involve all the members of the frame, with interactions between the members resulting from continuity at the joints connecting them. Some members will act as if positively restrained by the adjacent members, in which case their behaviour may be similar to that described in Chapters 6 and 8 for restrained columns and beams. However, the restraining actions between frame members at buckling are no less difficult to predict than are those between the adjacent segments of braced and continuous beams discussed in Chapter 10, and it may not be sufficient to consider only the behaviour of individual members.

The prediction of the flexural-torsional buckling behaviour of plane frames is further complicated because the interactions between adjacent members depend very much on the details of the joints connecting them and their restraints. At a right-angle joint, warping deformations in one member are compatible with cross-section distortions in the perpendicular member, as shown in Figure 12.1,

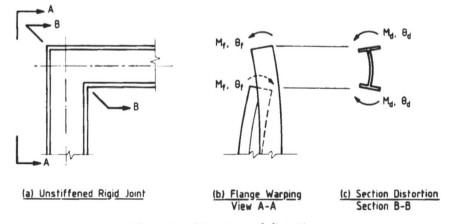

(a) Unstiffened Rigid Joint   (b) Flange Warping   (c) Section Distortion
View A-A   Section B-B

Figure 12.1 Warping and distortion.

while bimoments in one member induce distortional moment pairs in the other. Thus distortional effects may need to be accounted for.

This chapter is concerned with the flexural-torsional buckling under in-plane loading of plane frames whose members' stiffer principal planes coincide with the plane of the frame. The nature of the joints between the members and their influence on joint compatibility and equilibrium are first considered, and then methods for the accurate and approximate analysis of the buckling of plane frames are discussed. Finally, selected results determined for portals and other frames are presented.

12.2 Joints

12.2.1 IN-PLANE COMPATIBILITY

In general, the member in-plane axes y, z at a joint j will be inclined α to the frame axes Y, Z, as shown in Figure 12.2. If the member is rigidly connected to the joint, then compatibility requires that its end in-plane pre-buckling deformations

$$\{\delta_i\} = \{v, v', w\}^{\mathrm{T}} \tag{12.1}$$

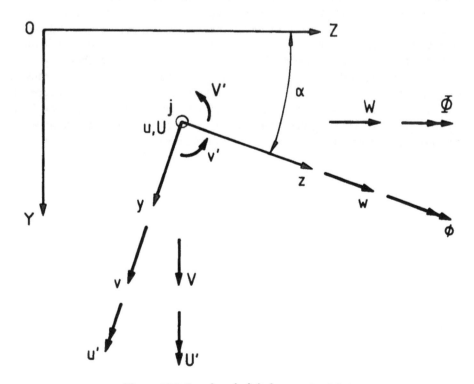

Figure 12.2 Local and global axes at a joint.

be related to the global joint in-plane deformations

$$\{\Delta_i\} = \{V, V', W\}^T \tag{12.2}$$

by

$$\{\delta_i\} = [T_i]\{\Delta_i\} \tag{12.3}$$

in which

$$[T_i] = \begin{bmatrix} \cos\alpha & 0 & -\sin\alpha \\ 0 & 1 & 0 \\ \sin\alpha & 0 & \cos\alpha \end{bmatrix}. \tag{12.4}$$

These in-plane compatibility conditions are used in the pre-buckling in-plane analysis of the frame to determine the distributions of axial forces and in-plane moments.

12.2.2 OUT-OF-PLANE DISPLACEMENT AND ROTATION COMPATIBILITY

Compatibility of a member m rigidly connected to the joint j shown in Figure 12.2 requires that its out-of-plane buckling deformations

$$\{\delta\} = \{u, u', \phi\}^T \tag{12.5}$$

be related to the joint out-of-plane deformations

$$\{\Delta\} = \{U, U', \Phi\}^T \tag{12.6}$$

by

$$\{\delta\} = [T]\{\Delta\} \tag{12.7}$$

in which

$$\{T\} = \begin{bmatrix} 1 & 0 & 0 \\ 0 & \cos\alpha & -\sin\alpha \\ 0 & \sin\alpha & \cos\alpha \end{bmatrix}. \tag{12.8}$$

These out-of-plane compatibility conditions are used in the out-of-plane buckling analysis of the frame.

12.2.3 WARPING COMPATIBILITY

12.2.3.1 General

The out-of-plane buckling analysis of a frame also requires consideration of the warping conditions at the ends of each member. In general, there may be different conditions for each member connected to a particular joint, and it is not always possible to associate each set of member warping displacements proportional to the twist ϕ' with a single deformation of the joint. For this reason, the member warping degrees of freedom ϕ' should remain as independent local degrees of freedom, instead of being associated with the global joint degrees of freedom. The following sub-sections deal with particular examples of member warping conditions at joints.

12.2.3.2 Warping free

In some cases, a member is connected to a joint in such a way that there is no restraint of warping, and the end warping stresses ($= E\omega\phi''$) are zero, in which case

$$\phi'' = 0. \tag{12.9}$$

Such a condition occurs for example, when an I-section member is connected through its web only, leaving its flanges free to warp. Studies of warping free joints between channel and zed-section members are discussed in [1, 2].

12.2.3.3 Warping prevented

If a member is rigidly connected to a joint which itself is sufficiently rigid effectively to prevent member end warping deformations $w(= \omega\phi')$, then member end warping may be assumed to be prevented, so that

$$\phi' = 0. \tag{12.10}$$

Such a condition may occur for example, when the flanges of an I-section member are rigidly connected to a stiffened joint.

12.2.3.4 Warping restrained

If a member is rigidly connected to a flexible joint, then end warping displacements $w(= \omega\phi')$ of the member will be elastically restrained by bimoments $B(= EI_w\phi'')$. This elastic restraint may be characterized by the relationship

$$EI_w\phi'' = -\alpha_w\phi' \tag{12.11}$$

in which α_w is the warping restraint stiffness and the negative sign is used to indicate that for positive warping restraint, the bimoment B opposes the warping deformations w. A zero value for α_w corresponds to warping free with $\phi'' = 0$, while very large values of α_w correspond to warping prevented with $\phi' = 0$.

Restrained warping at I-section joints such as those shown in Figure 12.3 has been studied in [3, 4]. For the unstiffened joints considered (Type A), the dimensionless warping restraint stiffness $\alpha_w/(EI_w GJ)$ is comparatively low, varying from 0.5 to 0.15 and back to 0.4 as the included angle α between the members varies from 60° to 120° to 150°, and may conservatively be taken as zero, so that the member can be considered to be free to warp.

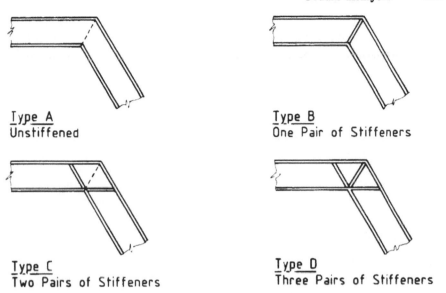

Type A
Unstiffened

Type B
One Pair of Stiffeners

Type C
Two Pairs of Stiffeners

Type D
Three Pairs of Stiffeners

Figure 12.3 Joint stiffening arrangements.

The value of $\alpha_w/\sqrt{(EI_wGJ)}$ is approximately 1 for joints with one stiffener (Type B), and varies between 1 and 10 for joints with 2 stiffeners (Type C). Joints with three stiffeners (Type D) effectively prevent end warping.

12.2.3.5 Warping continuous

When two collinear members of the same cross-section are rigidly connected together at a joint, the warping end displacements $w(=\omega\phi')$ are the same for each member, and are continuous through the joint. In this case each of the two member warping degrees of freedom ϕ' can be transformed to a global degree of freedom Φ' by using

$$\phi' = \Phi'. \tag{12.12}$$

12.3 Frame analysis

12.3.1 IN-PLANE PRE-BUCKLING ANALYSIS

Since an out-of-plane elastic buckling analysis requires a knowledge of the distributions of the axial forces and in-plane moments, a pre-buckling analysis of the frame is required. This can often be done by using a first-order linear elastic analysis (section 4.5 and [5]) for a set of initial loads. The out-of-plane buckling analysis then determines the lowest factor by which these initial loads must be multiplied to cause out-of-plane buckling.

Sometimes the frame may have significant elastic non-linear behaviour in-plane before out-of-plane buckling occurs. In this case a second-order non-linear elastic analysis [6] may be made, which allows for the effects of the second-order moments caused by the products of the applied loads and member forces with the structure and member displacements. Such an analysis must be carried out iteratively for a particular set of loads. The out-of-plane buckling analysis then determines the lowest load factor by which this load set and all axial forces and moments must be multiplied to cause out-of-plane buckling. Because there is a non-linear relationship between the loads and the axial forces and moments, the in-plane analysis must be repeated for the new set of loads predicted by the buckling analysis, and a new out-of-plane buckling analysis must then be made. This process should be repeated until a satisfactory convergence is obtained.

12.3.2 OUT-OF-PLANE BUCKLING ANALYSIS

There are a number of methods used when carrying out elastic buckling analyses to model flexible joints between members which induce elastic warping restraints. In these cases, warping of one member at the joint is elastically restrained by the resistance to distortion of a perpendicular member at the joint (section 12.2.3.4).

The most rigorous method of elastic buckling analysis requires the inclusion of distortional degrees of freedom corresponding to the warping degrees of freedom in the perpendicular members, and includes the effects of distortional moments (M_d in Figure 12.1) which are in equilibrium with the flange moments (M_f) equivalent to the bimoments in the perpendicular members. While the effects of web distortion on the flexural-torsional buckling of beams and beam-columns have been studied (see section 16.5), they do not appear to have been considered in frame buckling.

An alternative approximate method is to incorporate elastic warping restraints at the member ends whose stiffnesses α_w are based on studies such as those of [3], whose results are approximated in section 12.2.3.4. Computer methods have been developed for this [7–9] which are based on the finite integral method [10]. A finite element method may be developed by extending the method of [11], discussed in Chapter 4, to include the effects of the concentrated elastic end warping restraints α_w, by using section 8.2.2.

A third and less precise method is to idealize the warping conditions at member ends as being either free to warp ($\phi'' = 0$), or continuous ($\phi' = \Phi'$), or prevented from warping ($\phi' = 0$).

12.4 Portal frames

There have been a number of studies of the flexural-torsional buckling of portal frames [7, 8, 12–16], and the scopes of these are summarized in Figure 12.4. They include studies of the combinations of loads which interact to cause frame

Frame and Loading	Theoretical (T) or Experimental (E)	Elastic (E) or Inelastic (I)	Warping Restraints	Comments	Reference
	T, E	E	Zero	Interaction	[7]
	T	E	Full	-	[8, 12]
	T	E	Zero, Cont, Zero	-	[8]
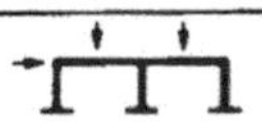	T	E	Full, Cont, Full	Interaction	[8]
	E, T	I	Full	Interaction	[13, 14]
	T	E	Full	Interaction	[8]
	E	I	Cont	Bracing	[15]
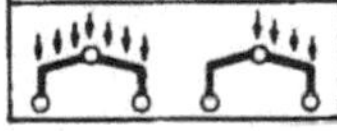	T	E	Cont	-	[16]

Figure 12.4 Studies of portal frames.

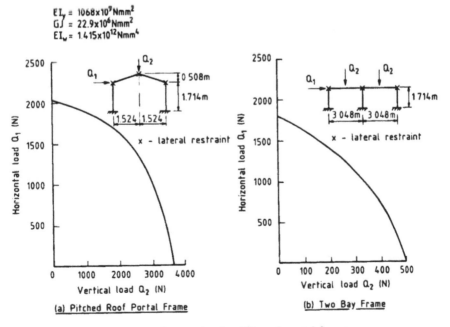

$EI_y = 1068 \times 10^9 \, Nmm^2$
$GJ = 22.9 \times 10^6 \, Nmm^2$
$EI_w = 1.415 \times 10^{12} \, Nmm^4$

(a) Pitched Roof Portal Frame

(b) Two Bay Frame

Figure 12.5 Interaction buckling of portal frames.

Frame and Loading	Theoretical (T) or Experimental (E)	Elastic (E) or Inelastic (I)	Warping Restraints	Comments	Reference
 *(frame diagram)*	T	E	$L_w = 0$	–	[17]
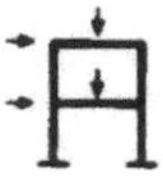 *(frame diagram)*	T	E	Full	Interaction	[8]
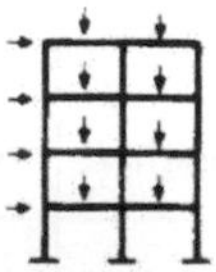 *(frame diagram)*	T	E	Full	–	[9]
 *(truss diagram)*	T	E	$L_w = 0$	–	[9]
 *(truss diagram)*	T	E	$L_w = 0$	–	[18]

Figure 12.6 Studies of other frames.

buckling. The load combinations shown in Figure 12.5 for the buckling of a pitched roof portal frame and of a two-bay portal frame under horizontal and vertical loads are typical, and indicate that as one load increases, the other load at buckling decreases.

12.5 Other frames

The scopes of some studies of other plane frames [8, 9, 17, 18], including multi-storey and triangulated frames, are summarized in Figure 12.6.

12.6 Problem

PROBLEM 12.1

Adapt the solution of Problem 4.12 so as to produce a computer program for analysing the elastic flexural-torsional buckling of plane frames whose member sections are symmetric about the plane of the frame.

12.7 References

1. Baigent, A.H. and Hancock, G.J. (1982) Structural analysis of assemblages of thin-walled members. *Engineering Structures*, **4**, 207–16.
2. Hancock, G.J. (1985) Portal frames composed of cold-formed channel and zed-

sections, in *Steel Framed Structures* (ed. R Narayanan), Applied Science Publishers, London, pp. 241–76.

3. Vacharajittiphan, P. and Trahair, N.S. (1974) Warping and distortion at I-section joints. *Journal of the Structural Division, ASCE*, 100(ST3), pp. 547–64.

4. Krenk, S. and Damkilde, L. (1991) Warping of joints in I-beam assemblages. *Journal of Engineering Mechanics, ASCE*, **117**, (11), 2457–74.

5. Harrison, H.B. (1990) *Structural Analysis and Design – Some Microcomputer Applications, Parts 1 and 2*, 2nd edn, Pergamon Press, Oxford.

6. Harrison, H.B. (1973) *Computer Methods in Structural Analysis*, Prentice-Hall, Englewood Cliffs, N.J.

7. Vacharajittiphan, P. and Trahair, N.S. (1973) Elastic lateral buckling of portal frames. *Journal of the Structural Division, ASCE*, **99** (ST5), 821–52.

8. Vacharajittiphan, P. and Trahair, N.S. (1973) Analysis of lateral buckling in plane frames. *Journal of the Structural Division, ASCE*, **101** (ST7), 1497–516.

9. Vacharajittiphan, P. and Trahair, N.S. (1974) Direct Stiffness Analysis of Lateral Buckling. *Journal of Structural Mechanics*, **3** (2), 107–37.

10. Brown, P.T. and Trahair, N.S. (1968) Finite integral solution of differential equations. *Civil Engineering Transactions*, Institution of Engineers, Australia, **CE10** (2), 193–6.

11. Hancock, G.J. and Trahair, N.S. (1978) Finite element analysis of the lateral buckling of continuously restrained beam-columns. *Civil Engineering Transactions*, Institution of Engineers, Australia, **CE20** (2), 120–7.

12. Hartmann, A.J. and Munse, W.H. (1966) Flexural-torsional buckling of planar frames. *Journal of the Engineering Mechanics Division, ASCE*, Vol. **92** (EM2), 37–59.

13. Cuk, P.E., Rogers, D.F. and Trahair, N.S. (1986) Inelastic buckling of continuous steel beam-columns. *Journal of Constructional Steel Research*, **6**, 21–52.

14. Bradford, M.A. and Trahair, N.S. (1986) Inelastic buckling tests on beam-columns. *Journal of Structural Engineering, ASCE*, **112** (3), 538–49.

15. Hancock, G.J. (1976) Tests of lightly restrained portal frames. *Steel Construction*, Australian Institute of Steel Construction, **10** (2), 10–6.

16. Krenk, S. (1990) Constrained lateral buckling of I-beam gable frames. *Journal of the Structural Division, ASCE*, **116** (12), 3268–84.

17. Argyris, J.H., Balmer, H., Doltsinis, J.St, Dunne, P.C., Haase, M., Kleiber, M., Malejannakis, G.A., Mlejnek, H.-P., Muller, M. and Scharpf, D.W. (1979) Finite Element method – The natural approach. *Computer Methods in Applied Mechanics and Engineering*, **17/18**, 1–106.

18. Kouhia, R. (1990) Nonlinear finite element analysis of space frames. Report 109, Department of Structural Engineering, Helsinki University of Technology.

13 Arches and rings

13.1 General

This chapter is concerned with the out-of-plane flexural-torsional buckling of planar arches, curved beams, and rings which are loaded in-plane. Each of these is curved in its plane, as shown in Figure 13.1. Constant curvature is common, but other profiles such as parabolic may be used. Arches are supported at both ends, usually in such a way that deflections of the ends away from each other are effectively prevented, either by the nature of the supports, or by tie members. Arches whose ends are free to move apart are better described as curved beams. Rings are closed arches with 360° included angles, so that there are no ends.

Arches whose ends are prevented from moving apart are often used structurally because of their high in-plane stiffness and strength, which result from their ability to transmit most of the applied loading by axial force actions, so that the bending actions are reduced. Curved beams whose ends are free to move apart have much more significant bending actions, and are less stiff and strong in-plane. Rings are self-reacting, and have high in-plane stiffness and strength with respect to uniform radial loading.

The resistances of arches and rings to out-of-plane flexural-torsional buckling depend on the rigidities EI_y for lateral bending, GJ for uniform torsion, and EI_w for warping torsion. The buckling resistance of an arch may be reduced by its in-plane curvature, and so it may require significant lateral bracing.

This chapter extends the energy method to the flexural-torsional buckling of arches and rings. Closed form solutions are developed for circular arches and rings in uniform compression or bending. The finite element method of buckling

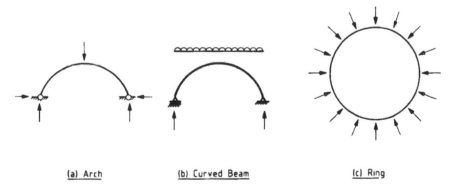

(a) Arch   (b) Curved Beam   (c) Ring

Figure 13.1 Arch, curved beam, and ring.

analysis developed in Chapter 4 for straight members is extended to arches, and used to obtain solutions for arches with concentrated loads.

13.2 In-plane behaviour

Under in-plane loading, the shear centre of a cross-section of a thin arch of radius R may deflect v radially and w_s tangentially as shown in Figure 13.2a, in which case the centroidal longitudinal strain ε and the curvature κ are approximated [1–3] by

$$\varepsilon = w'_s - v/R + y_0(v'' + w'_s/R),\tag{13.1}$$
$$\kappa = v'' + w'_s/R\tag{13.2}$$

in which $'$ indicates differentiation with respect to the shear centre distance s around the arch. The longitudinal strain ε_P at a point P at a distance y from the centroid is approximated by

$$\varepsilon_P = \varepsilon - y\kappa.\tag{13.3}$$

The corresponding longitudinal stresses

$$\sigma_P = E\varepsilon_P\tag{13.4}$$

have stress resultants

$$N = \int_A \sigma_P \, dA\tag{13.5}$$

and

$$M_x = \int_A \sigma_P y \, dA,\tag{13.6}$$

whence

$$N = EA[(w'_s - v/R) + y_0(v'' + w'_s/R)]\tag{13.7}$$

and

$$M_x = -EI_x(v'' + w'_s/R).\tag{13.8}$$

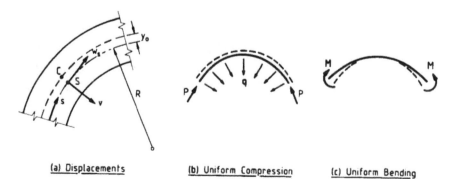

(a) Displacements   (b) Uniform Compression   (c) Uniform Bending

Figure 13.2 Displacements and uniform compression and bending.

Uniform compression $P = -N$ occurs in an arch when uniformly distributed radial loads $q = P/R$ cause constant radial displacements v only ($w_* = 0$), as shown in Figure 13.2b, so that $M_x = 0$. It can be seen that the ends of the arch move slightly towards each other. Arches whose ends are prevented from moving together have moments induced by horizontal end reactions.

Uniform bending occurs in a curved beam when only equal and opposite end moments M act, as shown in Figure 13.2c. It can be seen that the ends of the arch move relative to each other. Arches whose ends are prevented from deflecting have axial compressions induced by horizontal end reactions.

The in-plane behaviour of arches under other loadings than those of uniform compression and uniform bending is best analysed by the finite element method discussed in section 13.8.1.

13.3 Energy equation for flexural-torsional buckling

The energy equation for flexural-torsional buckling u, ϕ for the monosymmetric section arch of developed length L and constant radius of curvature R shown in Figure 13.3 can be written as [2, 3]

$$\frac{1}{2}\int_0^L EI_y\{(u'' + \phi/R)^2 + EI_w(\phi'' - u''/R)^2 + GJ(\phi' - u'/R)^2\}\,ds$$

$$-\frac{1}{2}\int_0^L P\{u'^2 + r_x^2\phi'^2 + r_y^2(\phi' - u'/R)^2 + y_0(2u'\phi' - \phi^2/R) + y_0^2\phi'^2\}\,ds$$

$$+\frac{1}{2}\int_0^L \{M_x(2\phi u'' + \phi^2/R + \beta_x\phi'^2) + q(y_q - y_0)\phi^2\}\,ds$$

$$+\frac{1}{2}\sum Q(y_Q - y_0)\phi^2 = 0 \tag{13.9}$$

in which EI_y is the section minor axis flexural rigidity, EI_w is the section warping rigidity, GJ is the section torsional rigidity, r_x and r_y are the major and minor axis radii of gyration, y_0 is the shear centre coordinate and β_x is the monosymmetry section constant. The arch has radial concentrated loads $Q_y = Q$ and distributed loads $q_y = q$ acting y_Q and y_q from the centroid and axial loads Q_z and in-plane moments M. These loads and moments have in-plane axial compression force resultants P and bending moments M_x. For zero values of q, Q, and $1/R$, equation 13.9 reduces to equation 2.107 for the buckling of straight monosymmetric beam-columns.

13.4 Differential equilibrium equations

The differential equilibrium equations for flexural-torsional buckling can be obtained from the energy equation by using the calculus of variations, according

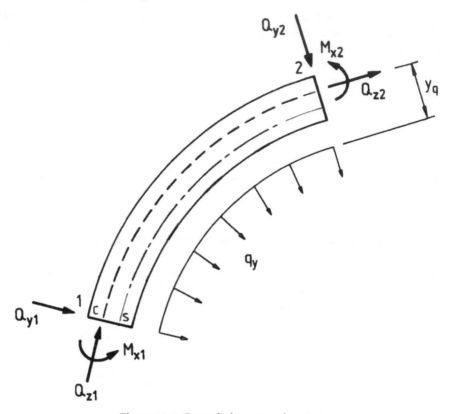

Figure 13.3 Curved element and actions.

to which the functions u, ϕ which make

$$\frac{1}{2}\delta^2 U_T = \int_0^L F(s, u', u'', \phi, \phi', \phi'')\,ds \tag{13.10}$$

stationary satisfy the conditions

$$\left.\begin{aligned}
-\frac{d}{ds}\left(\frac{\partial F}{\partial u'}\right) + \frac{d^2}{ds^2}\left(\frac{\partial F}{\partial u''}\right) &= 0, \\[2mm]
\frac{\partial F}{\partial \phi} - \frac{d}{ds}\left(\frac{\partial F}{\partial \phi'}\right) + \frac{d^2}{ds^2}\left(\frac{\partial F}{\partial \phi''}\right) &= 0.
\end{aligned}\right\} \tag{13.11}$$

Substituting the energy equation (equation 13.9) into equation 13.10 and using equation 13.11 leads to

$$\{EI_y(u'' + \phi/R)\}'' - \{EI_w(\phi'' - u''/R)/R\}'' + \{GJ(\phi' - u'/R)/R\}'$$
$$+ \{P(u' + y_0\phi' - r_y^2(\phi' - u'/R)/R)\}' + \{M_x\phi\}'' = 0, \tag{13.12}$$

$$\{EI_w(\phi'' - u''/R)\}'' - \{GJ(\phi' - u'/R)\}' + EI_y\{u'' + \phi/R\}/R$$

$$+ \{Py_0u'\}' + \{P(r_x^2 + r_y^2 + r_0^2)\phi'\}' - \{Pr_y^2u'/R\}' + \{Py_0\phi/R\}$$

$$+ M_xu'' - \{M_x\beta_x\phi'\}' + M_x\phi/R + q(y_q - y_0)\phi = 0, \tag{13.13}$$

which are the differential equilibrium equations for minor axis bending and torsion for the buckled position u, ϕ.

Slightly different differential equilibrium equations for arches have been obtained in [4–7].

13.5 Arch buckling under uniform compression and bending

13.5.1 GENERAL

Arches which are simply supported laterally so that $u_0 = u_L = 0, u_0'' = u_L'' = 0$, $\phi_0 = \phi_L = 0, \phi_0'' = \phi_L'' = 0$ under constant $P, M_x = M, q$ have possible buckled shapes defined by

$$u/\delta = \phi/\theta = \sin n\pi s/L \tag{13.14}$$

which correspond to n buckled half waves around the arch length L. Substituting these into equation 13.9 leads to

$$\frac{1}{2}\frac{n^2\pi^2}{L^2}\frac{L}{2}\left\{\begin{matrix}\delta\\0\end{matrix}\right\}^T\begin{bmatrix}k_{11} & k_{12}\\k_{21} & k_{22}\end{bmatrix}\left\{\begin{matrix}\delta\\0\end{matrix}\right\} = 0 \tag{13.15}$$

and

$$\left.\begin{matrix}k_{11} = P_{yn} - P + r_1^2 P_{zn}/R^2,\\[4pt] k_{12} = k_{21} = -(Py_0 + M) - (EI_y + r_1^2 P_{zn})/R,\\[4pt] k_{22} = r_1^2(P_{zn} - P) + M\beta_x + q(y_q - y_0)L^2/n^2\pi^2\\[4pt] \quad + (Py_0 + M)L^2/n^2\pi^2 R + EI_yL^2/n^2\pi^2 R^2,\end{matrix}\right\} \tag{13.16}$$

after making the approximations

$$\left.\begin{matrix}1 + r_y^2/R^2 \approx 1,\\[4pt] 1 - Pr_y^2/EI_y \approx 1.\end{matrix}\right\} \tag{13.17}$$

In these equations,

$$\left.\begin{matrix}P_{yn} = n^2\pi^2 EI_y/L^2,\\[4pt] P_{zn} = (GJ + n^2\pi^2 EI_w/L^2)/r_1^2,\\[4pt] r_1^2 = r_x^2 + r_y^2 + y_0^2.\end{matrix}\right\} \tag{13.18}$$

13.5.2 DOUBLY SYMMETRIC ARCHES IN UNIFORM COMPRESSION

Uniform compression P is produced in an arch by uniform radial loading $q = P/R$, as shown in Figure 13.2b. In this case, equation 13.16 can be written for doubly symmetric sections ($y_0 = 0$) as

$$\left.\begin{array}{l} k_{11}/P_{yn} = 1 - P/P_{yn} + a_n^2 b_n^2, \\ k_{12}/M_{yzn} = -(a_n/b_n + a_n b_n), \\ k_{22}/(r_0^2 P_{zn}) = 1 - P/P_{zn} + a_n^2/b_n^2, \end{array}\right\} \qquad (13.19)$$

in which

$$\left.\begin{array}{l} a_n = L/n\pi R, \\ b_n = n\pi M_{yzn}/P_{yn}L, \\ M_{yzn}^2 = r_0^2 P_{yn} P_{zn}, \\ r_0^2 = r_x^2 + r_y^2. \end{array}\right\} \qquad (13.20)$$

Equation 13.15 is satisfied when

$$k_{12}^2 = k_{11} k_{22} \qquad (13.21)$$

which leads to

$$(1 - P/P_{yn} + a_n^2 b_n^2)(1 - P/P_{zn} + a_n^2/b_n^2) = (a_n/b_n + a_n b_n)^2. \qquad (13.22)$$

The solutions P of this equation depend on the number n of half waves, and so the lowest solution must be determined by a trial and error process, by increasing n until the solution increases. The lowest solution usually occurs for $n = 1$.

Solutions of equation 13.22 for the particular case for which

$$P_{z1}/P_{y1} = 2.5 \qquad (13.23)$$

are shown in Figure 13.4 by the variations of P/P_{yn} with b_n and a_n. For $a_n = 0$, the arch degenerates into a straight line, and buckles at the lowest compression member flexural buckling load P_{y1}. For $a_n = 1.0$ and $n = 1$, the arch is semi-circular, and buckles under zero load in a rigid body mode, because the torsional end restraints which ensure $\phi_{0.L} = 0$ do not prevent rotation about the diameter joining the arch ends.

13.5.3 DOUBLY SYMMETRIC CURVED BEAMS IN UNIFORM BENDING

Uniform bending is produced in a curved beam by equal and opposite end moments M when the ends of the arch are free to move together or apart, as shown in Figure 13.2c. In this case, equation 13.16 can be written for doubly symmetric sections ($\beta_x = 0$) as

$$\left.\begin{array}{l} k_{11}/P_{yn} = 1 + a_n^2 b_n^2, \\ k_{12}/M_{yzn} = -M/M_{yzn} - a_n/b_n - a_n b_n, \\ k_{22}/(r_0^2 P_{zn}) = 1 + (M/M_{yzn})a_n/b_n + a_n^2/b_n^2. \end{array}\right\} \qquad (13.24)$$

Once again, equation 13.15 is satisfied by equation 13.21, which in this case leads to

$$\{1 + a_n^2/b_n^2 + (a_n/b_n) M/M_{yzn}\}(1 + a_n^2 b_n^2) = (a_n/b_n + a_n b_n + M/M_{yzn})^2. \qquad (13.25)$$

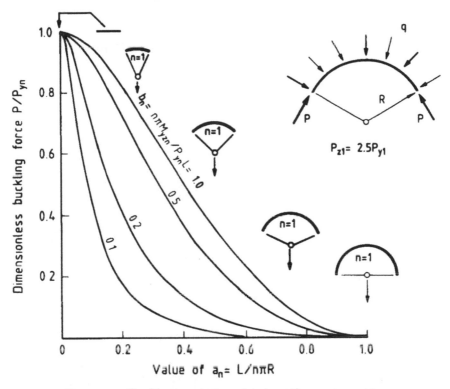

Figure 13.4 Doubly symmetric arches in uniform compression.

The solutions M of this equation depend on the number n of half waves, and so the lowest solution must be determined by trial and error. The lowest solution usually occurs for $n = 1$.

Solutions of equation 13.25 are shown in Figure 13.5 by the variations of M/M_{yzn} with b_n and a_n. For $a_n = 1$ and $n = 1$, the curved beam is semi-circular, and buckles under zero moment in a rigid body mode, because the torsional end restraints do not prevent rotation about the diameter joining the ends. For $a_n = 0$, the curved beam degenerates into a straight beam, and buckles at the lowest beam buckling moment M_{yz}. For negative values of a_n, the arch is inverted. The resistance to buckling generally increases as a_n decreases, and is higher for 'negative' curved beams whose outer (convex) surface is in tension than it is for 'positive' curved beams whose outer surface is in compression.

13.5.4 MONOSYMMETRIC ARCHES

The flexural-torsional buckling of monosymmetric arches in uniform compression or uniform bending has been studied in [3] and results for monosymmetric sections formed by reducing one flange of an I-section member are shown in Figures 13.6 and 13.7.

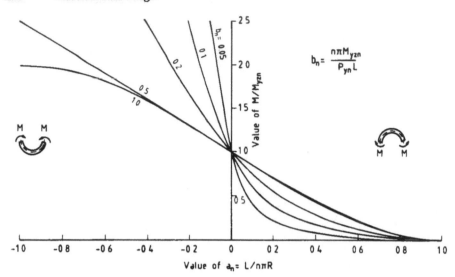

Figure 13.5 Doubly symmetric curved beams in uniform bending.

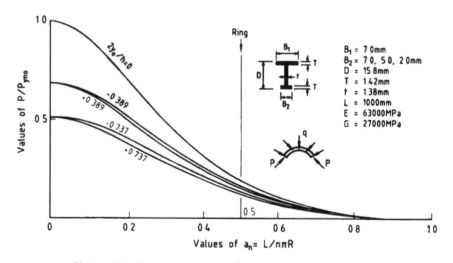

Figure 13.6 Monosymmetric arches in uniform compression.

For the arches in uniform compression, the variation of the dimensionless buckling force P/P_{yn0} with the arch parameter $a_n = L/n\pi R$ and the monosymmetry ratio $2y_0/h$ is shown in Figure 13.6. In this figure, P_{yn0} is the value of $n^2\pi^2 EI_y/L^2$ calculated for the doubly symmetric cross-section and h is the distance between the flange centroids. Positive values of $2y_0/h$ correspond to the outer (convex) flange being the smaller. It can be seen in Figure 13.6 that while

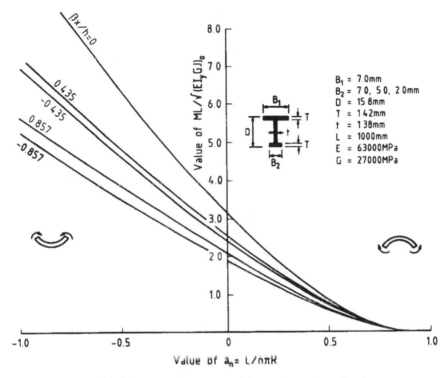

Figure 13.7 Monosymmetric curved beams in uniform bending.

there are significant reductions in the buckling resistance as the section is made more monosymmetric by reducing one flange, the resistance when the outer (convex) flange is the larger (negative $2y_0/h$) is only slightly greater than when the inner flange is the larger (positive $2y_0/h$).

For curved beams in uniform bending, the variation of the dimensionless buckling moment $ML/\sqrt{(EI_y GJ)_0}$ with the parameter $a_n = L/n\pi R$ and the monosymmetry parameter β_x/h is shown in Figure 13.7. In this figure, $(EI_y GJ)_0$ is the value calculated for the doubly symmetric cross-section. It can be seen that the resistance when the compression flange is the larger (positive β_x/h) is greater than when the tension flange is the larger (negative β_x/h).

13.6 Ring buckling under uniform compression

13.6.1 UNRESTRAINED RINGS

In the case of complete rings ($L = 2\pi R$), solutions for doubly symmetric sections can be obtained from equation 13.22 by noting that continuity conditions require there to be an even number n of half waves. The value of $n = 2$ corresponds to $a_n = L/n\pi R = 1.0$ and the rigid body mode already noted in section 13.5.2. If this rigid body mode is prevented, then the lowest solution corresponds to $n = 4$, so that $a_n = 0.5$. Solutions for $a_n = 0.5$ are shown in Figure 13.4.

13.6.2 RESTRAINED RINGS

Rings are often connected continuously to cylindrical, conical, or spherical shells, which may effectively prevent transverse displacement of the rings at the point of attachment, as shown in Figure 13.8 [8,9]. The ring then buckles with an enforced centre of rotation at the point of attachment y_r, so that the shear centre deflection magnitude is

$$\delta = (y_r - y_0)\theta, \tag{13.26}$$

in which case equation 13.15 becomes

$$(n^2\pi^2/4L)\{(y_r - y_0)^2 k_{11} + 2(y_r - y_0)k_{12} + k_{22}\}\theta = 0. \tag{13.27}$$

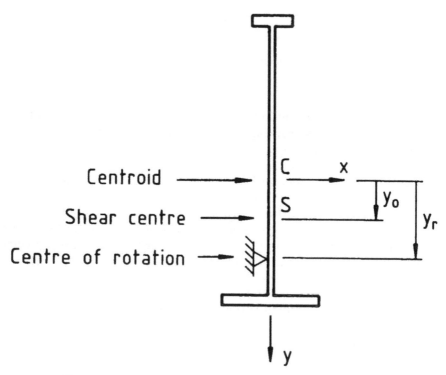

Figure 13.8 Ring buckling with an enforced centre of rotation.

In the special case of doubly symmetric sections ($y_0 = 0$) with centroidal loading ($y_q = 0$), equation 13.27 can be written in the form

$$P = n^2 P_a/4 + P_b + 4P_c/n^2 \qquad (13.28)$$

in which

$$\left.\begin{array}{l} r_P^2 P_a = EI_y y_r^2/R^2 + (EI_w/R^2)(1 - y_r/R)^2, \\ r_P^2 P_b = GJ(1 - y_r/R)^2 - 2EI_y y_r/R, \\ r_P^2 P_c = EI_y, \end{array}\right\} \qquad (13.29)$$

and

$$r_P^2 = r_0^2 + y_r^2. \qquad (13.30)$$

When $P_c < P_a$, the minimum solution occurs for $n = 2$, so that

$$P = P_a + P_b + P_c. \qquad (13.31)$$

When $P_c > P_a$, the minimum solution depends on the number of (even) half waves n, and can be approximated by

$$P = 2\sqrt{(P_a P_c)} + P_b. \qquad (13.32)$$

If the connected shell which enforces a centre of rotation during buckling also exerts a continuous torsional restraint of stiffness α_{rz}, then $r_P^2 P_c$ increases to

$$r_P^2 P_c = EI_y + \alpha_{rz} R^2. \qquad (13.33)$$

13.7 Effect of load height on arch buckling

The effects of load height y_Q on the flexural-torsional buckling of simply supported curved beams with central concentrated load Q are shown in Figure 13.9 [10]. It can be seen that the buckling resistance increases as the distance of the load below the curved beam increases, until a limiting value is approached at which the curved beam buckling mode changes from a single half wave to two half waves. Semi-circular curved beams ($L/\pi R = 1.0$) buckle in a rigid body mode under zero load when the load acts on the line joining the supports ($y_Q/L = 1/\pi$).

The effects of load height y_Q on the buckling of arches in uniform compression are shown in Figure 13.10 [10]. Again, the buckling resistance increases as the distance of the load below the arch increases.

The results of experiments [11, 12] on the buckling of doubly symmetric curved beams and monosymmetric arches under central concentrated loads are shown in Figures 13.11 and 13.12. These show very good agreement with the theoretical predictions obtained by finite element analysis, and provide general support for the theoretical findings summarized in this chapter.

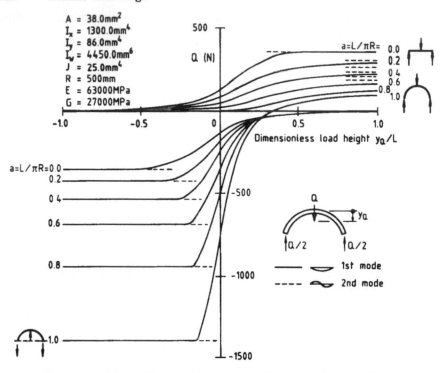

Figure 13.9 Effect of load height on curved beams with central loads.

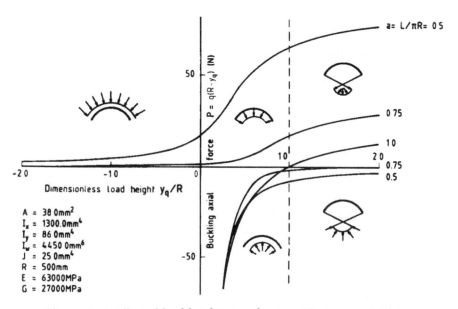

Figure 13.10 Effect of load height on arches in uniform compression.

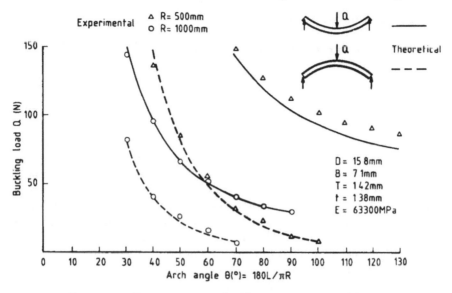

Figure 13.11 Experiments on doubly symmetric curved beams.

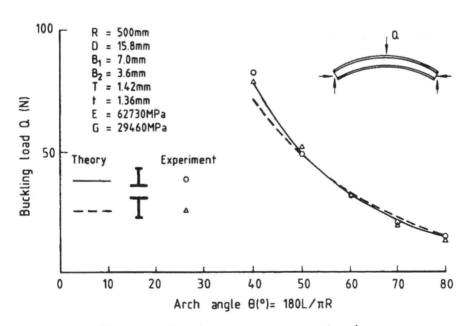

Figure 13.12 Experiments on monosymmetric arches.

13.8 Finite element analysis

13.8.1 IN-PLANE PRE-BUCKLING ANALYSIS

Many arches are statically indeterminate, and the distributions of the in-plane stress resultants P, M_x depend on the deflections v, w_s. In this case, these are best determined by using a computer method of analysis, such as the finite element method [1, 10, 12, 13].

The finite element method for the in-plane analysis of straight elements described in section 4.5 uses linear displacement fields for the in-plane longitudinal displacement w_s. It has been found [14] that linear fields may cause significant errors in the analysis of arches, but that improved accuracy can be achieved by using higher order elements. One study [1] has suggested that where 10 quintic elements may give an accuracy of within 1%, 60 cubic elements may be required for the same accuracy. The use of higher order elements is simplified by the use of Gaussian numerical integration [13] instead of formal integration.

The use of higher order elements leads to higher order nodal deformations which are generally associated with stress resultants. These are additional to those deformations used to ensure geometric continuity between adjacent elements. The enforcement during assembly of the elements of fictitious 'compatibility' conditions in which the higher order nodal stress resultant deformations for one element are equated to the corresponding ones for the other element at the node may introduce errors, especially when there are slope or area changes at a node between elements.

One method of avoiding these errors is to condense out the higher order deformations before the assembly. Another method [1] is to use short length elements at the node, so that the errors are reduced.

13.8.2 OUT-OF-PLANE BUCKLING ANALYSIS

The finite element method described in sections 4.2 and 4.4 for the elastic flexural-torsional buckling analysis of structures composed of straight elements has been extended to the curved elements of arches [10, 12]. For this, the energy equation of equation 13.9 was substituted for the straight element equations 4.58 and 4.62, and quintic deformation fields were used for the out-of-plane buckling deformations u, ϕ. The higher order nodal deformations u'', ϕ'' were not condensed out, but retained, so that the eigenvalue problem remained linear. This procedure is similar to that recommended in section 4.3.6 for internal hinges.

Studies [10, 12] have suggested that convergence is much slower for arches with in-plane bending than for uniform compression, and that up to 16 elements may be required to obtain an accuracy within 1%.

13.9 Problem

PROBLEM 13.1

Adapt the solution of Problem 4.12 so as to produce a computer program for analysing the elastic flexural-torsional buckling of an arch whose cross-sections are symmetric about the plane of the arch.

13.10 References

1. Papangelis, J.P. and Trahair, N.S. (1986) In-plane finite element analysis of arches. *Proceedings*, Pacific Structural Steel Conference, Auckland, August, Volume 4, pp. 333–50.
2. Papangelis, J.P. and Trahair, N.S. (1987) Flexural-torsional buckling of arches. *Journal of Structural Engineering, ASCE*, 113(4), 889–906.
3. Trahair, N.S. and Papangelis, J.P. (1987) Flexural-torsional buckling of monosymmetric arches. *Journal of Structural Engineering, ASCE*, 113(10), 2271–88.
4. Yang, Y.-B. and Kuo, S.-R. (1986) Static stability of curved thin-walled beams. *Journal of Engineering Mechanics, ASCE*, 112 (8), 821–41.
5. Rajasekaran, S. and Padmanabhan, S. (1989) Equations of curved beams. *Journal of Engineering Mechanics, ASCE*, 115 (5) 1094–111.
6. Yang, Y.-B., Kuo, S.-R. and Yan, J.-D. (1991) Use of straight-beam approach to study buckling of curved beams. *Journal of Structural Engineering, ASCE*, 117 (7) 1963–78.
7. Kuo, S.-R. and Yang, Y.-B. (1991) New theory on buckling of curved beams. *Journal of Engineering Mechanics, ASCE*, 117 (8), 1698–717.
8. Teng, J.G. and Rotter, J.M. (1988) Buckling of restrained monosymmetric rings. *Journal of Engineering Mechanics, ASCE*, 114 (EM10), 1651–71.
9. Teng, J.G. and Rotter, J.M. (1989) Buckling of rings in column-supported bins and tanks. *Thin-Walled Structures*, 7 (3–4), 257–80.
10. Papangelis, J.P. and Trahair, N.S. (1987) Finite element analysis of arch lateral buckling. *Civil Engineering Transactions*, Institution of Engineers, Australia, CE29 (1), 34–9.
11. Papangelis, J.P. and Trahair, N.S. (1987) Flexural-torsional buckling tests on arches, *Journal of Structural Engineering, ASCE*, 113 (7), 1433–43.
12. Papangelis, J.P. and Trahair, N.S. (1988) Buckling of monosymmetric arches under point loads. *Engineering Structures*, 10 (4), 257–64.
13. Zienkiewiez, O.C. and Taylor, R.L. (1989) *The Finite Element Method, Volume 1 – Basic Formulation and Linear Problems; (1991) Volume 2 – Solid and Fluid Mechanics, Dynamics, and Non-Linearity*, 4th edn, McGraw-Hill, London.
14. Dawe, D.J. (1974) Numerical studies using circular arch finite elements. *Computers and Structures*, 4 (4), 729–40.

14 Inelastic buckling

14.1 General

A slender beam or beam-column which has low resistances to lateral bending and torsion may buckle while still elastic by deflecting and twisting out of the plane of loading. This elastic flexural-torsional buckling was discussed in Chapters 7–11.

The resistance of a member to elastic buckling increases as its slenderness decreases, and a steel member of moderate stiffness may yield before its elastic buckling load is reached. Yielding is caused by a combination of the stresses induced by the applied loads with any residual stresses which remain after the manufacturing process is completed. Yielding reduces the effective out-of-plane rigidities, and decreases the buckling resistance below the elastic value, as shown in Figure 7.4.

A theoretical model of inelastic flexural-torsional buckling is developed in this chapter, and a computer method of analysis is presented. Comparisons are made of the theoretical predictions with the results of inelastic buckling experiments. Following this, the effects of the support conditions and the loading arrangements on the inelastic buckling of beams and beam-columns are discussed.

The members considered in this chapter are steel I-section members which are assumed to be perfectly straight and untwisted before loading. Real members have initial curvatures and twists, which further reduce their strengths. Design methods which account for the reductions caused by these geometrical imperfections are discussed in Chapter 15.

14.2 Tangent modulus theory of inelastic buckling

14.2.1 FLEXURAL BUCKLING OF COLUMNS

Most inelastic buckling predictions are based on the tangent modulus theory, which was originally developed to predict the flexural buckling of inelastic columns [1, 2]. Strictly speaking, this theory applies only to members whose non-linear stress–strain curves are elastic so that the loading and unloading paths are identical, as shown in Figure 14.1a. In this case, the tangent modulus E_t appropriate to the average stress level in the column may be substituted for the initial modulus E in the expression $\pi^2 E I_y/L^2$ for the elastic flexural buckling load P_y of a simply supported column to obtain the tangent modulus buckling load

$$P_t = \pi^2 E_t I_y/L^2. \tag{14.1}$$

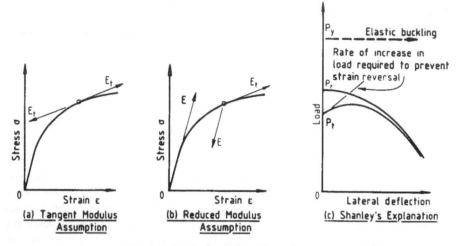

Figure 14.1 Inelastic buckling theories.

The application of this theory to inelastic materials is apparently invalid, since these have unloading paths which are parallel to the initial slopes E of the stress–strain curves, as shown in Figure 14.1b. This observation led to the development of the reduced modulus theory of inelastic buckling, for which E_t is used only in the loading region of the column (where compressive bending stresses increase the total strain during buckling), and the initial modulus E is used for the unloading region (where tensile bending stresses decrease the total strain). Under these circumstances, a reduced modulus E_r can be determined from E, E_t, and the geometry of the cross-section, and used to calculate a reduced modulus inelastic buckling load

$$P_r = \pi^2 E_r I_y/L^2. \tag{14.2}$$

The reduced modulus E_r always lies between E and E_t, so that

$$P_y > P_r > P_t \tag{14.3}$$

as shown in Figure 14.1c.

Although the tangent modulus theory appears to be invalid for inelastic materials, careful experiments have shown that it leads to more accurate predictions than the apparently rigorous reduced modulus theory. This paradox was resolved by Shanley [3], who reasoned that the tangent modulus theory is valid when buckling deflections are accompanied by simultaneous increases in the applied load of sufficient magnitude to prevent strain reversal, as shown in Figure 14.1c. When this happens, all the stress and strain increments are related by the tangent modulus E_t, and so the buckling load is equal to the tangent modulus load P_t. Thus lateral deflection initiates at P_t, and increases with increasing load (and decreasing E_t) until a maximum load is reached, as shown in Figure 14.1c.

The tangent modulus load P_t is the lowest load at which lateral deflection of a straight column can commence. It is theoretically possible for buckling to begin at higher loads, the limiting case being at the reduced modulus load P_r at which buckling initiates without any change in the applied load, as shown in Figure 14.1c. However, real members have small imperfections which lower the load deflection path below the tangent modulus path, and the maximum load capacity is often quite close to the tangent modulus prediction. Because of this, and because of its simplicity, the tangent modulus theory of inelastic buckling has gained wide acceptance.

14.2.2 APPLICATION TO STEEL COLUMNS

The application of the tangent modulus theory to the inelastic buckling of steel columns with stress–strain curves of the type shown in Figure 14.2 led at first to anomalous results, as demonstrated in Figure 14.3a. This figure shows that the predicted buckling loads for slender columns increase as the slenderness decreases, until the squash load $P_Y = AF_Y$ is reached, when the buckling load drops suddenly to zero, corresponding to the sudden change in E_t from E to zero when general yielding occurs. If, however, the general stress level reaches the strain-hardening range, there is a sudden increase from zero corresponding to the sudden change in E_t from zero to E_s.

That these anomalous predictions are not realized in practice is due to the fact that the sudden and general yielding which led to $P_t = 0$ does not occur. There are two reasons for this, the first being associated with the residual stresses σ_r such as those shown in Figure 14.4 which are induced in most structural steel members during their manufacture. The variations of these residual stresses across the section cause yielding to initiate locally, at points where the resultant compressive

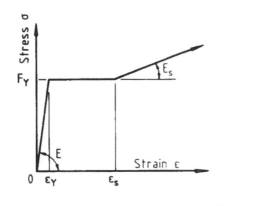

$F_Y = 250$ N/mm^2
$E = 200\ 000$ N/mm^2
$E_s = E/33$
$\varepsilon_Y = 1250 \times 10^{-6}$
$\varepsilon_s = 11\varepsilon_Y$

$\nu = 0\ 3$
$G = 76\ 920$ N/mm^2
$G_s = 20\ 940$ N/mm^2

(a) Tensile Stress-Strain Relationship   (b) Material Properties

Figure 14.2 Idealized properties of structural steel.

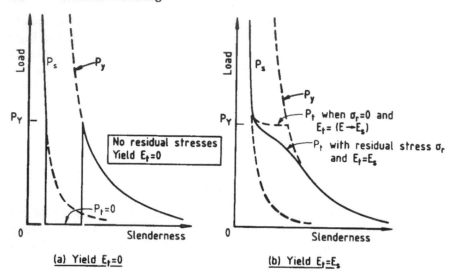

Figure 14.3 Tangent modulus predictions for steel columns.

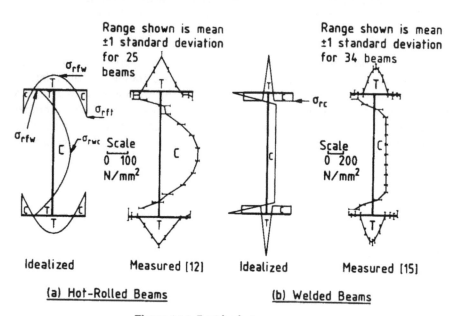

Figure 14.4 Residual stress patterns.

stresses are greatest. As the load increases further, there are corresponding increases in the yielded areas, and decreases in the remaining elastic core where $E_t = E$. The decreasing elastic core leads to decreases in the tangent modulus rigidity

$$(EI)_t \approx EI_e \tag{14.4}$$

where I_e is computed for the elastic core, and consequent decreases in the tangent modulus buckling load. Thus the residual stresses cause early yielding as shown in Figure 14.3b, and the tangent modulus load diverges increasingly from the elastic buckling load as yielding progresses through the column.

The second reason why yielding does not occur suddenly and generally is associated with the actual yielding process of steel. First it should be noted that the curve shown in Figure 14.2 does not represent the local variations between stress and strain, but rather the mean variations averaged over the instrumented volume of the test piece for which it was determined. Locally, yielding progresses as a series of sudden discontinuous slips in the matrix between the grains [4] from the elastic condition to the strain-hardened condition. Thus when the average stress–strain curve indicates that yielding is taking place, some of the material is elastic ($E_t = E$), and the rest is strain-hardened ($E_t = E_s$). Thus the average E_t representative of the whole yielded region decreases steadily from E to E_s as the strain increases from yielding at ε_Y to strain-hardening at ε_s. This has led to the conservative proposal that E_s should be used for E_t in the yielded regions.

The approximation of E_s for E_t in the yielded regions explains why short length columns which have had their residual stresses removed by annealing do not buckle suddenly when they first become fully yielded. The approximation is also in agreement with observations of other buckling phenomena. For example, many beams are capable of reaching the full plastic moment M_P and maintaining it over a considerable range of deformations before local buckling occurs. Since the flanges are fully yielded before M_P is reached, it can be concluded that the local buckling strength is not reduced to zero by yielding.

It can be seen that the tangent modulus buckling theory for steel columns should allow for the presence of residual stresses when determining the elastic, yielded, and strain-hardened regions of the column, and should use E for the elastic regions and E_s for the yielded and strain-hardened regions. Thus the tangent modulus flexural rigidity $(EI)_t$ will be equal to the elastic rigidity of an equivalent section obtained by transforming the yielded and strain-hardened regions according to the modular ratio E_s/E.

14.2.3 FLEXURAL-TORSIONAL BUCKLING

The application of the tangent modulus theory to the inelastic flexural-torsional buckling of a steel member requires only one further extension of the theory for the flexural buckling of columns. This extension is needed to determine appropriate values of the tangent shear modulus G_t to be used when evaluating the contributions of the yielded and strain-hardened regions to the effective torsional rigidity $(GJ)_t$.

Although the incremental theory of perfectly plastic solids suggests that the shear modulus of any yielded material is initially equal to the elastic modulus G, experiments on yielded steel members [5, 6] indicate that it decreases rapidly with even quite small strains. An examination [7] of the theoretical and experimental evidence led to the conclusion that the shear modulus of yielded structural steel could best be approximated by

$$\frac{G_t}{G} = \frac{2}{1 + (E/E_s)/4(1 + v)}$$

(14.5)

where v is the elastic Poisson's ratio.

14.3 Pre-buckling analysis of in-plane bending

14.3.1 GENERAL

Before an inelastic out-of-plane buckling prediction can be made, the in-plane bending must be analysed so that the distributions of the elastic, yielded, and strain-hardened regions throughout the member can be determined. The effective out-of-plane rigidities which contribute to the inelastic buckling resistance can then be evaluated using these distributions.

When the member is statically determinate, the in-plane analysis can be made in two separate stages. First, the variation of the axial force and bending moment along the member can be determined from statics. Following this, the locations of the boundaries of the elastic, yielded, and strain-hardened regions within selected cross-sections can be determined, using the cross-section geometry, material properties, residual stresses, and the axial force and bending moment.

When an inelastic member is statically indeterminate, the two stages cannot be separated, because the material non-linearity closes the chain of dependence of yielding on stress resultants, on redundant actions, on deflections, on stiffnesses, and on yielding. In addition, it may be necessary to consider the effects of geometric non-linearity, as for example when there are significant in-plane instability effects in the bending of some sway frames.

A finite element computer method of analysing the in-plane behaviour of steel frames is described in [8]. This method allows for the effects of residual stresses, yielding, strain-hardening, and finite deflections. The following sub-sections discuss the data required for this method, and show how the cross-section and member analyses are performed, and how the yielded and strain-hardened boundaries are determined.

14.3.2 SECTION GEOMETRY AND MATERIAL PROPERTIES

The geometry of an idealized doubly symmetric I-section member is defined in Figure 14.5a. For simplicity, all radii and fillets are ignored, as is any flange tapering.

An idealized tensile stress–strain relationship for structural steel is shown in Figure 14.2, with nominal values for E, E_s, F_Y, and ε_s. The existence of an upper

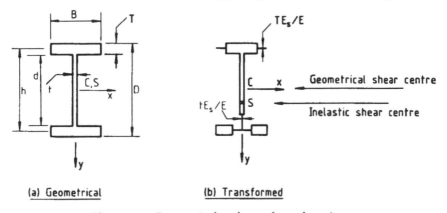

(a) Geometrical   (b) Transformed

Figure 14.5 Geometrical and transformed sections.

yield stress is ignored. This stress-strain curve is usually assumed to apply to all of the material in the cross-section, even though the web yield stress is usually significantly greater than the flange yield stress.

14.3.3 RESIDUAL STRESSES

14.3.3.1 Hot-rolled beams

Longitudinal residual stresses are induced in a hot-rolled steel member during cooling after rolling, and as a result of any mechanical processes used to straighten it [9]. During cooling, the more highly exposed regions of the cross-section at the flange tips and web centre cool more rapidly. These early cooling regions shrink, inducing matching plastic flows in the high temperature late-cooling regions at the flange-web junctions, which have correspondingly low yield stresses. Subsequent shrinkages during final cooling of the high temperature regions at the flange-web junctions are resisted by the regions already cooled which have developed high yield stresses, and induce residual compressions in them, as shown in Figure 14.4a. Equilibrating residual tensions are induced in the late-cooling flange-web junctions. Cold-working by mechanical straightening causes local yielding, and further modifies the residual stress pattern.

The magnitudes and distributions of residual stress vary considerably with the cross-section geometry and with the cooling and straightening processes. Idealized and measured distributions of the residual stresses in hot-rolled beams are shown in Figure 14.4a. The distributions across the flanges and web are often approximately parabolic, being compressive at the early-cooling flange tips and web centre, and tensile at the late-cooling flange-web junctions. It has been suggested [10] that the maximum residual stresses in hot-rolled beams may be approximated by

$$\left.\begin{array}{l} \sigma_{rft} = -137.5(2.2 - A/2BT)\,\mathrm{N\,mm^{-2}}, \\ \sigma_{rfw} = 100(-0.3 + A/2BT)\,\mathrm{N\,mm^{-2}}, \\ \sigma_{rwc} = -83.3(0.8 + A/2BT)\,\mathrm{N\,mm^{-2}}. \end{array}\right\} \tag{14.6}$$

While the flange and web distributions are approximately parabolic, it has been suggested [11] that quartic distributions of the form

$$\left.\begin{array}{l} \sigma_{rf}/F_Y = a_1 + a_2(2x/B)^2 + a_3(2x/B)^4, \\ \sigma_{rw}/F_Y = a_4 + a_5(2y/d)^2 + a_6(2y/d)^4 \end{array}\right\} \tag{14.7}$$

should be used, so that the conditions

$$\int_A \sigma_r \, dA = 0, \tag{14.8}$$

$$\int_A \sigma_r(x^2 + y^2) \, dA = 0 \tag{14.9}$$

are satisfied. The first of these conditions ensures that the residual stresses have a zero axial force resultant (zero moment resultants about the x and y axes are automatically ensured by symmetry). The second condition was suggested to ensure that the residual stresses have a zero axial torque effect when the member is twisted. If this condition is not satisfied, then the effective elastic torsional rigidity of the cross-section changes to $(GJ - \int_A \sigma_r(x^2 + y^2) \, dA)$.

In practice, some residual stress distributions may differ significantly from the idealized pattern discussed above, as can be seen from the measured distributions shown in Figure 14.4a [12].

It has been suggested [13] that comparative numerical analyses should be carried out using $a_3 = 0$ and

$$\left.\begin{array}{l} \sigma_{rft} = -0.35F_Y, \\ \sigma_{rfw} = 0.50F_Y, \end{array}\right\} \tag{14.10}$$

so that

$$\begin{Bmatrix} a_1 \\ a_2 \\ a_3 \\ a_4 \\ a_5 \\ a_6 \end{Bmatrix} = \frac{1}{16} \begin{bmatrix} 8 & 0 & 0 \\ -13.6 & 0 & 0 \\ 0 & 0 & 0 \\ 3 & 45 & -105 \\ -30 & -150 & 630 \\ -35 & 105 & -525 \end{bmatrix} \begin{Bmatrix} 1 \\ -\left(\dfrac{13}{60}\right)\dfrac{A_F}{A_W} \\ \dfrac{A_F}{A_W}\left(\dfrac{B^2 - 65h^2}{300d^2}\right) \end{Bmatrix} \tag{14.11}$$

where

$$A_F/A_W = 1/(A/2BT - 1). \tag{14.12}$$

14.3.3.2 Welded beams

Residual stresses are induced in welded beams by uneven heating and cooling as a result of flame-cutting and welding the flanges and webs. The process is similar to that described above for hot-rolled beams, with large residual tensions being induced in the late-cooling flange–web junctions, and equilibrating compressions in the flanges and web, as shown in Figure 14.4b.

The magnitudes and distributions of residual stress again vary with the cross-section geometry and with the cutting, welding, cooling and straightening processes. In the idealized residual stress distribution shown in Figure 14.4b, the compressive stresses are uniform across the flanges and web, except near the flange–web junctions. It has been suggested [14] that the maximum tensile stress can be assumed to be equal to the parent metal yield stress F_Y, and that the compressive stresses at the flange tip can be determined from

$$\sigma_{rc} = 0.1 B_w \sum (A_w/A) \tag{14.13}$$

in which A_w is the area of weld metal added in the biggest single pass at the weld site, by using their suggested values of B_w, the energy supplied per unit volume of electrode wire.

Again, practical residual stress distributions may differ significantly from this idealized pattern, as can be seen from the measured distributions shown in Figure 14.4b [15].

14.3.4 CROSS-SECTION ANALYSIS

The in-plane displacements v_P, w_P of any point $P(x, y)$ in the doubly symmetric cross-section shown in Figure 14.6 can be expressed in terms of the displacements

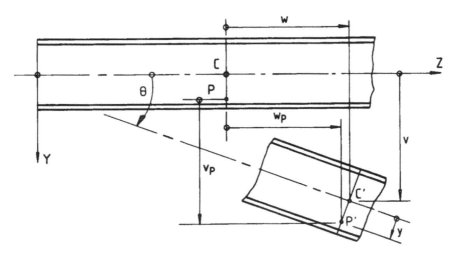

Figure 14.6 In-plane displacements.

v, w of the elastic centroid C as

$$v_P = v - y(1 - \cos \theta) \tag{14.14}$$

$$w_P = w - y \sin \theta \tag{14.15}$$

where

$$\sin \theta = v'. \tag{14.16}$$

The effects of finite rotations are included in these equations by the use of $\sin \theta$ and $\cos \theta$ instead of the usual small rotation approximations of θ and 1.

The total normal strain ε_P is sum of the residual strain $\varepsilon_r = \sigma_r/E$ and the second-order strains associated with the displacements v_P, w_P as shown in Figure 14.7, and can be obtained [8] from

$$\varepsilon_P = \varepsilon_r + w'_P + (v'^2_P + w'^2_P)/2 \tag{14.17}$$

as

$$\varepsilon_P = \varepsilon_r + (w' - yv'') + \{(v' - y\tan\theta v'')^2 + (w' - yv'')^2\}/2. \tag{14.18}$$

The inclusion of the second-order terms $(v'^2_P + w'^2_P)/2$ allows the effects of in-plane instability to be incorporated into the analysis.

The corresponding normal stresses σ_P are obtained from

$$\sigma_P = \int_0^{\varepsilon_P} \left(\frac{d\sigma}{d\varepsilon_P}\right) d\varepsilon_P \tag{14.19}$$

where $(d\sigma_P/d\varepsilon_P)$ is the slope of the stress–strain curve (Figure 14.2a).

14.3.5 MEMBER ANALYSIS

It is assumed that the displacements v, w of the elastic centroidal axis of an element of length L can be approximated by

$$\{v_i\} = \{v, w\}^T = [M]\{a\} = L\begin{bmatrix} \{Z\}^T & \{0\} \\ \{0\} & \{Z\}^T \end{bmatrix}\{a\} \tag{14.20}$$

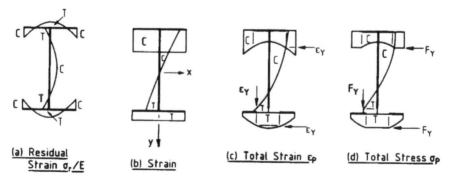

(a) Residual Strain σ_r/E (b) Strain   (c) Total Strain ε_P (d) Total Stress σ_P

Figure 14.7 Strain and stress distributions.

where

$$[M] = L \begin{bmatrix} \{Z\}^{\mathrm{T}} & \{0\} \\ \{0\} & \{Z^{\mathrm{T}}\} \end{bmatrix}, \qquad (14.21)$$

$\{Z\}^{\mathrm{T}} = \{1, z/L, z^2/L^2, z^3/L^3\}$, and $\{a\}$ contains eight arbitrary coefficients. The eight element nodal displacements

$$\{\delta_{ie}\} = \{v_0, v_L, v'_0, v'_L, w_0, w_L, w'_0, w'_L\}^{\mathrm{T}} \qquad (14.22)$$

can be obtained from equation 14.20 in the form of

$$\{\delta_{ie}\} = [C_i]\{a\} \qquad (14.23)$$

which enables the coefficients $\{a\}$ to be eliminated from equation 14.20, whence

$$\{v_i\} = [M][C_i]^{-1}\{\delta_{ie}\}. \qquad (14.24)$$

The second-order strains $\varepsilon_{\mathbf{P}}$ given by equation 14.18 are non-linear functions of the derivatives v', v'', w', which can be obtained from equation 14.24 in the form of

$$\{v', v'', w'\}^{\mathrm{T}} = L[B_i][C_i]^{-1}\{\delta_{ie}\} \qquad (14.25)$$

which enables the stresses $\sigma_{\mathbf{P}}$ given by equation 14.19 to be expressed in terms of the nodal displacements $\{\delta_{ie}\}$.

For equilibrium of the element, the virtual work principle requires

$$\int_V \delta\varepsilon_{\mathbf{P}}\sigma_{\mathbf{P}}\,\mathrm{d}V \quad \{\delta\delta_{ie}\}^{\mathrm{T}}\{Q_{ie}\} = 0 \qquad (14.26)$$

where the symbol δ indicates a virtual displacement, and the generalized in-plane nodal forces $\{Q_{ie}\}$ corresponding to $\{\delta_{ie}\}$ are the static equilibrants of the stresses $\sigma_{\mathbf{P}}$. This can be rewritten as

$$\int_L \{\delta\delta_{ie}\}^{\mathrm{T}}[C_i]^{-\mathrm{T}}[B_i]^{\mathrm{T}}\{b\}\,L\,\mathrm{d}z - \{\delta\delta_{ie}\}^{\mathrm{T}}\{Q_{ie}\} = 0 \qquad (14.27)$$

in which $\{b\}$ are non-linear functions of $\{\delta_{ie}\}$ and of the generalized stress resultants

$$
\left.
\begin{aligned}
\sigma_{i1} &= \int_A \sigma_{\mathbf{P}}\,\mathrm{d}A = -P, \\[4pt]
\sigma_{i2} &= \int_A \sigma_{\mathbf{P}}\,y\,\mathrm{d}A = M_x, \\[4pt]
\sigma_{i3} &= \int_A \sigma_{\mathbf{P}}\,y^2\,\mathrm{d}A, \\[4pt]
\sigma_{i4} &= \int_A \sigma_{\mathbf{P}}\,x^2\,\mathrm{d}A.
\end{aligned}
\right\}
\qquad (14.28)
$$

Performing the variation in equation 14.27 and noting that $\{\delta\delta_{ie}\}$ is arbitrary leads to a set of eight simultaneous non-linear equations of the form of

$$Q_{ie} = g_i(\delta_{ie1}, \delta_{ie2}, \ldots, \delta_{ie8}). \tag{14.29}$$

The terms in these equations are formed by integrations over the volume of the element, as indicated by equations 14.27 and 14.28. If Gaussian integration [16, 17] is used with respect to z, then the integrations over the area A need only be carried out at the Gauss points.

The tangent stiffness equations for the element can be expressed as

$$[k_{iet}]\{\Delta\delta_{ie}\} = \{\Delta Q_{ie}\} \tag{14.30}$$

in which the symbol Δ indicates an increment, and $\{\Delta\delta_{ie}\}$, $\{\Delta Q_{ie}\}$ are the increments from the equilibrium values $\{\delta_{ie}\}$, $\{Q_{ie}\}$, and $[k_{iet}]$ is the element tangent stiffness matrix. An approximation for $[k_{iet}]$ may be obtained from equation 14.29 by assuming that there are no unloading and no changes in the yielded and strain-hardened zones. These assumptions enable a quasi-elastic transformed element to be established for the stress levels corresponding to $\{\delta_{ie}\}$ whose tangent stiffness can be obtained from equation 14.29 as

$$\Delta Q_{ie} = \left\{\frac{\partial g_i}{\partial\delta_{ie1}}, \frac{\partial g_i}{\partial\delta_{ie2}}, \ldots, \frac{\partial g_i}{\partial\delta_{ie8}}\right\}^{\mathsf{T}}\{\Delta\delta_{ie}\}. \tag{14.31}$$

Further details of this process are given in [8].

The tangent stiffness equations for each element can be transformed and combined to form the global tangent stiffness equations

$$[K_{it}]\{\Delta\Delta_i\} = \{\Delta Q_i\} \tag{14.32}$$

in which $\{\Delta\Delta_i\}$, $\{\Delta Q_i\}$ are the increments in the structure's nodal displacements and forces and $[K_{it}]$ is the global stiffness matrix.

The load–deflection behaviour of the structure is non-linear in general, but the solution for a particular set of applied nodal forces $\{Q_i\}$ can be obtained iteratively by using the Newton–Raphson method [16, 17], starting from an initial set of global displacements $\{\Delta_{i1}\}$. This initial set may be arbitrary, but faster convergence can be obtained if it is calculated from a conventional linear elastic analysis, or extrapolated from a converged analysis corresponding to a lower set of nodal forces, as indicated diagrammatically in Figure 14.8a.

A typical cycle of the solution process is illustrated diagrammatically in Figure 14.8b. The cycle starts with the global displacements $\{\Delta_{in}\}$, which are transformed to the element displacements $\{\delta_{in}\}$. These are used in equation 14.29 to obtain the element forces $\{Q_{ien}\}$, which are transformed and combined into the global forces $\{Q_{in}\}$. If these are sufficiently close to the applied forces $\{Q_i\}$, then the process has converged, and the solution has been obtained.

When $\{Q_{in}\}$ is not sufficiently close to $\{Q_i\}$, the incremental global forces

$$\{\Delta Q_{in}\} = \{Q_i\} - \{Q_{in}\} \tag{14.33}$$

can be substituted into equation 14.32, which can be solved for the incremental

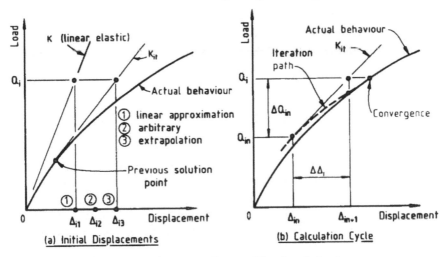

Figure 14.8 Iterative solution of in-plane behaviour.

global displacements $\{\Delta\Delta_i\}$. The new and more accurate set of global displacements

$$\{\Delta_{in+1}\} = \{\Delta_{in}\} + \{\Delta\Delta_i\} \qquad (14.34)$$

can then be used to begin a new cycle of calculations. The process is repeated until convergence is obtained.

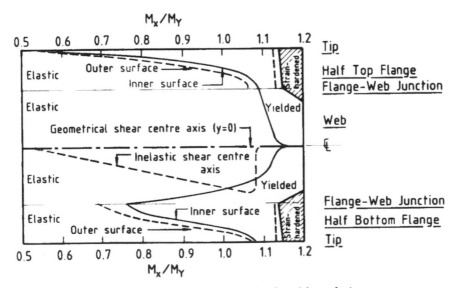

Figure 14.9 Yielded and strain-hardened boundaries.

14.3.6 YIELDED AND STRAIN-HARDENED BOUNDARIES

The converged solution for the element nodal displacements $\{\delta_{le}\}$ can be used in equations 14.25 and 14.18 to find the cross-sectional variations of the strain ε_p at the Gauss points. The positions of the yielded and strain-hardened boundaries where $\varepsilon_p = \varepsilon_Y, \varepsilon_s$ can then be determined.

Typical variations of these boundaries with the major axis moment M_x for beams ($P = 0$) are shown in Figure 14.9 [18]. The top flange commences yielding at the flange tips at 0.5 M_Y (M_Y is the nominal first yield moment $F_Y Z_x$) since the residual compressive stress is 0.5 F_Y in this case, and is fully plastic at 1.09 M_Y. The bottom flange commences yielding at 0.7 M_Y at its centre (where the residual tensile stress was 0.3 F_Y), and is fully plastic at 1.08 M_Y. Yielding of the web commences at the bottom at 0.76 M_Y, and progresses slowly until the flanges are fully yielded at 1.09 M_Y. The process then accelerates, and the web is virtually fully yielded at 1.14 M_Y, when strain-hardening commences in both flanges.

14.4 Inelastic buckling analysis

14.4.1 METHODS OF ANALYSIS

A number of methods have been used to analyse the inelastic flexural-torsional buckling of steel I-section members, including the finite difference, finite integral, transfer matrix and finite element methods. These are summarized in [19]. In this chapter, only the inelastic finite element method is described. This method is an adaptation of the method presented in Chapter 4 for elastic buckling.

14.4.2 ELEMENT DEFORMATIONS

The elastic finite element method of Chapter 4 models the deformations of an element in terms of the deformations of a one-dimensional line. For an elastic element, it is convenient to use the shear centre axis for this line, since this allows the first-order relationships between the stress resultants and the curvatures and twist to be decoupled, and simplifies the buckling equations.

For inelastic elements, there are two shear centre axes. The first is the geometrical shear centre axis, which passes through the shear centre of the geometrical cross-section, as shown in Figure 14.5a. The second is the inelastic shear centre axis, which is the locus of the shear centres of sections transformed from the geometrical section by using the tangent modulus ratio $E_t/E = E_s/E$ for the yielded and strain-hardened regions, as shown in Figure 14.5b. The position of the inelastic shear centre axis varies with the yielding of the section, as shown in Figure 14.9. For this example, the shear centre axis moves towards the bottom flange as M_x increases from 0.5 M_Y to 1.08 M_Y. This happens because the bottom flange is the more effective as a result of the significant reductions in the effective flexural rigidity of the top flange caused by early yielding near the flange tips (yielding at the centre of the bottom flange causes only minor reductions in its rigidity). For $M_x > 1.08\ M_Y$, both flanges are completely yielded or strain-hardened, and so the effective section is doubly symmetric again, and the inelastic shear centre moves back to the geometrical axis.

While the inelastic shear centre axis seems the natural one to use when describing the resistances of an inelastic element to lateral bending and torsion, there are significant disadvantages. First of all, the axis is often inclined to the geometrical axis, to which the stress resultants are usually referred, which requires the variable position of the axis to be included in the formulation. Secondly, the first-order decoupling between the stress resultants and the curvatures and twist for elastic elements is largely destroyed for inelastic elements, even when the second-order flexural-torsional coupling is unimportant, and the significance of the elastic shear centre axis as the axis of pure torsion disappears.

On the other hand, while the use of the geometrical axis requires some care when formulating the lateral bending and torsional resistances, its position does not change with the amount of yielding, but remains aligned with the stress resultant axes. For these reasons, the inelastic finite element method is most simply developed from the displacements u, ϕ of the geometrical shear centre axis.

14.4.3 STRAIN ENERGY

The flexural strain energy stored in the flanges during inelastic buckling can be written as

$$\frac{1}{2}\delta^2 U_{fb} = \frac{1}{2}\int_0^L \{(EI_T)_t(u'' + h\phi''/2)^2 + (EI_B)_t(u'' - h\phi''/2)^2\}\,dz \quad (14.35)$$

where $(EI_T)_t$, $(EI_B)_t$ are the tangent flexural rigidities of the top and bottom flanges, and h is the distance between the flange centroids. The tangent rigidities may be determined by transforming the yielded and strain-hardened regions of the cross-section using the tangent modulus ratio E_t/E, as shown in Figure 14.5b. For this purpose, the positions of the boundary planes between the elastic and yielded zones are obtained from the in-plane pre-buckling analysis, as described in section 14.3.6. For sections with residual stresses, these boundary planes are almost perpendicular to the flange mid-plane, and so for convenience they are assumed to be perpendicular.

The tangent torsional rigidity $(GJ)_t$ can be obtained by using G_t for the yielded and strain-hardened regions so that

$$(GJ)_t = \sum G_t bt^3/3 + C \quad (14.36)$$

where b and t are the actual width and thickness of each zone (elastic, yielded, or strain-hardened) of the cross-section, and C is a correction to allow for end, junction, and finite thickness effects. For simplicity, C is usually taken as the elastic value given in [20]. The uniform torsional strain energy stored during inelastic buckling can then be written as

$$\frac{1}{2}\delta^2 U_t = \frac{1}{2}\int_0^L (GJ)_t \phi'^2 \, dz. \tag{14.37}$$

The total strain energy stored in the element during buckling can therefore be obtained from equations 14.35 and 14.37 as

$$\frac{1}{2}\delta^2 U_e = \frac{1}{2}\int_0^L \{\varepsilon_u\}^T [D_u] \{\varepsilon_u\} \, dz \tag{14.38}$$

where

$$\{\varepsilon_u\}^T = \{u'', \phi', -\phi''\} \tag{14.39}$$

and

$$[D_u] = \begin{bmatrix} (EI_T + EI_B)_t & 0 & -(EI_T - EI_B)_t h/2 \\ 0 & (GJ)_t & 0 \\ -(EI_T - EI_B)_t h/2 & 0 & (EI_T + EI_B)_t h^2/4 \end{bmatrix}. \tag{14.40}$$

The off-diagonal terms $-(EI_T - EI_B)_t h/2$ allow for monosymmetry effects caused by unequal yielding of the flanges. The use of equation 14.38 in which tangent rigidities vary along the element automatically allows for the non-uniformity of the inelastic element.

Typical variations of $(EI_T + EI_B)_t$, and $(GJ)_t$ with moment M_x are shown non-dimensionally in Figure 14.10 for the case of a beam $(P = 0)$. These remain at their elastic values until yielding commences at $M_x = 0.5 \, M_Y$, and then decrease towards their strain-hardened values as yielding extends through the cross-section. The initial rate of decrease of the flange flexural rigidities is fastest for the top flange, where the residual stresses cause early yielding at the flange tips, and consequent large reductions in $(EI_T)_t$. On the other hand, early yielding at the centre of the bottom flange has very little early effect on $(EI_B)_t$.

14.4.4 WORK DONE

The work done on the applied forces acting on a monosymmetric element $(x_0 = 0)$ can be determined from the second-order normal strain (equation 17.163)

$$\varepsilon_{PQ} = \frac{1}{2}(u'^2 + y_0^2\phi'^2) + y_0 u'\phi' + y(-y_0\phi'^2 + \phi u'') + \frac{1}{2}(x^2 + y^2)\phi'^2 \tag{14.41}$$

as

$$\frac{1}{2}\delta^2 V_e = \frac{1}{2}\int_0^L \int_0^A \sigma_P \{u'^2 + 2yu''\phi + 2y_0 u'\phi' + [x^2 + (y - y_0)^2]\phi'^2\} \, dA \, dz$$

$$+ \frac{1}{2}\int_0^L \{q(y_q - y_0)\phi^2\} \, dz \tag{14.42}$$

where σ_P is the longitudinal normal stress (tension positive). This equation is independent of the elastic moduli, and is valid for both elastic and inelastic

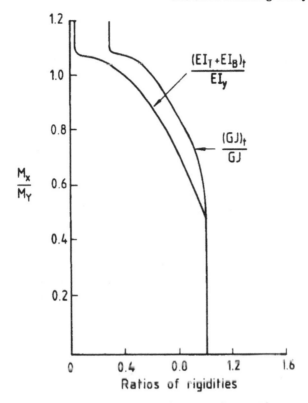

Figure 14.10 Variations of the inelastic rigidities with moment.

elements, since the work done by a set of forces acting on an element depends only on the forces and their displacements.

If the integrations across the section are carried out, then equation 14.41 becomes

$$\frac{1}{2}\delta^2 V_e = \frac{1}{2}\int_0^L \{\sigma_{11}(u'^2 + 2y_0 u'\phi' + y_0^2\phi'^2) + \sigma_{12}(2u''\phi - 2y_0\phi'^2) + (\sigma_{13} + \sigma_{14})\phi'^2$$

$$+ q(y_q - y_0)\phi^2\} dz \tag{14.43}$$

where $\sigma_{11}, \sigma_{12}, \sigma_{13}$, and σ_{14} are the stress resultants defined by equation 14.28.

For doubly symmetric sections ($y_0 = 0$), equation 14.42 can be expressed as

$$\frac{1}{2}\delta^2 V_e = \frac{1}{2}\int_0^L \{\varepsilon_v\}^T [D_v] \{\varepsilon_v\} dz \tag{14.44}$$

where

$$\{\varepsilon_v\}^T = \{u', u'', \phi, \phi'\} \tag{14.45}$$

and

$$[D_v] = \begin{bmatrix} \sigma_{11} & 0 & 0 & \sigma_{11}y_0 \\ 0 & 0 & \sigma_{12} & 0 \\ 0 & \sigma_{12} & q(y_q - y_0) & 0 \\ \sigma_{11}y_0 & 0 & 0 & (\sigma_{11}y_0^2 - 2\sigma_{12}y_0 + \sigma_{13} + \sigma_{14}) \end{bmatrix}. \quad (14.46)$$

14.4.5 SOLUTION METHOD

The element stiffness and stability matrices $[k_e]$, $[g_e]$ which correspond to the element strain energy $\frac{1}{2}\delta^2 U_e$ and work $\frac{1}{2}\delta^2 V_e$ may be obtained for a particular load set by using the results of the in-plane analysis in equations 14.38 and 14.44. These equations involve integrations along the length of the element, and if these are done by Gaussian integration [16, 17], then the quantities in equations 14.40 and 14.46 need only be evaluated at the Gauss points. It should be noted that the stress resultants $\sigma_{11}, \sigma_{12}, \sigma_{13}$, and σ_{14} are automatically obtained in the in-plane analysis.

The element matrices $[k_e]$, $[g_e]$ may be transformed and assembled to form global stiffness equations

$$([K] + \lambda[G])\{\Delta\} = \{0\} \quad (14.47)$$

in which G is an initial stability matrix corresponding to an initial load set, and λ is a load factor. In elastic buckling problems, $[K]$ and $[G]$ are independent of the load level λ, and need only be evaluated for the initial load set ($\lambda = 1$). In this case, equation 14.47 is a generalized linear eigenvalue equation, for which efficient methods have been developed for the extraction of the buckling load factors λ and the eigenvectors $\{\Delta\}$ which define the buckled shapes. These methods are referred to in Chapter 4.

However, in inelastic buckling problems, the stiffness and stability matrices $[K]$ and $\lambda[G]$ must be recalculated for each load level because of the changes in the inelastic section properties and the stress distributions, and so these eigenvalue extraction methods lose their efficiency. In general, a more reasonable computation procedure is to iterate through a series of load levels towards a solution. At each load level the in-plane analysis is performed, and the results are then used to establish the matrix $[K + \lambda G]$ and the value of its determinant is calculated. This is repeated for different load levels until a zero value for the determinant is found, which defines a buckling load set, and enables a corresponding buckling mode $\{\Delta\}$ to be determined. Some care must be taken with this method to ensure that the lowest buckling load is not missed. This can usually be done by steadily increasing the load level from first yield by reasonably small increments.

14.5 Inelastic buckling predictions

14.5.1 EFFECTS OF RESIDUAL STRESSES

The effects of residual stresses on the inelastic lateral buckling of steel beams
($\sigma_{i1} = 0$) may be conveniently studied by considering simply supported doubly
symmetric ($x_0 = y_0 = 0$) beams in uniform in-plane bending ($q = 0$), firstly because in this case the modifying effects of moment gradient and non-uniform
yielding are excluded, and secondly because the conditions at inelastic buckling
can be expressed by the comparatively simple equation

$$\{\sigma_{i2} - \pi^2(EI_T - EI_B)_t h/2L^2\}^2$$
$$= \frac{\pi^2(EI_T + EI_B)_t}{L^2}\left\{(GJ)_t + \frac{\pi^2(EI_T + EI_B)_t}{L^2}\frac{h^2}{4} + (\sigma_{i3} + \sigma_{i4})\right\} \qquad (14.48)$$

which is obtained from equations 14.38 and 14.44 by assuming that the buckled
shapes are half sine waves, as they are for elastic buckling. This equation is most
easily solved for the span length L for which inelastic buckling occurs at a
specified moment $M_1 = \sigma_{i2}$, since the tangent modulus rigidities and the stress
resultants σ_{i3}, σ_{i4} can be determined directly for this moment. The solutions are
conveniently displayed in plots of the type shown in Figure 14.11, in which the
dimensionless inelastic buckling resistance M_I/M_Y is plotted against a modified
slenderness $\sqrt{(M_Y/M_F)}$, in which $M_F = M_{yz}$ is the elastic buckling resistance, and

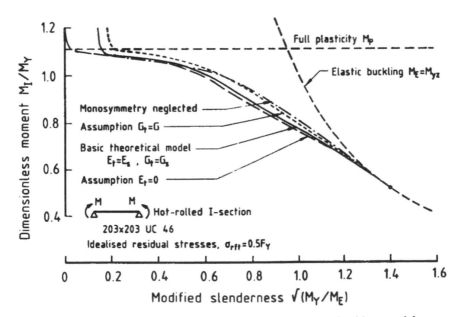

Figure 14.11 Effects of assumptions made for inelastic buckling model.

$M_Y = F_Y Z_x$ is the nominal first yield moment of the section. This method of plotting largely removes the influences of cross-section and yield stress.

The effects of the residual stresses, cross-section, yield stress, and tangent moduli assumptions have been studied by many investigators [19], and some of these are demonstrated in Figures 14.11 and 14.12. The influences of the assumptions made concerning the inelastic rigidities are illustrated in Figure 14.11 for a hot-rolled I-section beam with idealized residual stresses [20]. The curve labelled 'basic theoretical model' was calculated using the assumptions discussed earlier in section 14.2. It shows an almost linear increase in the inelastic buckling moment with decreasing slenderness, commencing at $M_I/M_Y = 0.5$ when the compression flange tips first yield, to $M_I/M_Y = 1.08$ when both flanges are nearly fully yielded. For $M_I/M_Y > 1.11$, the beam acts as if completely strain-hardened. The other curves of Figure 14.11 show that the assumption of $G_t = G$ in the yielded regions lead to moderate increases in the buckling moment, and that the assumption that $E_t = 0$ in the yielded regions leads to slight decreases. More important are the overestimates of buckling moment caused by neglecting the effects of monosymmetry (associated with $(EI_T - EI_B)_t$) in the yielded cross-section.

The effects of the magnitude and the shape of the residual stress distribution are demonstrated in Figure 14.12. The most important factor is the magnitude of the residual stress at the flange tips. High compressive residual stresses lead to early yielding at the top (compression) flange tips, with significant reductions in $(EI_T)_t$. It can be seen from the form of equation 14.48 that even moderate decreases in

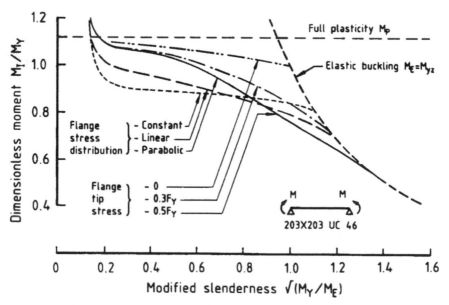

Figure 14.12 Effects of residual stress on inelastic buckling.

$(EI_T)_t$ will cause substantial decreases in M_t, since most terms are affected adversely. On the other hand, the small decreases that occur in $(EI_B)_t$ caused by tensile yielding in the bottom flange have very little effect on M_t because there are compensating monosymmetry terms.

Related explanations for the adverse effects of yielding at the tips of the top flange can be obtained by considering either the position of the centre of buckling rotation, or the effects of monosymmetry. The centre of buckling rotation is often close to the bottom flange, as shown in Figure 7.6, so that it makes only a small contribution to the buckling resistance. Thus the top flange is the more important, and so reductions in its rigidity will cause substantial reductions in the buckling resistance. Alternatively, it can be reasoned that the larger reductions in the top flange rigidity move the shear centre downwards and create an unfavourable monosymmetry effect, so that the reductions in the buckling resistance are accentuated, as shown in Figures 7.9 and 7.11.

Residual stresses may increase the resistance to inelastic buckling, as in the case of welded beams where flame-cutting of the flanges induces high residual tensile stresses at the flange tips. This is because the tensile residual stresses delay yielding of the top (compression) flange tips, thereby maintaining its flexural rigidity at higher moments. On the other hand, early yielding at the tips of the bottom (tension) flange has a smaller effect because the tension flange deflects less during buckling and makes a smaller contribution to the buckling resistance. The effect of the tensile residual stresses at the flange tips is therefore to change the sense of monosymmetry, causing the shear centre to move upwards, and increasing the buckling resistance.

Also important in the effects of residual stresses is their distribution across the flanges, as the more nearly constant is the residual stress, then the more dramatic is the decrease in the top flange flexural rigidity $(EI_T)_t$ once yielding commences. Thus, while the peak residual compressive stresses are often higher in hot-rolled beams than in welded beams (Figure 14.4), causing earlier reductions in the buckling strengths of intermediate slenderness beams, yet the more constant residual compressive stresses in welded beams cause greater reductions for low slenderness beams, as can be seen in Figure 14.12.

14.5.2 SIMPLY SUPPORTED BEAMS

14.5.2.1 General

Many theoretical studies [19] have shown that the effect of variations in the major axis moment distribution along a beam on its inelastic buckling resistance is very important. This is because yielding takes place only in the high moment regions, and so the consequent reductions in the section rigidities are localized. The inelastic buckling resistance depends markedly on the location and extent of these regions of reduced rigidity. The following sub-sections summarize some of these theoretical findings.

14.5.2.2 Moment gradients

Inelastic buckling predictions for hot-rolled beams with unequal end moments M, βM are shown in Figure 14.13 [21]. The effect of the end moment ratio β is very important, as the inelastic buckling moments of beams in uniform bending ($\beta = -1$) are reduced substantially below the elastic values, while those for beams with high moment gradient ($\beta \to 1$) are only slightly reduced, at least until the full plastic moment $M_P = F_Y S_x$ is reached. This is because a high moment gradient ensures that yielding is confined to short regions at the ends of the beam. In this case the central region, which is mostly responsible for buckling resistance, remains elastic, and there is only a small reduction in buckling strength.

Also shown in Figure 14.13 are simple approximations given by

$$\frac{M_1}{M_P} = 0.70 + \frac{0.30(1 - 0.70\,M_P/M_E)}{(0.61 - 0.30\beta + 0.07\beta^2)} \tag{14.49}$$

in which $M_E = \alpha_m M_{yz}$ is the elastic buckling moment for the same value of β (Chapter 7). These approximations, which are valid in the range $0.7 \leqslant M_1/M_P \leqslant 1.0$, are generally conservative, except for the small discrepancies which occur for uniform bending ($\beta = -1$). They can be used to determine approximate limiting modified slendernesses at which the inelastic buckling moment M_1 is equal to the full plastic moment M_P. These approximate limiting values are

$$\sqrt{(M_P/M_E)_P} = \sqrt{\{(0.39 + 0.30\beta - 0.07\beta^2)/0.70\}} \tag{14.50}$$

and vary from 0.17 to 0.94 as β varies from -1 to $+1$.

Experimental results for beams with central concentrated braced loads (which are equivalent to half-length beams with $\beta = 0$) are also shown in Figure 14.13 [22]. It can be seen that these are reasonably close to the theoretical predictions.

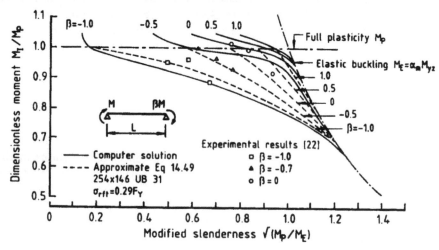

Figure 14.13 Hot-rolled beams with unequal end moments.

14.5.2.3 Transverse loads

Inelastic buckling predictions for beams with unbraced central concentrated loads Q acting at the geometrical axis are shown by the curve $\gamma = 0$ in Figure 14.14 [21]. These predictions are only a little higher than those for uniform bending ($\beta = -1$), because while yielding is confined to the mid-span region, it is here that the reductions in the rigidities have their greatest effect. A reasonably accurate approximation for the inelastic buckling moment

$$M_1 = QL/4 \tag{14.51}$$

is obtained by using $\beta = -0.70$ in equation 14.49.

Predictions have also been made for beams with uniformly distributed load, or with two loads, each placed a distance a off-centre. In each case, the maximum moment at inelastic buckling may be approximated by using equation 14.49 with $\beta = -0.90$ for uniformly distributed load, or

$$\beta = -0.7 - 0.6a/L \tag{14.52}$$

for off-centre loading.

It has been suggested [21] that the effects of transverse loads acting away from the beam axis could also be approximated by using equation 14.49, provided that the moment M_E used in equation 14.49 includes the effect of load height on elastic buckling (section 7.6). Some limited experimental evidence for this is shown in Figure 14.14 [23].

14.5.2.4 End moments and transverse loads

The effects of equal in-plane end moments $\gamma QL/8$ on the inelastic buckling of hot-rolled beams with central concentrated loads Q are shown in Figure 14.14

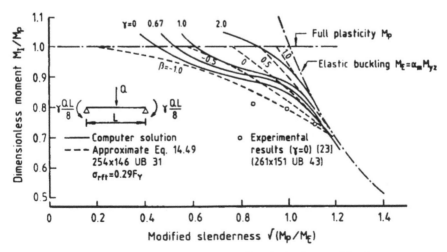

Figure 14.14 Hot-rolled beams with unbraced central loads.

[21]. For high end moments ($\gamma \rightarrow 2.0$), the behaviour is similar to that of beams with high moment gradient ($\beta \rightarrow 0.5$), since yielding occurs only at the ends where it is relatively unimportant. These loading arrangements may be regarded as equivalent unequal end moments, so that equation 14.49 may be used with

$$\beta = -0.7 + 0.3\gamma^2. \tag{14.53}$$

The effects of unequal end moments $\gamma_L QL/8$ and $\gamma_R QL/8$ on the inelastic buckling of hot-rolled beams with central concentrated loads Q were investigated in [24]. The results of this study are summarized by the equivalent end moment ratios β shown in Figure 14.15.

14.5.3 CANTILEVERS

The inelastic buckling of hot-rolled cantilevers with concentrated end loads or uniformly distributed loads has been investigated in [25]. This study includes the effect of the height of the loading relative to the geometrical axis of the cantilever. In all cases, yielding was confined to the support region, and caused substantial reductions from the elastic buckling resistance which were similar to those for simply supported beams with central concentrated loads.

14.5.4 DETERMINATE BRACED BEAMS

Approximate methods have been developed for analysing the inelastic buckling of determinate beams which are prevented from deflecting and twisting at their

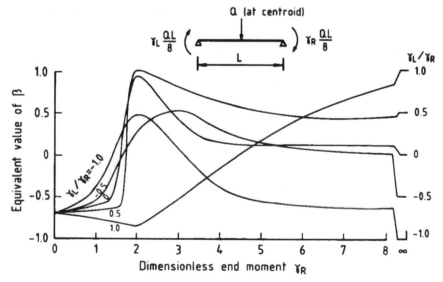

Figure 14.15 Equivalent β values for hot-rolled beams.

supports and brace points. The simplest method is to ignore the buckling interactions which take place between adjacent segments during buckling, which are similar to those which occur in elastic braced beams (Chapter 10). A lower bound may then be obtained from the lowest of the inelastic buckling load factors calculated for the segments of the beam. The information given in section 14.5.2 may be used to assess these load factors. The accuracy of this method is demonstrated in Figure 14.13 by the braced beam experimental results of [22] for $\beta = -1.0, -0.7$.

When a more accurate prediction is required, then the effects of interactions between adjacent segments may be approximated by using an extension [21, 24] of the elastic method described in Chapter 10. The accuracy of the extension is demonstrated in Figure 14.16. A more elaborate extension is developed in [26].

14.5.5 CONTINUOUS BEAMS

A continuous beam is statically indeterminate, and as yielding progresses in the beam, its in-plane bending moments are redistributed from the elastic distribution towards that of the plastic collapse mechanism. (However, this redistribution may not be as rapid as is suggested by simple plastic theory because of the effects of strain-hardening.) The redistribution allows the loads to increase to levels which are often significantly higher than the nominal loads which cause first yield in beams without residual stresses. Because of this, there are usually several areas of the beam with appreciable yielding. The changes in both the moment and yield region distributions may be favourable or unfavourable with respect to the inelastic buckling resistance.

Theoretical predictions of the inelastic buckling loads of two-span continuous

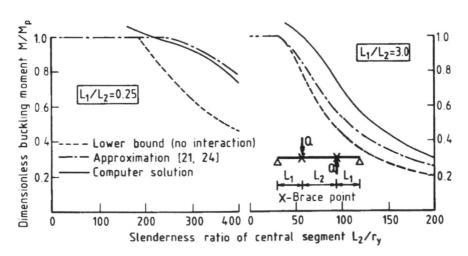

Figure 14.16 Inelastic buckling of braced beams.

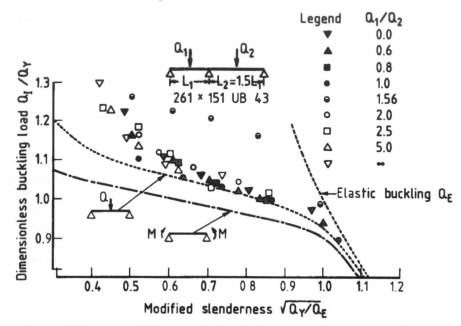

Figure 14.17 Inelastic buckling predictions for continuous beams.

hot-rolled beams are shown in Figure 14.17 [27], together with the corresponding predictions for simply supported beams with central concentrated loads or uniform bending. All of the plotted points (with the exception of those for the load ratio $Q_1/Q_2 = 1.56$) are very closely grouped, and for modified slendernesses $\sqrt{(Q_Y/Q_E)} > 0.6$ are quite close to the curve for simply supported beams with central concentrated loads. This is because all of these continuous beams yield first at one of the load points at mid-span. For decreasing slenderness, the plotted points rise more rapidly than the curve for simply supported beams, principally because yielding spreads more slowly from the load points as a result of the higher moment gradients, particularly towards the interior support.

For continuous beams with load ratios of $Q_1/Q_2 = 1.56$, yielding occurs first at the interior support, where it causes comparatively small reductions from the elastic buckling resistance, in much the same way as do the end moments of simple beams with high moment gradient (Figure 14.13). Thus the inelastic buckling resistances of these beams are noticeably higher than those of the other continuous beams shown in Figure 14.17.

Some of the theoretical predictions for two-span continuous beams (for $L_1 = 2.44\,\text{m}$) are also shown in Figure 14.18, together with the experimental results of [28]. The experimental results are surprisingly low, even though the corresponding results shown in Figure 14.14 for simple beams of the same cross-section [23] are in reasonable agreement with theory. No satisfactory explanation has yet been advanced for these discrepancies.

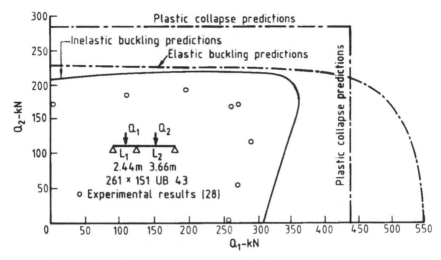

Figure 14.18 Experimental maximum loads for continuous beams.

Approximate theoretical predictions for the inelastic buckling of continuous beams with central concentrated loads may be made by using the methods for determinate braced beams described in section 14.5.4, provided that the in-plane moment distribution can be determined. It is suggested that two approximations should be obtained, one for the elastic moment distribution, and one for the distribution at plastic collapse. If necessary, the final approximation may be determined by interpolation.

14.5.6 BEAM-COLUMNS

14.5.6.1 Isolated braced beam-columns

The inelastic buckling of isolated hot-rolled beam-columns with unequal end moments M, βM was studied theoretically in [13]. The predictions were compared with approximations obtained from the linear interaction equation

$$\frac{P}{P_1} + \frac{C_m}{(1 - P/P_x)}\frac{M}{M_{10}} \leqslant 1 \tag{14.54}$$

with

$$C_m = 0.6 - 0.4\beta \geqslant 0.4. \tag{14.55}$$

In these equations, P_1 is the inelastic buckling load of a simply supported column, and M_{10} is the uniform inelastic buckling moment of a simply supported beam. Approximations for these were developed from theoretical predictions for a wide range of hot-rolled I-sections as

$$P_1/P_Y = 1.035 - 0.181\sqrt{(P_Y/P_y)} - 0.128\,P_Y/P_y \leqslant 1.0, \tag{14.56}$$

$$M_{10}/M_P = 1.008 - 0.245\, M_P/M_{yz} \leqslant 1.0 \tag{14.57}$$

in which P_y, P_Y are the elastic flexural buckling and squash loads of the column, and M_{yz}, M_P are the uniform elastic buckling and full plastic moments of the beam. Equation 14.57 yields slightly different results from those obtained from equation 14.49 for $\beta = -1.0$.

It was found that these approximations were generally conservative, and especially so for high moment gradients ($\beta \geqslant 0.5$). The reasons for this conservatism were attributed to the use of a linear interaction equation instead of a parabolic one (see equation 11.13), and to the use of a C_m factor which was independent of the axial load P (see equation 11.31).

Because of this it was decided to use a modification of the elastic parabolic interaction equation (equation 11.30) to

$$\left(\frac{M}{C_{bc}M_{10}}\right)^2 = \left(1 - \frac{P}{P_1}\right)\left(1 - \frac{P}{P_z}\right) \tag{14.58}$$

where

$$\frac{1}{C_{bc}} \approx \left(\frac{1-\beta}{2}\right) + \left(0.40 - 0.23\frac{P}{P_1}\right)\left(\frac{1+\beta}{2}\right)^3. \tag{14.59}$$

These equations proved to be of high accuracy, as is demonstrated in Figure 14.19.

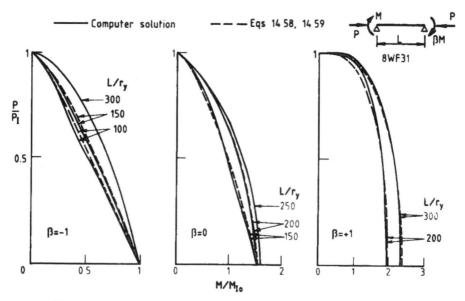

Figure 14.19 Improved interaction equation for inelastic buckling.

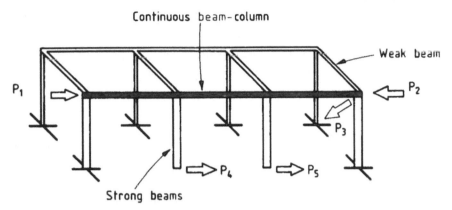

Figure 14.20 Arrangement of a three-span beam-column.

14.5.6.2 Continuous beam-columns

The inelastic buckling of beam-columns continuous over three spans has been studied experimentally [29] and theoretically [30, 31]. The beam-columns were of hot-rolled I-section and were loaded by the end forces P_1, P_2, the brace force P_3, and concentrated in-plane moments developed by the forces P_4, P_5 shown in Figure 14.20. These actions caused significant in-plane yielding of the continuous beam-column, reducing its resistance to out-of-plane buckling. Because of this, restraining actions developed by the weak out-of-plane beams played more important roles in determining the beam-column strength.

The purpose of the experiments was to obtain data which could be used to test inelastic buckling theories. The values of the ratios P_T/P_F of the theoretical predictions obtained from [30] with the experimental failure results are shown in Figure 14.21, which indicates extremely close agreement.

14.6 Problems

PROBLEM 14.1

A simply supported I-section beam whose properties are given in Figure 7.23 has a span of 6.0 m and equal and opposite end moments, as shown in Figure 14.22a, and a yield stress of 250 MPa. Determine the inelastic buckling moment.

PROBLEM 14.2

A simply supported I-section beam whose properties are given in Figure 7.23 has a span of 10.0 m and unequal end moments M and $0.4 M$ which causes double

Load Set	Nominal Load Configuration	Specimen Number	$(P_5/P_1)_n$	(P_T/P_F)
1	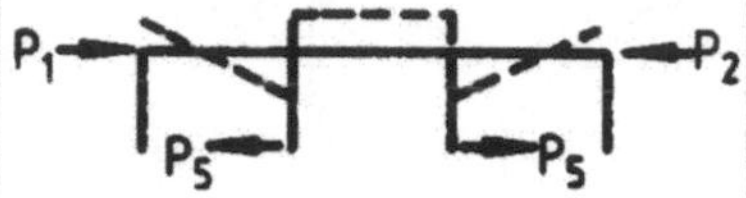 P_1 ... P_2 ; P_S, P_S	1 2	0.082 0.221	0.96 0.96
2	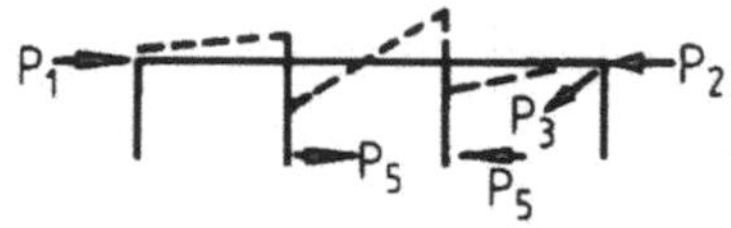 P_1 ... P_2 ; P_3, P_S, P_S	3 5A	0.066 0.032	0.95 1.03
3	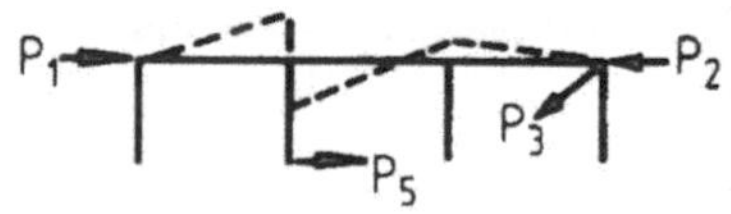 P_1 ... P_2 ; P_3, P_S	4 6A 7A	0.116 0.045 0.045	1.00 1.00 0.99
4	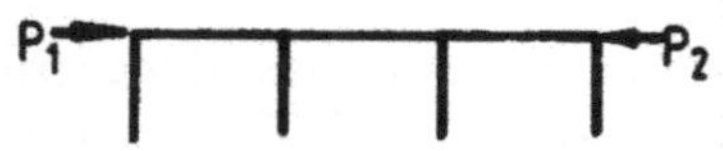 P_1 ... P_2	3A 4A 5	0 0 0	1.06 1.00 0.98
5	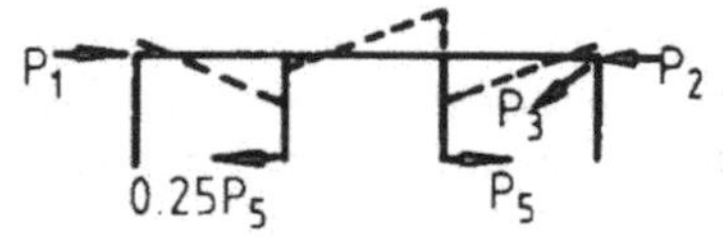 P_1 ... P_2 ; P_3, $0.25P_S$, P_S	6 7	0.122 0.048	1.01 1.03
6	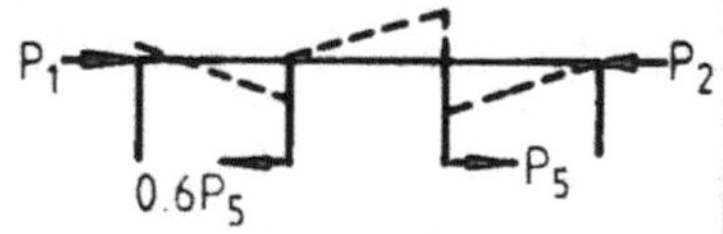 P_1 ... P_2 ; $0.6P_S$, P_S	8 9	0.055 0.136	1.02 0.95

Note: 'A' indicates specimen previously tested to failure

Figure 14.21 Comparison of continuous beam-column failure loads.

curvature bending, as shown in Figure 14.22b, and a yield stress of 250 MPa. Determine the value of M at inelastic buckling.

PROBLEM 14.3

A simply supported I-section beam whose properties are given in Figure 7.23 has a concentrated load at the centre of its 20.0 m span where bracing prevents lateral deflection and twist, as shown in Figure 14.22c, and a yield stress of 250 MPa. Determine the maximum moment at inelastic buckling.

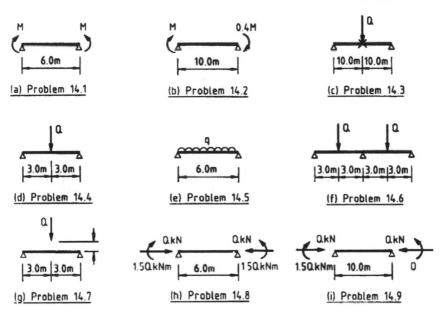

Figure 14.22 Problems 14.1–14.9.

PROBLEM 14.4

A simply supported I-section beam whose properties are given in Figure 7.23 has a concentrated load at the shear centre of its 6.0 m span which is unrestrained, as shown in Figure 14.22d, and a yield stress of 250 MPa. Determine the maximum moment at inelastic buckling.

PROBLEM 14.5

A simply supported I-section beam whose properties are given in Figure 7.23 has uniformly distributed shear centre loading, as shown in Figure 14.22e, and a yield stress of 250 MPa. Determine the maximum moment at inelastic buckling.

PROBLEM 14.6

A two-span continuous I-section beam whose properties are given in Figure 7.23 has equal concentrated loads at the shear centre at the centre of both 6.0 m spans which are unrestrained, as shown in Figure 14.22f, and a yield stress of 250 MPa. Determine the maximum moment at inelastic buckling

(a) by assuming an elastic distribution of the in-plane bending moment;
(b) by assuming a plastic collapse mechanism distribution of the in-plane bending moment.

PROBLEM 14.7

A simply supported I-section beam whose properties are given in Figure 7.23 has a top flange concentrated load at the centre of its 6.0 m span which is unrestrained, as shown in Figure 14.22g, and a yield stress of 250 MPa. Determine the maximum moment at inelastic buckling.

PROBLEM 14.8

A simply supported I-section beam-column whose properties are given in Figure 7.23 has a span of 6.0 m, axial compressions Q kN, and equal opposite end moments $1.5Q$ kN m, as shown in Figure 14.22h, and a yield stress of 250 MPa. Determine the value of Q at inelastic buckling.

PROBLEM 14.9

A simply supported I-section beam-column whose properties are given in Figure 7.23 has a span of 10.0 m, axial compressions Q kN, and end moments 1.5 Q kN m and 0 kN m, as shown in Figure 14.22i, and a yield stress of 250 MPa. Determine the value of Q at inelastic buckling.

14.7 References

1. Galambos, T.V. (ed.) (1988) *Guide to Stability Design Criteria for Metal Structures*, 4th edn, John Wiley and Sons, New York.
2. Trahair, N.S. and Bradford, M.A. (1991) *The Behaviour and Design of Steel Structures*, revised 2nd edn, Chapman and Hall, London
3. Shanley, F.R. (1947) Inelastic column theory. *Journal of Aeronautical Sciences*, **14**(5), 261–8.
4. Lay, M.G. (1965) Yielding of uniformly loaded steel members. *Journal of the Structural Division, ASCE*, **91**(ST6), 49–66.
5. Haaijer, G. (1957) Plate buckling in the strain-hardening range. *Journal of the Engineering Mechanics Division, ASCE*, **83**(EM2), 1212.1–47.
6. Massey, P.C. (1963) The torsional rigidity of steel I-beams. *Civil Engineering and Public Works Review*, **58**(680), 367–71; **58**(681), 488–92.
7. Lay M.G. (1965) Flange local buckling in wide-flange shapes. *Journal of the Structural Division, ASCE*, **91**(ST6), 95–116.
8. El-Zanaty, M.H. and Murray, D.W. (1983) Nonlinear finite element analysis of steel frames. *Journal of Structural Engineering, ASCE*, **109**(2), 353–68.
9. Lay, M.G. and Ward, R. (1969) Residual stresses in steel structures. *Steel Construction*, **3**(3), 2–21.
10. Young, B.W. (1975) Residual stresses in hot-rolled sections, in *Proceedings*, International Colloquium on Column Strength, IABSE, Vol. 23, pp. 25–38.
11. Lee, G.C., Fine, E.S. and Hastreiter, W.R. (1967) Inelastic torsional buckling of H-columns. *Journal of the Structural Division, ASCE*, **93**(ST5), 295–307.

12. Fukumoto, Y., Itoh, Y. and Kubo, M. (1980) Strength variation of laterally unsupported beams. *Journal of the Structural Division, ASCE*, 106(ST1), 165–81.
13. Bradford, M.A. and Trahair, N.S. (1985) Inelastic buckling of beam-columns with unequal end moments. *Journal of Constructional Steel Research*, 5(3), 195–212.
14. Dwight, J.B. and White, J.D. (1977) Prediction of weld shrinkage stresses in plated structures. *Preliminary Report*, 2nd International Colloquium on Stability of Steel Structures, ECCS-IABSE, Liege, pp. 31–7.
15. Fukumoto, Y. and Itoh, Y. (1981) Statistical study of experiments on welded beams. *Journal of the Structural Division, ASCE*, 107(ST1), 89–103.
16. Bathe, K.-J. (1982) *Finite Element Procedures in Engineering Analysis*, Prentice-Hall, Englewood Cliffs, NJ.
17. Zienkiewicz, O.C., and Taylor, R.L. (1989) *The Finite Element Method, Volume 1 – Basic Formulation and Linear Problems*; (1991) *Volume 2 – Solid Mechanics, Dynamics, and Non-Linearity*, 4th edn, McGraw-Hill, New York.
18. Kitipornchai, S., and Trahair, N.S. (1975) Buckling of Inelastic I-Beams Under Moment Gradient', *Journal of the Structural Division, ASCE*, 105(ST5), 991–1004.
19. Trahair, N.S. (1983) Inelastic lateral buckling of beams, in *Beams and Beam Columns* (ed. R. Narayanan), Applied Science Publishers, Barking, pp 35–69.
20. Trahair, N.S. and Kitipornchai, S. (1972) Buckling of inelastic I-beams under uniform moment. *Journal of the Structural Division, ASCE*, 98(ST11), 2551–66.
21. Nethercot, D.A. and Trahair, N.S. (1976) Inelastic lateral buckling of determinate beams. *Journal of the Structural Division, ASCE*, 102(ST4), 701–17.
22. Dux, P.F. and Kitipornchai, S. (1983) Inelastic beam buckling experiments. *Journal of Constructional Steel Research*, 3(1), 3–9.
23. Kitipornchai, S. and Trahair, N.S. (1975) Inelastic buckling of simply supported steel I-beams. *Journal of the Structural Division, ASCE*, 101(ST7), 1333–47.
24. Nethercot, D.A. and Trahair, N.S. (1977) Lateral buckling calculations for braced beams, *Civil Engineering Transactions*, Institution of Engineers, Australia, CE19(2), 211–4.
25. Nethercot, D.A. (1975) Inelastic buckling of steel beams under non-uniform moment. *The Structural Engineer*, 53(2), 73–8.
26. Dux, P.F. and Kitipornchai, S. (1984) Buckling approximations for inelastic beams. *Journal of Structural Engineering, ASCE*, 110(ST3), 559–74.
27. Yoshida, H., Nethercot, D.A. and Trahair, N.S. (1977) Analysis of inelastic buckling of continuous beams. *Proceedings*, IABSE, No. P-3/77, pp. 1-14.
28. Poowannachaikul, T. and Trahair, N.S. (1976) Inelastic buckling of continuous steel I-beams. *Civil Engineering Transactions*, Institution of Engineers, Australia, CE18(2), 134–9.
29. Cuk, P.E., Rogers, D.F. and Trahair, N.S. (1986) Inelastic buckling of continuous steel beam-columns. *Journal of Constructional Steel Research*, 6(1), 21–52.
30. Bradford, M.A., Cuk, P.E., Gizejowski, M.A. and Trahair, N.S. (1987) Inelastic lateral buckling of beam-columns. *Journal of Structural Engineering, ASCE*, 113(11), 2259–77.
31. Bradford, M.A. and Trahair, N.S. (1986) Inelastic buckling tests on beam-columns. *Journal of Structural Engineering, ASCE*, 112(3) 539–49.

15 Strength and design of steel members

15.1 General

Steel members which are loaded in a principal plane may fail in that plane as a result of excessive yielding or local buckling, or by excessive bending and twisting out of the plane of loading caused by flexural-torsional buckling effects.

The out-of-plane strength M_u is significantly influenced by elastic or inelastic buckling, but is reduced below the buckling strength M_E or M_I as a result of geometrical imperfections such as initial crookedness or twist, as shown in Figure 15.1. Perfectly straight members do not deflect laterally or twist until the elastic or inelastic buckling load is reached, and failure occurs then or soon after. On the other hand, an elastic member with realistic initial crookedness or twist deflects non-linearly, with the deflection asymptoting towards the elastic buckling load. This elastic behaviour continues until the first yield occurs, as a result of the combination of stresses due to in-plane bending and compression, residual stresses, and out-of-plane deflection and twist. Thereafter, a lower load–deformation curve is followed until a maximum moment M_u is reached which is equal to the out-of-plane strength.

Designers may avoid flexural-torsional buckling completely, either by using sections which are not susceptible to this form of buckling, or by providing sufficient bracing to prevent buckling. In these cases, the member's design capacity will be governed by in-plane considerations, such as excessive plasticity or local buckling.

Designers who do not avoid flexural-torsional buckling altogether must design by estimating the member's capacity to resist this mode of failure, and by ensuring that this exceeds the design action.

The out-of-plane strengths of beams which are susceptible to flexural-torsional buckling failure are considered in section 15.2, while code rules for designing against the flexural-torsional buckling of beams are discussed in sections 15.3 and 15.4. The strengths and design of columns which fail in a flexural-torsional mode are considered in section 15.5, and of beam-columns in section 15.6.

15.2 Beam strength

15.2.1 ELASTIC BUCKLING

The elastic flexural-torsional buckling behaviour of straight beams is discussed in Chapter 7 (simply supported beams), Chapter 8 (restrained beams), Chapter 9

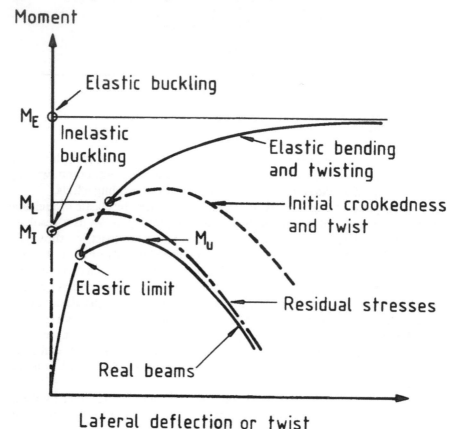

Figure 15.1 Behaviour of real beams.

(cantilevers), and Chapter 10 (braced and continuous beams). The resistance to elastic buckling depends principally on:

(a) the beam rigidities EI_y, GJ, and EI_w;
(b) the beam span or segment length between braces L;
(c) the bending moment distribution M_x;
(d) the load height y_Q;
(e) the restraints against lateral deflection u, lateral rotation du/dz, twist rotation ϕ, and warping proportional to $d\phi/dz$.

The maximum moment M_m in a beam segment at elastic flexural-torsional buckling provides an upper bound to the beam strength.

15.2.2 INELASTIC BUCKLING

A straight steel beam of intermediate slenderness fails before its elastic buckling load is reached, as a result of premature yielding caused by the in-plane actions and the residual stresses induced in it by the method of manufacture. The influence of residual stresses on inelastic buckling is discussed in section 14.5.1. For simply supported beams in uniform bending which are of hot-rolled I-section, the inelastic buckling moment M_1 is approximated by (see also equation 14.49)

$$M_1/M_P = 0.70 + 0.30(1 - 0.70M_P/M_{yz})/0.98 \qquad (15.1)$$

according to which M_1 decreases from M_P to $0.7M_P$ as the modified slenderness $\sqrt{(M_P/M_{yz})}$ increases from 0.169 to 1.195, as shown in Figure 14.13 by the curve for $\beta = -1$. At $\sqrt{(M_P/M_{yz})} = 1.195$, the approximate inelastic buckling moment M_1 is equal to the elastic buckling moment M_{yz}, and so elastic buckling at M_{yz} controls for higher slendernesses.

It was shown in section 14.5.2 that the effect of the bending moment distribution on inelastic bending is very important. Beams in uniform bending have their resistances reduced substantially below their elastic buckling resistances because yielding takes place uniformly along the beam. On the other hand, yielding is confined to the end regions of beams in double curvature bending, and the reductions from their elastic buckling resistances are much smaller, as shown in Figure 14.13 by the curve for $\beta = 1$.

For simply supported beams with unequal end moments M and βM, the inelastic buckling moment M_1 was approximated by

$$\frac{M_1}{M_P} = 0.70 + \frac{0.30(1 - 0.70M_P/M_E)}{(0.61 - 0.30\beta + 0.07\beta^2)} \qquad (15.2)$$

in which

$$M_E = \alpha_m M_{yz} \qquad (15.3)$$

is the elastic buckling moment and

$$\alpha_m = 1.75 + 1.05\beta + 0.3\beta^2 \leqslant 2.5 \qquad (15.4)$$

is the moment modification factor of section 7.3. These approximations, which are valid in the range $0.7 < M_1/M_P < 1.0$, are shown in Figure 14.13.

15.2.3 BENDING AND TWISTING OF CROOKED BEAMS

Real beams are not perfectly straight, but have small initial crookednesses and twists which cause them to deflect laterally and twist at the beginning of loading. If an elastic simply supported beam with equal and opposite end moments M has initial crookedness and twist given by

$$\frac{u_0}{\delta_0} = \frac{\phi_0}{\theta_0} = \sin\frac{\pi z}{L} \qquad (15.5)$$

in which δ_0 is the central crookedness and θ_0 is the central twist, then its deformation can be analysed by considering the minor axis bending and torsion equations [1, 2].

$$EI_y\frac{d^2u}{dz^2} = -M(\phi + \phi_0) \qquad (15.6)$$

and

$$EI_w \frac{d^3\phi}{dz^3} - GJ\frac{d\phi}{dz} = -M\left(\frac{du}{dz} + \frac{du_0}{dz}\right). \tag{15.7}$$

If the central crookedness δ_0 and central twist θ_0 are related by

$$\delta_0/\theta_0 = M_{yz}/P_y \tag{15.8}$$

in which

$$M_{yz} = \sqrt{\{(\pi^2 EI_y/L^2)(GJ + \pi^2 EI_w/L^2)\}} \tag{15.9}$$

is the elastic beam buckling moment, and

$$P_y = \pi^2 EI_y/L^2 \tag{15.10}$$

is the elastic column flexural buckling load, then the solution of equations 15.6 and 15.7 which satisfies the simply supported boundary conditions of

$$\left. \begin{array}{r} u_0 = u_L = 0, \\ \phi_0 = \phi_L = 0, \\ (d^2\phi/dz^2)_0 = (d^2\phi/dz^2)_L = 0 \end{array} \right\} \tag{15.11}$$

is given by

$$\frac{u}{\delta} = \frac{\phi}{\theta} = \sin\frac{\pi z}{L}, \tag{15.12}$$

in which

$$\frac{\delta}{\delta_0} = \frac{\theta}{\theta_0} = \frac{M/M_{yz}}{-M/M_{yz}}. \tag{15.13}$$

The variations of the dimensionless central deflection δ/δ_0 and twist θ/θ_0 are shown in Figure 15.2, and it can be seen that deformations begin at the commencement of loading, and increase rapidly as the elastic buckling moment M_{yz} is approached.

The bending and twisting of beams with central concentrated loads which have either initial crookedness and twist or are loaded eccentrically to the plane of the web have been analysed in [1], and similar results obtained to those above for beams in uniform bending. The results shown in Figure 15.2 can therefore be expected to be indicative of a wide range of situations.

As the deformations increase with the applied moments M, so do the stresses. The maximum longitudinal stress in the beam in the absence of residual stresses is the sum of the stresses due to in-plane bending, out-of-plane bending, and warping, and is equal to

$$\sigma_m = \frac{M}{Z_x} - \frac{EI_y}{Z_y}\left(\frac{d^2(u + h\phi/2)}{dz^2}\right)_{L/2}. \tag{15.14}$$

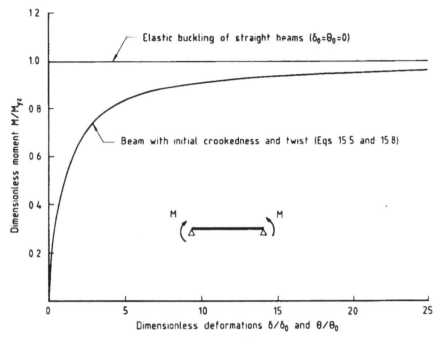

Figure 15.2 Deformations of beams with initial crookedness and twist.

The limiting moment

$$M_L = \sigma_L Z_x \qquad (15.15)$$

at which the maximum stress reaches first yield so that

$$\sigma_m = F_Y \qquad (15.16)$$

is given by

$$M_L = M_Y - \frac{P_y \delta_0}{M_{yz}}\left(1 + \frac{P_y h/2}{M_{yz}}\right)\frac{Z_x}{Z_y}\frac{M_L}{(1 - M_L/M_{yz})} \qquad (15.17)$$

in which

$$M_Y = F_Y Z_x. \qquad (15.18)$$

When the central crookedness and twist are given by

$$\frac{P_y \delta_0}{M_{yz}} = \theta_0 = \frac{Z_y/Z_x}{(1 + hP_y/2M_{yz})}\frac{M_Y}{4M_{yz}} \qquad (15.19)$$

then equation 15.17 can be solved for the dimensionless limiting moment

$$\frac{M_L}{M_Y} = \left(\frac{1.25 + M_{yz}/M_Y}{2}\right) - \sqrt{\left\{\left(\frac{1.25 + M_{yz}/M_Y}{2}\right)^2 - \frac{M_{yz}}{M_Y}\right\}}. \qquad (15.20)$$

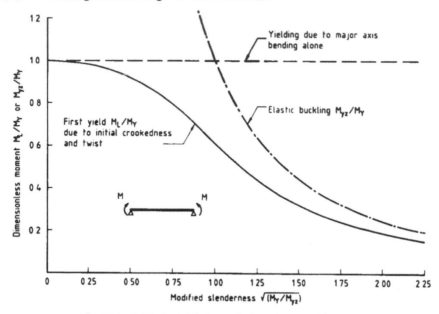

Figure 15.3 First yield of crooked and twisted beams.

The variation of the dimensionless limiting moment M_L/M_Y is shown in Figure 15.3, in which the quantity $\sqrt{(M_Y/M_{yz})}$ plotted on the horizontal axis is a modified slenderness of the beam. It can be seen that the limiting moments of short beams approach the nominal yield moment M_Y, and underestimate the capacities of real beams which can be expected to reach the full plastic moment M_P, provided local buckling does not cause premature failure. While the limiting moments M_L of long beams shown in Figure 15.3 approach the elastic buckling moment M_{yz}, the values for beams of intermediate slenderness may overestimate the strengths of real beams, whose residual stresses cause premature yielding.

15.2.4 EXPERIMENTAL STUDIES

More than 450 flexural-torsional buckling tests of commercial hot-rolled and welded steel I-beams have been reported in six studies reviewed in [3]. The results of some of these tests (on beams in uniform or near uniform bending) are shown in Figures 15.4 and 15.5, where variations in the elastic range due to beam geometry, support and loading conditions, and restraints have been eliminated by plotting the non-dimensional maximum moment M_u/M_P against the modified slenderness $\sqrt{(M_P/M_E)}$. There is considerable scatter in each figure, demonstrating not only the effects of variable initial crookedness and twist and load eccentricity on strength, but also those of variable residual stress, moment distribution and lateral continuity between adjacent segments on inelastic buckling.

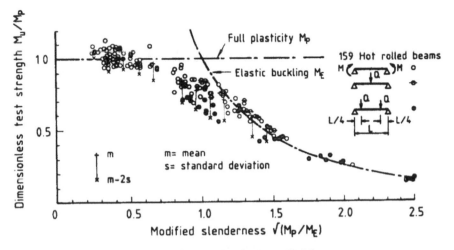

Figure 15.4 Test results for hot-rolled beams.

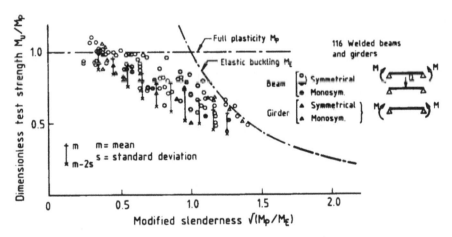

Figure 15.5 Test results for welded beams and girders.

Points representing the mean minus two standard deviations of the hot-rolled test results of Figure 15.4 are shown in Figure 15.6. Also shown in Figure 15.6 is the uniform bending inelastic buckling approximation of equation 15.1, and a modification of equation 15.20 for the first yield of crooked and twisted beams, for which the first yield moment M_Y has been replaced by the full plastic moment M_p.

It can be seen that neither the inelastic buckling approximation nor the modified first yield curve is completely successful in representing the lower bounds of the test results. The modified first yield curve is a little low for high and

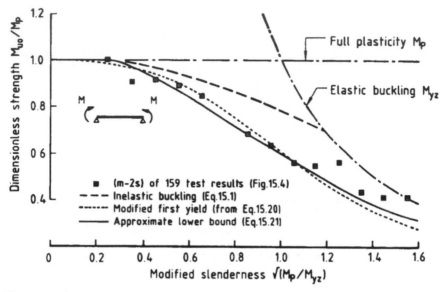

Figure 15.6 Comparison of strength approximations for hot-rolled beams in bending.

low modified slendernesses $\sqrt{(M_P/M_{yz})}$, and the inelastic buckling curve is too high for $\sqrt{(M_P/M_{yz})} > 0.4$. On the other hand, a close fit to the lower bounds is obtained by using

$$\frac{M_{u0}}{M_P} = 0.6\left\{\sqrt{\left[\left(\frac{M_P}{M_{yz}}\right)^2 + 3\right]} - \frac{M_P}{M_{yz}}\right\} \leqslant 1.0 \qquad (15.21)$$

as shown in Figure 15.6.

15.3 Working stress design of steel beams

In the working (or allowable) stress method of designing steel beams, the inequality

$$\sigma \leqslant \sigma_a \qquad (15.22)$$

must be satisfied, in which σ is the maximum stress determined by analysing the structure for its working loads, and σ_a is the maximum allowable stress given in the code. The values given for the allowable stress σ_a are usually obtained by dividing the nominal maximum stress at failure $\sigma_u = M_u/Z_x$ by a factor of safety, which is often taken as 1/0.6.

Working stress design methods are somewhat illogical, in that design against failure is carried out at loads approximately equal to 60% of those which cause failure, and are being replaced by the limit states design methods discussed in sections 15.4–15.6 following. The working stress design of steel structures according to the American [4], British [5], and Australian [6] design codes is discussed in detail in [7].

15.4 Limit states strength design of steel beams

15.4.1 GENERAL

The limit states (or load and resistance factor) strength design of steel beams [2, 8, 9] requires the inequality

$$M^* \leqslant \phi M_u \tag{15.23}$$

to be satisfied for all moment failure states and under all loading sets. In this equation, M^* is the design moment, M_u is the nominal moment capacity, and ϕ is the capacity factor.

The design moment M^* is determined from the analysis of the beam under the relevant design load combination obtained from the nominal gravity, live, wind and other loads multiplied by appropriate load factors which take into account the possible overload situations that may occur. The nominal moment capacities M_u are specified in the design codes for the various failure states which may occur, including failure by in-plane plasticity, local buckling or flexural-torsional buckling, and represent the compromises made by the code writers between the conflicting requirements of accuracy and simplicity. The capacity factor ϕ (which is often close to 0.9) is chosen by the code writers so that it will lead to a satisfactory level of safety when used with M^* and M_u in equation 15.23. It can be thought of as taking account of the possibility that the structure may be understrength.

The moment failure states that must be considered for the limit states strength design of steel beams include yielding and local buckling of the cross-section, and plastic collapse or flexural-torsional buckling of the beam as a whole.

For the cross-section limit states, the design moment M^* is the value for the cross-section under consideration, while the nominal moment capacity M_s depends on both yielding and local buckling considerations. Beams whose compression flanges have low width–thickness ratios do not buckle locally, and are often described as compact if they can reach the full plastic moment M_P, which is then taken as the section capacity. Beams whose compression flanges have high width–thickness ratios buckle locally, and are described as slender. Their moment capacities are reduced below the nominal first yield moment M_Y. Beams whose flanges have intermediate width–thickness ratios may be described as non-compact, and their capacities are often interpolated between M_P and M_Y.

The member limit state of plastic collapse is restricted to beams which do not fail prematurely by local or flexural-torsional buckling, and which can reach and sustain the full plastic moment at the early-forming hinge locations. The member limit states for other beams are taken as those of section failure due to yielding or local buckling, provided there is sufficient bracing to prevent premature failure by flexural-torsional buckling.

For the flexural-torsional buckling limit state, the design moment M^* is taken as the maximum moment in the beam or beam segment under consideration, while the nominal member capacity M_b depends on the beam slenderness, the bending moment distribution, the load height above the shear centre, and the restraints. The formulations given by a number of design codes [10–14] for the effects of these on the nominal member capacity are discussed in the following sub-sections.

15.4.2 SLENDERNESS

The effects of slenderness on the nominal moment capacities M_{bu} of compact beams in uniform bending according to different design codes [10–14] are compared in Figure 15.7 with the lower bound approximation of equation 15.21 to the test results shown in Figures 15.4 and 15.6.

The Canadian nominal moment capacity [10] is given by

$$\left. \begin{aligned} 0 \leqslant \sqrt{(M_P/M_{yz})} < 0.683, \quad & M_{bu}/M_P = 1.0, \\ 0.683 \leqslant \sqrt{(M_P/M_{yz})} < 1.224, \quad & M_{bu}/M_P = 1.15(1 - 0.28\,M_P/M_{yz}), \\ 1.224 < \sqrt{(M_P/M_{yz})}, \quad & M_{bu}/M_P = M_{yz}/M_P. \end{aligned} \right\} \quad (15.24)$$

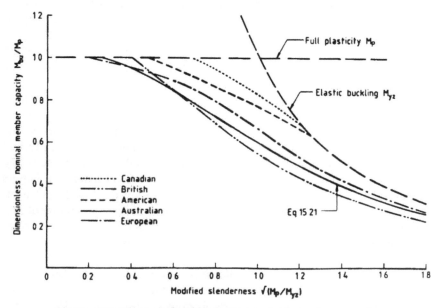

Figure 15.7 Effect of slenderness on nominal member capacity.

The second of equation 15.24 provides a parabolic transition from the full plastic moment M_P to the elastic buckling moment M_{yz}.

The British nominal moment capacity [11] is given by

$$0 \leqslant \sqrt{(M_P/M_{yz})} \leqslant 0.4, \quad M_{bu}/M_P = 1.0,$$
$$\left. 0.4 \leqslant \sqrt{(M_P/M_{yz})}, \quad \frac{M_{bu}}{M_P} = \frac{M_{yz}/M_P}{\phi_B/M_P + \sqrt{(\phi_B^2/M_P^2 - M_{yz}/M_P)}} \right\} \quad (15.25)$$

in which

$$\phi_B/M_P = \{1 + (1 + \eta) M_{yz}/M_P\}/2 \quad (15.26)$$

and

$$\eta = 0.007 \sqrt{(\pi^2 E/F_Y)}(\sqrt{(M_P/M_{yz})} - 0.4). \quad (15.27)$$

The second of equation 15.25 and equations 15.26 and 15.27 correspond to a modification of the first yield condition of equation 15.20 with M_Y replaced by M_P and 1.25 by $(1 + \eta M_{yz}/M_P)$. Equations 15.25–15.27 are shown in Figure 15.7 for the case where $E = 200\,000$ MPa and $F_Y = 250$ MPa.

The American nominal moment capacity [12] is given by

$$0 \leqslant L \leqslant L_p, \quad M_{bu}/M_P = 1.0,$$
$$\left. L_p \leqslant L \leqslant L_r, \quad M_{bu}/M_r = 1.0 - (1 - M_r/M_P)(L - L_p)/(L_r - L_p), \right\} \quad (15.28)$$
$$L_r \leqslant L, \quad M_{bu}/M_P = M_{yz}/M_P,$$

in which

$$M_r/M_P = M_Y(1 - 10/F_Y)/M_P, \quad (15.29)$$

$$L_p = 50r_y/\sqrt{(F_Y/36)}. \quad (15.30)$$

L_r is the member length for which $M_r = M_{yz}$, and F_Y is in kips/in². The second of equation 15.28 provides a linear transition with L from the full plastic moment M_P to the elastic buckling moment M_{yz}. Equations 15.28–15.30 are shown in Figure 15.7 for the case of an 18WF50 beam with $E = 29\,000$ kips/in² and $F_Y = 36$ kips/in².

The Australian nominal moment capacity [13] is given by the lower bound approximation of equation 15.21 to the test results shown in Figure 15.4 and 15.6 while $0.26 < \sqrt{(M_P/M_{yz})}$.

The European nominal moment capacity [14] is given by

$$0 \leqslant \sqrt{(M_P/M_{yz})} \leqslant 0.2, \quad M_{bu}/M_P = 1.0,$$
$$\left. 0.2 \leqslant \sqrt{(M_P/M_{yz})}, \quad \frac{M_{bu}}{M_P} = \frac{M_{yz}/M_P}{\phi_B/M_P + \sqrt{(\phi_B^2/M_P^2 - M_{yz}/M_P)}}, \right\} \quad (15.31)$$

in which ϕ_B/M_P is given by equation 15.26

and

$$\eta = 0.021(\sqrt{(M_P/M_{yz})} - 0.2).\tag{15.32}$$

Equation 15.31 is the same modification as that of the British equation 15.25 of the first yield condition of equation 15.20 with M_Y replaced by M_P and 1.25 by $(1 + \eta M_{yz}/M_P)$.

The comparison shown in Figure 15.7 of the different nominal moment capacities shows that the highest capacities are predicted by the Canadian and American codes, which appear to ignore any effects of initial crookedness and twist, and to be based entirely on the elastic and inelastic buckling strengths. The Canadian capacities are close to upper bounds of the test results of Figure 15.4, while the American capacities are closer to the means of the test results.

The British capacities shown in Figure 15.7 appear to be the lowest, but this is deceptive because of the use of a high capacity factor of $\phi = 1.0$ compared with the values of $\phi = 0.9$ and $1/1.1$ used by the other codes. The Australian capacities are close to the lower bounds of the test results of Figure 15.4, while the European capacities lie between the lower bounds and means of the test results.

15.4.3 MOMENT DISTRIBUTION

The effects of moment distribution on the flexural-torsional buckling of beams with unequal end moments $M, \beta M$ are accounted for in the Canadian, American and Australian codes [10, 12, 13] by using the moment modification factor

$$\alpha_m = 1.75 + 1.05\beta + 0.30\beta^2\tag{15.33}$$

for elastic buckling presented in section 7.3, with maximum values of 2.5, 2.3 and 2.5 respectively. In the American and Australian codes, α_m is applied directly to the uniform bending nominal capacity M_{bu}, whereas its application in the Canadian code is to the elastic buckling resistance M_{yz}. This latter method is conservative, and does not take account of the important effect of localized yielding on inelastic buckling, which was discussed in section 14.5.2.2.

The effects of unequal end moments are accounted for in the British code [11] by using a moment modification factor

$$1/m = 1/(0.57 - 0.33\beta + 0.10\beta^2) \leqslant 1/0.43\tag{15.34}$$

whose values are close to the corresponding values of α_m. The factor is applied directly to the uniform bending nominal capacity M_{bu} of equal flanged beams of uniform section.

The effects of unequal end moments are accounted for in the European code [14] by using

$$\alpha_m = 1.88 + 1.40\beta + 0.52\beta^2 \leqslant 2.70\tag{15.35}$$

which is somewhat higher than equation 15.33. However, equation 15.35 is conservatively applied to the uniform bending elastic buckling resistance M_{yz} instead of the uniform moment capacity M_{bu}.

The effects of non-uniform moment distributions caused by transverse loads are allowed for approximately in the British, Australian, and European codes, principally by using values for the factor α_m or its equivalent which depend on the bending moment distribution. For the Australian code, a simple approximation can also be obtained by using

$$\alpha_m = \frac{1.7\,M_m}{\sqrt{(M_2^2 + M_3^2 + M_4^2)}} \tag{15.36}$$

in which M_m is the maximum moment, and M_2, M_3, and M_4 are the moments at the quarter, mid-, and three-quarter points of the segment length. The British and European codes apply these factors conservatively to the uniform elastic buckling resistance M_{yz}, but the Australian code is more realistic in applying them to M_{bu}.

15.4.4 LOAD HEIGHT

The destabilizing effects of gravity loads which act at the top flange (see section 7.6) are allowed for approximately in the British and Australian codes [11, 13] by calculating increased lengths

$$L_e = k_1 L \tag{15.37}$$

for use in calculating the elastic buckling resistance M_{yz}, in which the load height factors k_1 are approximately 1.2 or 1.4 for beams, and higher for cantilevers.

A more accurate method is permitted in both the Australian and European codes [13, 14], in which the effect of load height on the elastic buckling resistance is calculated more accurately by using approximations such as those of equation 7.65.

No account of load height is taken in the Canadian and American codes [10, 12].

15.4.5 RESTRAINTS

The effects of end restraints against lateral rotation du/dz out of the plane of bending are allowed for in the British and Australian codes [11, 13] by calculating decreased effective lengths

$$L_e = k_{Ry} L \tag{15.38}$$

in which the lateral rotation restraint factor k_{Ry} is 0.7 or 0.85 depending on whether the restraint is full or partial or whether there are restraints at both ends or only one end. The European code [14] decreases the 0.7 factor to 0.5, the theoretical value for full restraint at both ends (section 8.5.1.1). The Canadian and American codes [10, 12] do not account for the effects of lateral rotation restraints.

The effects of partial torsional restraints which allow some end twisting are accounted for approximately in the British and Australian codes by increasing the effective length L_e. In the British code the increase is twice the beam depth for partial restraints at both ends, while the increase given in the Australian code also depends on the ratio t/T of the web and flange thicknesses. The Australian code also allows more accurate approximations to be made for partial torsional end restraints which are based on those of section 8.5.1.1. The Canadian and American codes do not account for the effects of partial torsional end restraints.

The effects of end warping restraint (section 8.5.1.1) are not generally accounted for, except in the European code, which allows the use of a reduced effective length

$$L_e = k_W L \tag{15.39}$$

without specifying values of k_W.

15.4.6 DESIGN BY BUCKLING ANALYSIS

A method of designing against flexural-torsional buckling by using the results of an elastic buckling analysis is allowed implicitly in the British and European codes [11, 14]. The implication is made through the use of expressions for the elastic buckling resistance for a range of loading, support and restraint conditions. Thus the maximum moment at elastic buckling determined by the analysis is used in place of the uniform bending buckling resistance M_{yz} in the design process.

Design by buckling analysis is allowed explicitly in the Australian code [13]. Because this code multiplies the elastic uniform bending buckling resistance M_{yz} by the moment modification factor α_m, the maximum moment at elastic buckling M_{ob} determined by buckling analysis is reduced to

$$M_{oa} = M_{ob}/\alpha_m$$

so that this value of M_{oa} can be used instead of M_{yz} in calculating the slenderness factor α_s ($= M_{uo}/M_P$) in equation 15.21. The value of α_m is then reintroduced in the final calculation of the nominal moment capacity

$$M_b = \alpha_m \alpha_s M_P \leqslant M_P. \tag{15.40}$$

15.5 Limit states strength design of steel columns

The American code [12] provides for the design of columns against torsional or flexural-torsional buckling. This is carried out by substituting the load P_E at elastic flexural-torsional buckling (sections 5.2–5.5) for the value of P_y at elastic flexural buckling in the formulations for the column strength P_u. This substitution is an example of design by buckling analysis, and is based on the method given in [15] of designing cold-formed columns against flexural-torsional buckling. Thus the dimensionless nominal design capacity P_u/P_Y is given by the

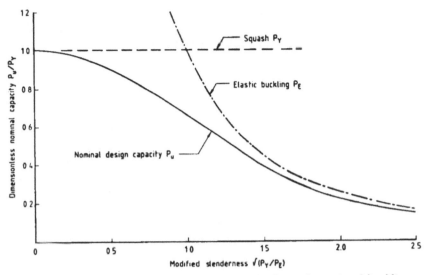

Figure 15.8 Nominal column design capacity for flexural–torsional buckling.

American code as

$$P_u/P_Y = 0.658^{(P_Y/P_E)} \tag{15.41}$$

while $\sqrt{(P_Y/P_E)} < 1.5$, in which $P_Y = AF_Y$ is the squash load, and by

$$P_u/P_Y = 0.877/(P_Y/P_E) \tag{15.42}$$

while $\sqrt{(P_Y/P_E)} > 1.5$, as shown in Figure 15.8.

The Canadian code [10] requires a rational analysis to be made, which may be taken to be the elastic buckling analysis method used for the American code. The European code [14] refers to flexural-torsional buckling of columns, but the British and Australian codes [11, 13] do not. These latter codes protect indirectly against the torsional buckling of angle and cruciform section columns through their rules against the local buckling of thin outstands.

15.6 Limit states strength design of steel beam-columns

Steel beam-columns must be designed against each of three modes of failure:

- (a) local failure at a cross-section due to yielding and/or local buckling;
- (b) in-plane failure due to excessive bending caused by the in-plane moments, including the second-order effects due to moment amplification by the axial compression;
- (c) out-of-plane failure due to flexural-torsional buckling under the in-plane bending and axial force actions.

For each of these failure modes, it is common to use interaction equations to represent the design inequality, and at the simplest level these usually take the form of the linear interaction equation (see Figure 15.9)

$$\frac{P^*}{\phi P_u} + \frac{M^*}{\phi M_u} \leqslant 1 \qquad (15.43)$$

in which P^* and M^* are the design axial force and maximum amplified moment, and P_u and M_u are the nominal capacities in axial compression and bending.

In some cases the same interaction equation is used for two or more of the three failure modes of cross-section, in-plane and out-of-plane failure, but in each case the nominal capacities are interpreted differently so that they reflect the capacities to resist the particular failure mode being considered. Thus when designing against out-of-plane flexural-torsional buckling, P_u is interpreted as the out-of-plane column buckling capacity, which is usually taken as the capacity P_{uy}

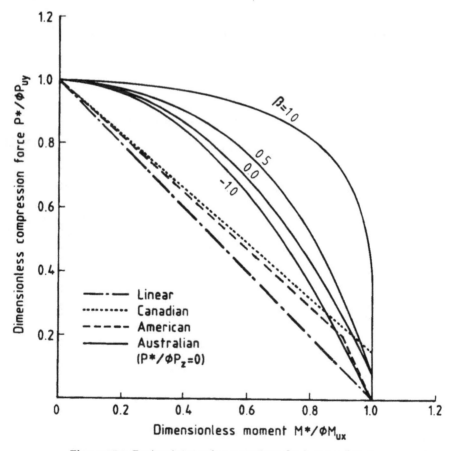

Figure 15.9 Design interaction equations for beam-columns.

associated with the load P_y for elastic flexural buckling out of the plane of bending, while M_u is interpreted as the beam nominal capacity M_{ux} associated with the elastic flexural-torsional buckling moment M_{yz}.

The design moments M^* used in equation 15.43 generally account for second-order effects on the in-plane bending moment distribution. They may be determined by amplifying the moments calculated from a first-order analysis, or in many codes [10, 12–14] by using a second-order in-plane analysis.

Non-linear interaction equations are used as a higher tier in some codes, usually for compact sections. The Canadian code [10] uses

$$\frac{P^*}{\phi P_{uy}} + \frac{0.85 M_x^*}{\phi M_{ux}} \leqslant 1.0 \qquad (15.44)$$

provided $M_x^* \leqslant \phi M_{ux}$, while the American code [12] uses

$$\frac{P^*}{\phi P_{uy}} + \frac{8 M_x^*}{9 \phi M_{ux}} \leqslant 1.0 \qquad (15.45)$$

while $P^*/\phi P_{uy} \geqslant 0.2$, and

$$\frac{P^*}{2\phi P_{uy}} + \frac{M_x^*}{\phi M_{ux}} \leqslant 1.0 \qquad (15.46)$$

while $P^*/\phi P_{uy} \leqslant 0.2$. The Australian code [13] uses

$$M_x^* \leqslant \phi M_{0x} \qquad (15.47)$$

in which

$$M_{0x} = \alpha_{bc} M_{bx0} \sqrt{\{(1 - P^*/\phi P_{uy})(1 - P^*/\phi P_z)\}} \qquad (15.48)$$

$$1/\alpha_{bc} = \left(\frac{1-\beta}{2}\right) + \left(\frac{1+\beta}{2}\right)^3 \left(0.4 - 0.23\frac{P^*}{\phi P_{uy}}\right) \qquad (15.49)$$

and M_{bx0} is the value of M_b for uniform bending. These formulations are based on experimental and theoretical studies of the flexural-torsional buckling of beam-columns with unequal end moments M, βM [16, 17]. The linear and non-linear interaction equations (equations 15.43–15.49) for the design of beam-columns against flexural-torsional buckling are compared in Figure 15.9.

Most design codes ignore the strengthening effects of axial tension on the resistance to out-of-plane flexural-torsional buckling (see sections 11.2.1 and 11.2.2), and only require the moment condition of equation 15.23 to be satisfied. The Australian code [13] does allow for this strengthening effect through the interaction equation

$$-\frac{P^*}{\phi P_t} + \frac{M^*}{\phi M_{ux}} \leqslant 1.0 \qquad (15.50)$$

in which P_t is the tension capacity, provided the section and in-plane capacities are not exceeded.

15.7 Problems

15.7.1 GENERAL

Problems for designing against flexural-torsional buckling may be based on many of the elastic and inelastic buckling problems given in other chapters. These design problems will require the determination of the design capacity according to an appropriate design code [10–14]. The design data will need to include an appropriate value of the yield stress F_Y (often equal to 250 MPa). The following sub-sections list suitable problems.

15.7.2 SIMPLY SUPPORTED BEAMS

Problems 7.1, 3, 4, 5, 6, 7, 8, 9, 10, 11, 13.

15.7.3 RESTRAINED BEAMS

Problems 8.1, 2, 3, 4, 5.

15.7.4 CANTILEVERS

Problems 9.1, 2, 3, 4, 5.

15.7.5 BRACED AND CONTINUOUS BEAMS

Problems 10.1, 2, 3.

15.7.6 BEAM-COLUMNS

Problems 11.1, 2, 4, 5.

15.7.7 INELASTIC BUCKLING
Problems 14.1, 2, 3, 4, 5, 6, 7, 8, 9.

15.7.8 STEPPED AND TAPERED MEMBERS

Problems 16.1, 2.

15.8 References

1. Trahair, N.S. (1969) Deformations of geometrically imperfect beams. *Journal of the Structural Division, ASCE*, **95** (ST7), 1475–96.
2. Trahair, N.S. and Bradford, M.A. (1991) *The Behaviour and Design of Steel Structures*, revised 2nd edn, Chapman and Hall, London.

3. Nethercot, D.A. and Trahair, N.S. (1983) Design of laterally unsupported beams, in *Developments in the Stability of Structures, Volume 2, Beams and Beam-Columns* (ed. R. Narayanan), Applied Science Publishers, Barking, pp. 70–94.
4. American Institute of Steel Construction, (1969) *Specification for the Design, Fabrication, and Erection of Structural Steel for Buildings*, AISC, New York.
5. British Standards Institution (1969) *BS449: Part 2: 1969, Specification for the Use of Structural Steel in Building*, BSI, London.
6. Standards Association of Australia (1975) *AS1250–1975, SAA Steel Structures Code*, SAA, Sydney.
7. Trahair, N.S. (1977) *The Behaviour and Design of Steel Structures*, 1st edn, Chapman and Hall, London.
8. Ravindra, M.K. and Galambos, T.V. (1978) Load and resistance factor design for steel. *Journal of the Structural Division, ASCE*, **104**(ST9), 1337–54.
9. Yura, J., Galambos, T.V. and Ravindra, M.K. (1978) The bending resistance of steel beams, *Journal of the Structural Division, ASCE*, **104**(ST9), 1355–70.
10. Canadian Standards Association (1989) *CAN/CSA-S16.1-M89 Limit States Design of Steel Structures*, CSA, Toronto.
11. British Standards Institution (1990) *BS5950: Part 1: 1990, Structural Use of Steelwork in Building, Part 1, Code of Practice for Design in Simple and Continuous Construction: Hot Rolled Sections*, BSI, London.
12. American Institute of Steel Construction (1986) *Load and Resistance Factor Design Specification for Structural Steel Buildings*, AISC, Chicago.
13. Standards Australia (1990) *AS4100, Steel Structures*, SA, Sydney.
14. Commission of the European Communities (1990) *Eurocode No. 3: Design of Steel Structures, Part 1: General Rules for Building*, CEC.
15. American Iron and Steel institute (1991) *Load and Resistance Factor Design Specification for Cold-Formed Steel Structural Members*, AISI, Washington, DC.
16. Cuk, P.E., Rogers, D.F., and Trahair, N.S. (1986) Inelastic buckling of continuous steel beam-columns. *Journal of Constructional Steel Research*, **6**(1), 21–52.
17. Bradford, M.A. and Trahair, N.S. (1985) Inelastic buckling of beam-columns with unequal end moments. *Journal of Constructional Steel Research*, **5**(3), 195–212.

16 Special topics

16.1 Stepped and tapered members

16.1.1 GENERAL

Non-uniform stepped and tapered members are often used because of the economies produced by reducing the cross-section in the low moment regions. While the cross-section reductions are usually determined by in-plane bending considerations, these reductions may significantly reduce the member's resistance to flexural-torsional buckling, by reducing the cross-sectional out-of-plane rigidities EI_y, GJ, and EI_w. The flexural-torsional buckling of stepped and tapered members has been reviewed in [1–4].

Stepped members (Figure 16.1a) have sudden changes in the cross-section, usually produced by increasing the flange thickness of a welded beam or by welding additional flange plates to a hot-rolled I-section. Sometimes, stepped members have sudden changes in flange width. Web depths are never changed suddenly, although short tapered transitions may be used to join sections of different depth.

Tapered members (Figure 16.1b) have gradual changes in the cross-section dimensions. Tapered web depths are very common, but less common are tapered flange widths. Flange or web thicknesses are never tapered in civil engineering steel structures, but sometimes are in aerospace structures, when savings in weight can produce significant economies.

The effects of reducing the cross-section of a member depend on the cross-section dimension reduced, and the location of the reduced cross-section along the member. The approximate effects of dimension changes are summarized in Table 16.1. It can be seen that changing the flange width B or thickness T causes moderate to large changes in the out-of-plane rigidities, that changing the web depth h has no effect on EI_y, small effect on GJ, and a large effect on EI_w, and that changing the web thickness t causes no changes, except for moderate changes to GJ.

The effects of reductions in the flexural rigidity EI_y are greatest in the region where the buckled beam has the greatest buckling curvature u'', which is usually at mid-span. The effects of reductions in the warping rigidity EI_w are usually greatest also at mid-span. However, the effects of reductions in the torsional rigidity GJ are greatest where the twist $d\phi/dz$ is greatest, which is usually at the supports.

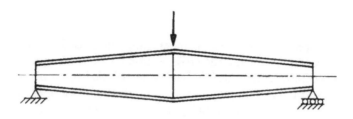

Original section   Heavier flanges   Additional flange plates

(a) Stepped Beam Sections

(b) Elevation of Tapered Beam

Figure 16.1 Stepped and tapered beams.

Table 16.1   Approximate effects of dimension changes on buckling rigidities

Dimension changed	Approximate Change in Rigidity		
	EI_y	GJ	EI_w
Flange width (B)	Large (B^3)	Moderate (B)	Large (B^3)
Flange thickness (T)	Moderate (T)	Large (T^3)	Moderate (T)
Wed depth (h)	0	Small (h)	Large (h^2)
Web thickness (t)	0	Moderate (t^3)	0

16.1.2 STEPPED BEAMS

The elastic flexural-torsional buckling of simply supported stepped I-beams under central concentrated loads was studied in [1]. The flanges were stepped in thickness as shown in Figure 16.1a or in width, and the step reductions were confined to the regions of length αL adjacent to the supports as indicated in Figure 16.2, where the in-plane moments were reduced. Tabulations were given of the variations in the dimensionless buckling loads $QL^2/\sqrt{(EI_yGJ)}$ with the flange width and thickness reduction ratios β and γ and the length ratio α for

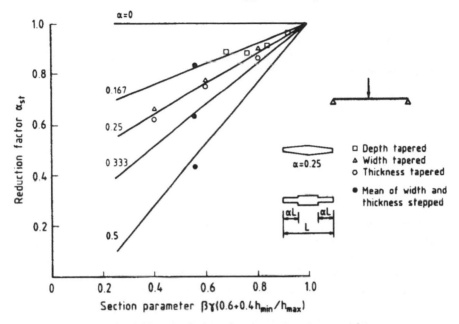

Figure 16.2 Reduction factors for stepped and tapered beams.

loads acting at the top flange, centroid, or bottom flange for a range of beam parameters $K = \sqrt{(\pi^2 EI_w/GJL^2)}$.

It was found that the buckling loads varied almost linearly with the length ratio α, and that reasonably close approximations could be determined by linear interpolation between the buckling loads of uniform members. Thus

$$Q \approx Q_{1,1} - 2\alpha(Q_{1,1} - Q_{\beta,\gamma}) \tag{16.1}$$

in which $Q_{1,1}$ is the buckling load for a uniform member having the full cross-section ($\beta = \gamma = 1$), and $Q_{\beta,\gamma}$ is the buckling load for a uniform member having a reduced cross-section corresponding to the values of the width and thickness reduction ratios β and γ.

For the elastic buckling moments M_{st} of stepped beams of constant depth h, [5] uses the approximation

$$M_{st} = \alpha_{st} M_{yz} \tag{16.2}$$

in which

$$\alpha_{st} = 1.0 - 2.4\alpha(1 - \beta\gamma) \tag{16.3}$$

and M_{yz} is the elastic buckling moment of a uniform member having the full cross-section. Equation 16.2 is compared in Figure 16.2 with the mean values for the width and thickness stepped beams of [1].

16.1.3 TAPERED MEMBERS

The elastic flexural-torsional buckling of tapered simply supported equal flange I-section members under central concentrated load was studied in [2]. The section flange width B, thickness T, or web depth h tapered linearly from maximum values at mid-span to minimum values at the supports, as shown in Figure 16.1b, following the corresponding reductions in the bending moments. It was predicted theoretically and confirmed experimentally that the elastic buckling resistance reduces almost linearly with the flange width and thickness reduction ratios β and γ, but is virtually independent of any reduction in the web depth h. Reference [5] adapts the approximation of equation 16.2 by redefining α_{st} as

$$\alpha_{st} = 1.0 - 0.6\{1 - \beta\gamma(0.6 + 0.4h_{min}/h_{max})\}. \tag{16.4}$$

The elastic buckling of cantilevers whose cross-section dimensions taper linearly from maximum values at the support to minimum values at the free end was studied in [6, 7], and tabulations were given for concentrated end and uniformly distributed loads acting at the top flange, centroid, or bottom flange. The elastic buckling of simply supported beams with equal and opposite end moments or uniformly distributed load was also studied.

The elastic buckling of simply supported beams whose flange width B and web depth h taper linearly from maximum values at one support to minimum values at the other was studied in [8], and a series of graphs was given for beams with unequal end moments M at the larger end and βM at the smaller end. Also studied were simply supported beams tapered from a maximum section at mid-span to minimum sections at the supports under uniformly distributed load.

The elastic buckling of simply supported beams which tapered linearly from one end to the other was also studied in [9], which provided the basis for the design rules of [10]. The effects of restraints on continuous beams were investigated in [11].

The elastic buckling of monosymmetric I-beams was studied in [12] for simply supported beams with central concentrated load and which tapered from mid-span to the supports. The buckling of tapered and haunched beams was discussed in [13], and test results reported in [14].

A finite element method of analysing the elastic buckling of tapered monosymmetric beam-columns was presented in [4], and extended to beam-columns with continuous restraints in [15], and to inelastic buckling in [16].

16.2 Optimum beams

It is always desirable to distribute the material in a beam so as to minimize its cost or maximize its efficiency. In many cases, current practice represents the results of a cost optimization process in which designs are successively adjusted so as to achieve the most economical solution. This usually depends on a very wide range of parameters which affect the design, manufacture, erection, and use of the beam, and not just on the weight of material used.

Optimization may be carried out by minimizing the weight of material used when this dominates in the total cost structure. In this case, the optimization should consider the distributions of material both within the cross-section and along the length of the member. Very often the cross-sectional shape is fixed, and its size is varied along the length.

The optimization must be carried out while ensuring that the member is able to resist all failure modes, including in-plane yielding and local flange or web buckling, as well as flexural-torsional buckling. Often the proportions of the cross-section will be fixed so that local buckling will not occur, in which case the optimization can be made with respect to in-plane yielding, and out-of-plane buckling. In-plane yielding is usually considered to be governed by the in-plane bending moment resistance, and the beam is very simply optimized so that its section moment capacity follows the bending moment distribution. In the case of a beam subjected to uniform bending, this will result in a constant section beam.

Optimization with respect to member buckling will distribute the material along the length of the member so as to maximize its buckling resistance. This will result in material being more concentrated in the regions where the local

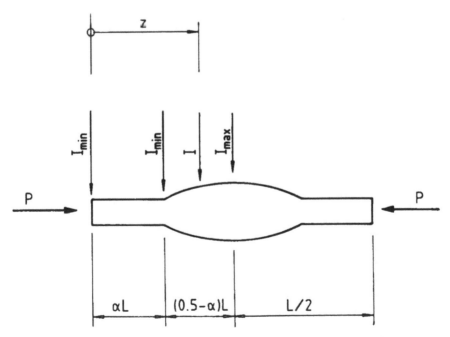

Figure 16.3 Column optimized against flexural buckling and yielding.

contribution to the buckling resistance is greatest. In the case of the flexural buckling of columns [17], the internal resistance to flexure $EI_y d^2 u/dz^2$ is greatest in the central region of a simply supported column of constant cross-section, and so the optimum column has minimum cross-sections in the end regions (to prevent premature yielding), and elsewhere has cross-sections which increase towards the mid-length, as shown in Figure 16.3. The volume ratio of the optimal column to the uniform column having the same strength varies from 0.866 for very long columns which are dominated by buckling to 1.0 for short columns which are dominated by yielding.

For a simply supported narrow rectangular section beam in uniform bending, the flexural-torsional buckling resistance depends on the internal resistances $EI_y d^2 u/dz^2$ to flexure and $GJ d\phi/dz$ to uniform torsion. For a uniform section beam, $EI_y d^2 u/dz^2$ is greatest at the beam centre and zero at the supports, but the reverse is true for $GJ d\phi/dz$. In this case the optimal resistance to flexural-torsional buckling is produced by a uniform section beam.

For a simply supported I-section beam in uniform bending, the internal resistance to warping torsion $- EI_w d^3\phi/dz^3$ is greatest at the centre of a uniform section beam, and least at the supports. For the case where $GJ d\phi/dz$ is negligible in comparison with $- EI_w d^3\phi/dz^3$, the constant depth optimal beam will have its flange material distributed as in Figure 16.3, with its flange volume ranging from 0.866 to 1.0 times the flange volume of a uniform beam of the same strength.

The optimization of narrow rectangular and I-section beams against flexural-torsional buckling has been studied in [18, 19], which consider cantilevers and simply supported beams under concentrated or uniformly distributed loads, or unequal end moments.

16.3 Secondary warping

Thin-walled open section members of rectangular, angle, tee, or cruciform section have zero warping rigidity EI_w, when calculated using the theory of thin-walled open sections (section 17.3). Each of these sections consists of a series of rectangular elements which radiate from a common point, as shown in Figure 16.4. Thus any warping shears in their mid-surface planes are concurrent, and so the warping torque $- EI_w d^3\phi/dz^3$, which is the torque resultant of these shears, is zero.

The thin-walled theory predicts zero warping displacements during the torsion of members of this type of section, and so boundary conditions associated with warping (such as free to warp, or prevented from warping) cannot be invoked when $I_w = 0$. Because of this, theoretical solutions for the twisting of the member show 'kinks' at points of concentrated torque or at built-in supports, as shown in Figure 16.5, which do not occur in practice.

While this does not cause any theoretical difficulties, a smoothing out of these kinks can be predicted by including the effects of secondary or thickness warping

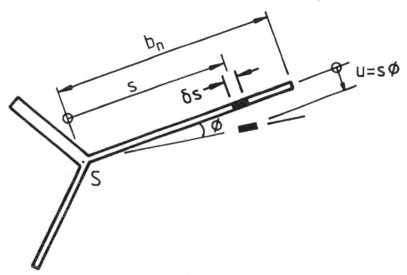

Figure 16.4 Section with radiating elements.

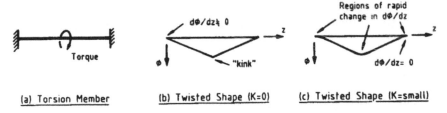

(a) Torsion Member   (b) Twisted Shape (K=0)   (c) Twisted Shape (K=small)

Figure 16.5 Smoothing effect of secondary warping.

[20, 21] which are neglected in thin-walled theory. The non-zero section warping constant I_{ws} can be obtained by considering the displacement

$$u = s\phi \qquad (16.5)$$

of an element $t_n \delta s$ of the typical rectangular element $b_n \times t_n$, as shown in Figure 16.4. This element will develop a shear component

$$\delta V = E(\delta s t_n^3/12)u''' \qquad (16.6)$$

which has a torque effect $s\delta V$ about the axis of twist through the common point S. The integrated effect of all such torque effects is equal to the secondary warping torque

$$M_{ws} = -EI_{ws}d^3\phi/dz^3. \qquad (16.7)$$

Thus

$$I_{ws} = \sum_n \int_0^{b_n} \left(\frac{t_n^3}{12}\right) s^2 \, ds, \qquad (16.8)$$

so that

$$I_{ws} = \sum_n b_n^3 t_n^3 / 36. \qquad (16.9)$$

Normally, the secondary warping section constant I_{ws} is so small that the torsion parameter $K = \sqrt{(\pi^2 E I_w / GJL^2)}$ is very small. Mathematically, this leads to rapid changes in the twist $d\phi/dz$ in the regions of concentrated torques, as shown in Figure 16.5c, which in the limit of $K = 0$ become the kinks shown in Figure 16.5b.

Doubly symmetric circular and hollow rectangular sections behave similarly to the sections with radiating elements discussed above, but for different reasons. Such sections do not warp during twisting, because the warping deflections due to twisting are exactly balanced by the warping deflections due to shear straining. The shear stresses produced by the shear strains are equivalent to a circulating shear flow which is very effective in resisting uniform torsion. Thus doubly symmetric hollow sections have very large torsion section constants J, and zero warping sections constants I_w. The kinks which appear at points of concentrated torque shown in Figure 16.5b are consistent with the sudden changes in the circulating shear strains which occur when there are sudden changes in the uniform torque, as at concentrated load points. Such kinks do not appear in practice because of rapid changes in the stress distribution, which can be predicted by using the theory of elasticity in place of thin-walled theory.

16.4 Interaction of local and flexural-torsional buckling

Elastic local buckling of a very thin compression flange will significantly reduce the resistance of a beam in uniform bending to flexural-torsional buckling [22]. In this case of uniform bending, local buckles appear along the whole length of the compression flange, and even though local failure is postponed by the flange post-local buckling resistance, the flexural and torsional stiffnesses of the flange are reduced, so that the effective out-of-plane rigidities EI_y, GJ, and EI_w of the beam are also reduced along the whole length of the beam [23]. The reduced resistance of the beam to flexural-torsional buckling can be predicted by transforming it into a monosymmetric beam which has these reduced rigidities.

The interaction between local and flexural-torsional buckling for other than uniform bending has been studied [24]. However, flange local buckling is then confined to the high moment regions, and its effect on flexural-torsional buckling is reduced.

When the elastic local buckling load of the compression flange is close to the elastic flexural-torsional buckling load, the actual strength may be reduced by imperfection sensitivity effects. Some small reductions were reported in [25] for thin-flange beams in uniform bending. In practice, very thin flanges are rarely used in beams in uniform bending, and it is unlikely that strength reductions will occur.

However, it is not uncommon to use welded beams with flanges of intermediate slenderness, so that the local buckling strength is of the same order as the inelastic flexural-torsional buckling capacity. An experimental study [26] of simply supported steel beams with central concentrated loads showed that their strengths could be closely predicted by substituting a reduced cross-section moment capacity for the full plastic moment capacity when determining the flexural-torsional buckling capacity (see section 15.4.2), the reduced section moment capacity allowing for the effects of yielding and local buckling and being based on test results.

16.5 Web distortion

It is usually assumed in the analysis of flexural-torsional buckling that there is no change in the shape of the cross-section during buckling. Web distortion (Figure 16.6b) may significantly reduce the flexural-torsional buckling resistance. Flexural-torsional buckling with web distortion is often described as distortional buckling. It should be noted that distortional buckling need not be associated with flexural-torsional buckling, as for example in the case of the buckling of columns with thin webs (see Figures 1.11 and 16.7), where the buckling mode is of intermediate wavelength between those of local and member buckling [27, 28].

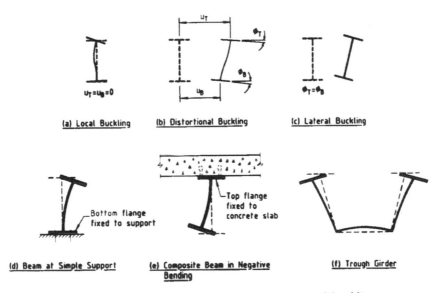

(a) Local Buckling   (b) Distortional Buckling   (c) Lateral Buckling

(d) Beam at Simple Support   (e) Composite Beam in Negative Bending   (f) Trough Girder

Figure 16.6 Effect of distortion of flexural-torsional buckling.

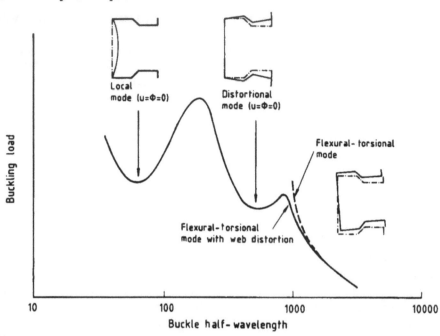

Figure 16.7 Column buckling modes.

Web distortion allows larger deflections of the critical flange than usual. The resistance to buckling is reduced through a corresponding reduction in the strain energy stored during buckling caused by reduced flange twists. Web distortion may occur in beams with flexible webs, due either to their thinness or to holes. It also occurs in beams whose compression flanges are unrestrained at the supports [29] as shown in Figure 16.6d, near the supports of continuous composite beams [30] as shown in Figure 16.6e, and in trough girders [31], as shown in Figure 16.6f.

The distortional buckling of beams in uniform bending has been analysed [27] by using the finite strip method [32]. A simple approximate solution has been obtained for I-section beam-columns in uniform bending by assuming that the web distorts in a cubic shape [33]. This assumption has also been used to analyse beam-columns of general, thin-walled open cross-section under general loading conditions [29, 34]. This method has been applied to a wide range of situations and extended to inelastic buckling problems. A recent study [35] has used a super finite element to include the effects of local, distortional, and lateral buckling. An extensive review is given in [36].

16.6 Pre-buckling deflections

The usual method of analysing flexural-torsional buckling neglects the effects of the pre-buckling deflections in the plane of bending, which increase the buckling resistance. For a simply supported beam in uniform bending, the increased elastic buckling resistance is given by

$$M = \frac{M_{yz}}{\sqrt{[(1 - EI_y/EI_x)\{1 - (GJ/2EI_x)(1 + \pi^2 EI_w/GJL^2)\}]}} \qquad (16.10)$$

in which

$$M_{yz} = \sqrt{\{(\pi^2 EI_y/L^2)(GJ + \pi^2 EI_w/L^2)\}} \qquad (16.11)$$

is the classical resistance (section 7.2.1).

The pre-buckling deflections transform the beam into a 'negative' arch. The concave curvature of the deflected beam **increases** its buckling resistance, just as the convex curvature of an arch (section 13.5.3) **decreases** its buckling resistance. The resistance of the beam in uniform bending increases with the ratio I_y/I_x, and theoretically becomes infinite when $I_y = I_x$. This is consistent with the common assumption that a beam with $I_y > I_x$ does not buckle, because it finds it easier to remain in the more flexible plane of bending than to buckle out-of-plane by deflecting in the stiffer plane. (Note that this common assumption does not always hold true, as for example when loads are applied well above the shear centre axis (see section 7.6), which may lead to buckling in a dominantly torsional mode.)

Early work on the effects of pre-buckling deflections was reviewed in [37] which considered simply supported beams in uniform bending. This was extended to a wide range of load, support, and restraint conditions in [38]. Finite element methods of analysis have been developed in [39–41].

16.7 Post-buckling behaviour

Slender statically determinate beams which remain elastic have slowly rising load–deformation paths after flexural-torsional buckling [42], as shown in Figure 16.8. This behaviour is similar to that of the behaviour of slender columns after flexural buckling [43]. The increased load carrying capacity over that predicted by the small deflection theory of elastic buckling is associated with disturbing effects at large deflections which are not as large as those predicted by the small deflection theory.

However, slender redundant beams [44–46] may have significant increases in load capacity above those predicted by the small deformation theory of flexural-torsional buckling. The twist rotations that occur after initial buckling reduce locally the effective bending rigidity in the plane of loading, which causes a redistribution of the bending moments. In continuous beams, the moments near the untwisted regions at the supports increase, and play a greater part in resisting the applied load.

The buckling resistance is generally affected by the bending moment distribution (see sections 7.3–7.5 and Figure 7.17), and in this case the redistribution is favourable, and increases the buckling resistance. The buckling resistance may

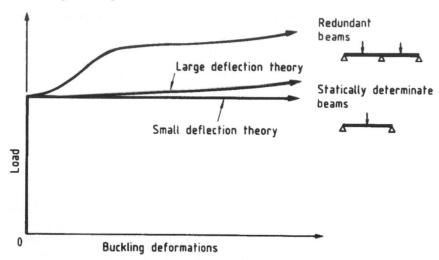

Figure 16.8 Post-buckling behaviour of beams.

asymptote towards a limiting value (Figure 16.8), but then may continue to increase, but only slowly, as in the case of statically determinate beams.

Significant increases in strength after post-buckling are only realized in slender rectangular beams or in very slender I-section beams. Yielding effects usually cause practical beams to fail before any significant post-buckling reserve can be achieved [46].

16.8 Viscoelastic beams

Materials which creep under sustained loading may be modelled as being linear-viscoelastic [47]. Perfectly straight viscoelastic members may buckle at some time after the application of load, while the deformations of members with initial crookedness or twist will increase with time, and may become excessive, or may lead to material non-linearity, and even higher deformations.

The stress (σ)–strain (ε)–time (t) relationship for the three-element material model shown in Figure 16.9a is given by

$$\sigma\left(1+\frac{E_1}{E_3}\right)+\frac{\eta_2}{E_3}\frac{d\sigma}{dt}=E_1\varepsilon+\eta_2\frac{d\varepsilon}{dt}. \tag{16.12}$$

The material strains $\varepsilon_0 = \sigma/E_3$ immediately a stress σ is applied, and then creeps viscously towards a final strain $\varepsilon_\infty = \sigma(E_1 + E_3)/E_1 E_3$, as shown in Figure 16.9b. When the material is under constant stress, it can be regarded as having an effective modulus of elasticity which decreases with the time from the initial application of load from an initial value $E_0 = E_3$ towards a long term value

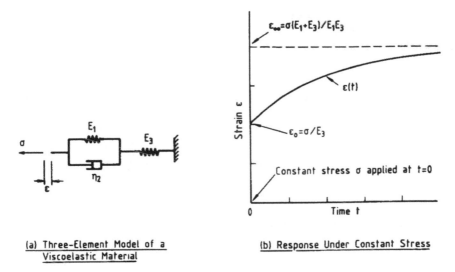

(a) Three-Element Model of a
     Viscoelastic Material

(b) Response Under Constant Stress

Figure 16.9 Viscoelastic material behaviour.

$E_\infty = E_1 E_3/(E_1 + E_3)$. Thus the possibility of creep buckling under constant stress can be investigated by first finding the modulus of elasticity for which elastic buckling would occur immediately under the applied load, and then determining the time at which the effective modulus would have reduced to this value.

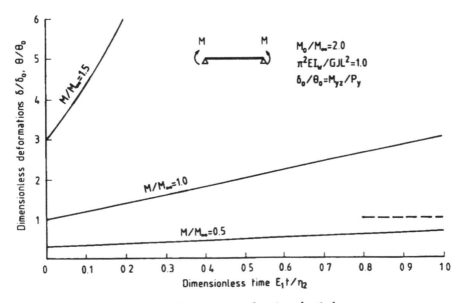

Figure 16.10 Deformations of a viscoelastic beam.

The deformations of viscoelastic members with initial crookedness and twist have been investigated in [48] by using the correspondence principle, which allows the viscoelastic problem to be transformed into a corresponding elastic problem. Some of the results of this study are shown in Figure 16.10 for a simply supported beam in uniform bending. When the applied moment M is equal to the long term buckling moment M_∞, the deformations increase linearly with time, and can be expected to cause the material to become non-linear at some later time. Higher values of M than M_∞ cause deformations that accelerate with time, but lower values cause the deformations to asymptote to those that would be predicted (see section 15.2.3) by using an elastic model with the long term moduli E_∞, G_∞.

16.9 Flexural-torsional vibrations

16.9.1 GENERAL

The vibrations of structural systems [49] are generally introduced by considering the behaviour of a single degree of freedom system (Figure 16.11a), for which the equation of motion which governs the variation of the displacement v with time t is

$$M\ddot{v} + \alpha_{T_y} v = Q(t) \tag{16.13}$$

in which M is the mass, α_{T_y} is the spring constant, $Q(t)$ is the driving force and $(\cdot) \equiv (d/dt)$. For free vibrations, $Q(t) = 0$, and the solution of equation 16.13 is

$$v = A \sin (\Omega t - \theta) \tag{16.14}$$

in which

$$\Omega = \sqrt{(\alpha_{T_y}/M)} \tag{16.15}$$

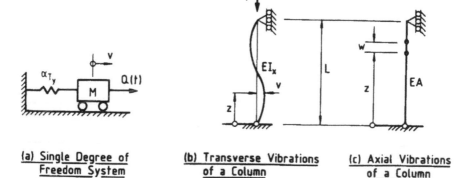

| (a) Single Degree of | (b) Transverse Vibrations | (c) Axial Vibrations |
| Freedom System | of a Column | of a Column |

Figure 16.11 Free vibrations.

is the natural frequency, and the initial conditions determine the amplitude A and the phase angle θ.

For continuous systems such as the simply supported column shown in Figure 16.11b, the equation of motion for free transverse flexural vibrations is

$$m\ddot{v} + EI_x v'''' + Pv'' = 0 \tag{16.16}$$

in which m is the mass per unit length. Free vibrations

$$v = A \sin (n\pi z/L) \sin (\Omega t - \theta) \tag{16.17}$$

have natural frequencies

$$\Omega = \frac{n^2\pi^2}{L^2} \sqrt{\left\{ \left(\frac{EI_x}{m}\right)\left(1 - \frac{P}{n^2\pi^2 EI_x/L^2}\right) \right\}}. \tag{16.18}$$

Continuous systems may be analysed by using the energy equation

$$\tfrac{1}{2}(\delta^2 U + \delta^2 V)_{max} = \tfrac{1}{2}(\delta^2 KE)_{max} \tag{16.19}$$

in which $\tfrac{1}{2}\delta^2 U$ is the change in the strain energy, $\tfrac{1}{2}\delta^2 V$ is the change in the potential energy of the applied loads, and $\tfrac{1}{2}\delta^2 KE$ is the change in the kinetic energy. This equation is a statement of the principle of conservation of energy during the free vibrations, according to which the total energy of the system does not vary with time. The total energy is equal to the sum of the maximum strain and potential energies when the kinetic energy is zero (zero velocity), and also equal to the maximum kinetic energy when the strain and potential energies are zero (zero displacement).

The energy equation can be used to develop a finite element method of analysing free vibrations, which can be expressed in the form of

$$\tfrac{1}{2}\{\Delta\}^T([K]) + \lambda[G] - \Omega^2[M]\}\{\Delta\} = 0 \tag{16.20}$$

in which $[K]$ is the stiffness matrix associated with the strain energy of the structure, $[G]$ is the stability matrix associated with the potential energy of an initial set of loads, $[M]$ is the mass matrix associated with the kinetic energy of the masses, $\{\Delta\}$ is the vector of nodal deformations of the structure, and λ is the load factor. The natural frequencies Ω of the structure satisfy

$$|K + \lambda G - \Omega^2 M| = 0. \tag{16.21}$$

16.9.2 STRUCTURAL VIBRATION MODES

Structural members may vibrate in axial, flexural, or torsional modes, or in combinations of these. Axial vibrations involve axial displacements w. For uniform members fixed at one end and free at the other (Figure 16.11c), the displacements are given by

$$w = A \sin \{(2n - 1)\pi z/2L - \theta\} \tag{16.22}$$

when the mass per unit length m is constant, and the natural frequencies Ω by

$$\Omega^2 = (2n-1)^2 \pi^2 EA/4mL^2 \tag{16.23}$$

in which n is an integer. These frequencies are usually quite high because of the high axial stiffness EA/L.

The free flexural vibrations v about the x axis of simply supported members with axial loads are governed by equations 16.16–16.18. Corresponding equations govern the free flexural vibrations u about the y axis.

Torsional vibrations ϕ of members of doubly symmetric cross-section which are prevented from twisting but free to warp at both ends take the form of

$$\phi = A \sin\frac{n\pi z}{L} \sin(\Omega t - \theta) \tag{16.24}$$

in which

$$\Omega^2 = \frac{n^2\pi^2}{I_p L^2}\left(GJ + \frac{n^2\pi^2 EI_w}{L^2}\right) \tag{16.25}$$

in which

$$I_p = \int_A \rho(x^2 + y^2)\,dA \tag{16.26}$$

is the rotary inertia, which depends on the distribution of material of density ρ through the mass area A attached to the member, and is uniform along the member.

Structural vibration modes may be coupled flexural-torsional modes. The coupling may be caused by the axial force or bending moment, in which case it is similar to the coupling leading to the flexural-torsional buckling of compression members (Chapter 5), beams (Chapter 7), and beam-columns (Chapter 11). Coupling may also be caused by the mass distribution, as for example when the centre of mass does not coincide with the shear centre.

16.9.3 EFFECTS OF COMPRESSION AND BENDING ON VIBRATION

Axial forces and bending moments change the natural frequencies of vibration of columns, beams, and beam-columns. This is caused by the change in the effective stiffness of the member from the elastic stiffness associated with the strain energy U (see equation 16.19) caused by the change in the potential energy V of the loading system. Thus in equation 16.20, the stability matrix $\lambda[G]$ indicates the change caused by the loading from the elastic stiffness $[K]$.

The natural frequencies Ω for transverse vibrations of a simply supported column decrease as the axial compression P increases, and become zero when P reaches the column flexural buckling load (see equation 16.18). These decreases are caused by the reduction in the effective flexural stiffness EI_x/L^2 caused by the destablizing effects of the axial compression P.

The natural frequencies of flexural-torsional vibrations of a simply supported monosymmetric ($x_0 = 0$) beam-column in uniform bending can be obtained by writing the vibration mode as

$$\begin{Bmatrix} u \\ \phi \end{Bmatrix} = \begin{Bmatrix} \delta \\ \theta \end{Bmatrix} \sin \frac{n\pi z}{L} \sin \Omega t \tag{16.27}$$

in which case the terms of the energy equation (equation 16.20) become

$$\{\Delta\} = \{\delta, \theta\}^T, \tag{16.28}$$

$$[K] = L \begin{bmatrix} (n^4\pi^4 EI_y/L^4) & 0 \\ 0 & (n^2\pi^2 GJ/L^2 + n^4\pi^4 EI_w/L^4) \end{bmatrix}, \tag{16.29}$$

$$-\lambda[G] = L \begin{bmatrix} (n^2\pi^2 P/L^2) & (n^2\pi^2 M/L^2 + n^2\pi^2 Py_0/L^2) \\ (n^2\pi^2 M/L^2 + n^2\pi^2 Py_0/L^2) & \{n^2\pi^2 P(r_0^2 + y_0^2) - n^2\pi^2 M\beta_x/L^2\} \end{bmatrix}, \tag{16.30}$$

$$\Omega^2[M] = \Omega^2 \begin{bmatrix} m & -I_{0y} \\ -I_{0y} & I_0 \end{bmatrix}, \tag{16.31}$$

in which

$$\left. \begin{aligned} m &= \int_A \rho \, dA \\ I_{0y} &= \int_A \rho(y - y_0) \, dA \\ I_0 &= \int_A \rho\{x^2 + (y - y_0)^2\} \, dA \end{aligned} \right\} \tag{16.32}$$

are mass integrals of the density (ρ) distribution over the mass area A connected to the member.

The frequency equation (equation 16.21) can then be expressed as

$$\alpha_{11}\alpha_{22} - \alpha_{12}^2 = 0 \tag{16.33}$$

in which

$$\left. \begin{aligned} \alpha_{11} &= n^4\pi^4 EI_y/L^4 - n^2\pi^2 P/L^2 - \Omega^2 m/L, \\ \alpha_{12} &= -n^2\pi^2 M/L^2 - n^2\pi^2 Py_0/L^2 + \Omega^2 I_{0y}/L, \\ \alpha_{22} &= (n^2\pi^2 GJ/L^2 + n^4\pi^4 EI_w/L^4) \\ &\quad - \{n^2\pi^2 P(r_0^2 + y_0^2) - n^2\pi^2 M\beta_x\}/L^2 - \Omega^2 I_0/L. \end{aligned} \right\} \tag{16.34}$$

This leads to a quadratic equation in Ω^2, which can be solved for the natural frequencies Ω.

The natural frequencies of cantilevered beam-columns have been investigated in [50].

16.10 Erection buckling

Spreader beams used for lifting during erection as shown in Figure 16.12a may buckle in a flexural-torsional mode. The inclined upper cable causes compression in the central portion of the beam, while the loads Q being lifted cause bending in the vertical plane.

There are no supports which prevent lateral deflection or twisting, but these buckling deformations are resisted by the loading system itself. Rigid body rotation about the vertical axis through the cable suspension point A is un-restrained, but this rigid body mode is of no structural significance. Rigid body horizontal deflections are restrained because the flexibilities of the upper and lower cables require each pair of attachment points B, B' and C, C' to the beam to remain in a vertical plane containing the suspension point A, as shown in Figure 16.12b. Because of this, torques resisting twist rotations are provided by the upper cable forces acting above the beam, and by the loads Q being lifted which act below the beam.

The elastic flexural-torsional buckling of spreader beams has been analysed in [51]. It was found that the optimum arrangement of the cable attachment points B, B' is close to that for which the cable line intersects the lifting beam axis at the load attachment points C, C', in which case there are no bending actions. The elastic buckling capacity is then close to that calculated for flexural buckling of a

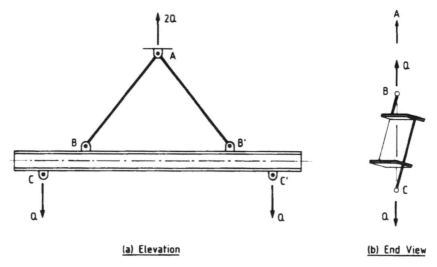

(a) Elevation   (b) End View

Figure 16.12 Spreader beam used during erection.

column simply supported between the load attachment points C, C′. The buckling capacity reduces rapidly as the cable points B, B′ move away from the
optimum positions.

The elastic flexural-torsional buckling of cambered I-section beams during
erection has also been analysed in [52].

16.11 Directed loading

Not all structural loads have the characteristic of gravity loads of remaining
parallel to their original lines of action as the structure deflects. Directed loads
exerted through tension cables or by articulated jacking systems act through a
fixed point, and their lines of action become inclined as the structure deflects
laterally. The resulting lateral components of the directed loads may exert
restoring or disturbing actions on the structure, and may increase or decrease the
buckling resistance.

The effect of directed loading on the flexural buckling of a cantilever column is
shown in Figure 16.13 [53]. The buckling load is given by

$$P/P_y = (\mu L/\pi)^2 \tag{16.35}$$

in which

$$P_y = \pi^2 EI_y/L^2 \tag{16.36}$$

is the flexural buckling load of a simply supported column under gravity loading,

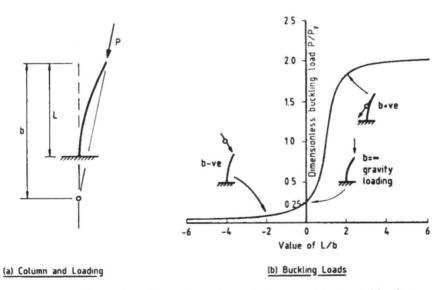

(a) Column and Loading   (b) Buckling Loads

Figure 16.13 Flexural buckling of cantilevered columns with directed loading.

and μL is the solution of

$$(\tan \mu L)/(\mu L) = (1 - b/L) \tag{16.37}$$

in which b defines the point through which the load is directed. For $L/b = 0$, the directed load is a gravity load, and $P/P_y = 0.25$. As L/b increases, the buckling load increases to $P = P_y$ for $L/b = 1.0$, and then approaches a limiting value of 2.045 P_y. For negative values of L/b, the directed load exerts a disturbing component, and the buckling load decreases from 0.25 P_y towards zero.

The effects of directed loading on the elastic flexural-torsional buckling of simply supported beams and cantilevers with concentrated loads have been investigated in [54]. For downwards loading, the buckling resistances are increased when the load direction points are below the load application points, and decreased when above. Beams and cantilevers with top flange loads directed from above may have very low buckling resistances, while there are often very significant increases in the buckling resistances when bottom flange loads are directed to lower points.

16.12 Follower loading

Some static loading methods cause structures to buckle suddenly in a dynamic manner, which cannot be predicted by using the concepts of equilibrium or conservation of energy based on strain and potential energy alone. These include follower loading, as shown in Figure 16.14, in which the applied load P acts tangentially at the end of the member [55, 56]. A simple demonstration of this dynamic buckling can be made by holding a garden hose some distance from its end, and increasing the water flow until instability occurs. In this case the follower loading is supplied by the reaction of the water as it leaves the hose. The same problem may be caused for aerospace vehicles by their rocket thrusts. Related problems may be caused by the thrusts of jet engines mounted on the cantilevered wings of aircraft.

Follower loads are examples of non-conservative loads, for which the work done during deformation is deformation path-dependent. They differ significantly from conservative loads, such as gravity loads for which the work done depends only on the vertical distance between the initial and final positions, and is independent of the deformation path. The principles of static equilibrium and of conservation of energy based on strain and potential energy alone can be applied to conservative load systems.

The path dependence of the work done by a follower load can be demonstrated by considering the two alternative deformation paths shown in Figure 16.14. The first path shown in Figure 16.14a consists of an initial end rotation (during which a small amount of work is done), followed by a final horizontal deflection in which substantial negative work is done by the horizontal component of the load. For the second path shown in Figure 16.14b, a small amount of work is done by

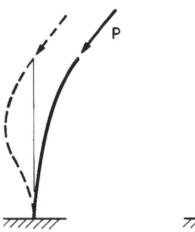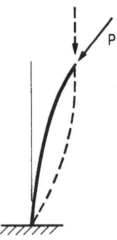

(a) Rotation and Translation   (b) Translation and Rotation

Figure 16.14 Deformation paths for follower loading.

the initial horizontal deflection, followed by another small amount of work during the end rotation.

The behaviour of a non-conservative load system can be analysed dynamically by considering the natural frequency of vibrations. If the follower load point of Figure 16.14 has a concentrated mass M, then it can be shown [55] that the natural frequency of vibration Ω is given by

$$\Omega^2 = \frac{(\mu L)^3 EI/ML^3}{\sin \mu L - \mu L \cos \mu L} \tag{16.38}$$

in which

$$(\mu L)^2 = PL^2/EI. \tag{16.39}$$

The natural frequency Ω increases with P from $\sqrt{(EI/3ML^3)}$ at $P = 0$ until it becomes infinite when

$$\sin \mu L - \mu L \cos \mu L = 0 \tag{16.40}$$

at

$$P = 20.19 EI/L^2. \tag{16.41}$$

Beyond this load, Ω^2 becomes negative, so that Ω becomes imaginary, and the vibration proportional to $\sin \Omega t$ changes from one of steady vibrations to one of exponential divergence proportional to $e^{-i\Omega t}$. This divergence is fed by kinetic energy extracted from the follower loading.

The load at divergence is approximately eight times the value of $\pi^2 EI/4L^2$ for cantilevers with gravity loads. It can be seen that the restoring component of the

320   Special topics

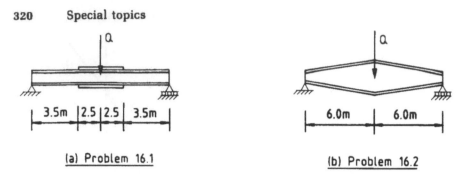

(a) Problem 16.1   (b) Problem 16.2

Figure 16.15 Problems 16.1–16.2.

follower load has a substantial effect on the buckling resistance of the cantilever. However, the application of a static equilibrium buckling analysis to this problem [55] predicts an infinite buckling resistance. Thus, while idealizing the follower load as a gravity load is overly conservative, using a conventional stability analysis is quite dangerous.

A survey of early work on follower loading is given in [55], which includes studies of the flexural-torsional behaviour of beams and cantilevers under concentrated loads. More recent work is reported in [50, 57].

16.13 Problems

PROBLEM 16.1

A simply supported beam whose properties are given in Figure 7.23 is strengthened by welding a plate 300 mm × 20 mm to each flange over the central 5.0 m portion of the 12.0 m span. The beam has a concentrated load at the shear centre at mid-span, which is unrestrained, as shown in Figure 16.15a. Determine the increased elastic flexural-torsional buckling load.

PROBLEM 16.2

A welded steel beam ($E = 200\,000$ MPa, $G = 80\,000$ MPa) has two 300 mm × 20 mm flanges welded to a 10 mm web which tapers so that the overall depth reduces from 800 mm at mid-span to 500 mm at the supports. The beam is simply supported over a span of 12.0 m, and has a central concentrated load at mid-span, which is unrestrained, as shown in Figure 16.15b. Determine the elastic flexural-torsional buckling load.

16.14 References

1. Trahair, N.S. and Kitipornchai, S. (1971) Elastic lateral buckling of stepped I-beams. *Journal of the Structural Division, ASCE,* **97** (ST10), 2535–48.

2. Kitipornchai, S. and Trahair, N.S. (1972) Elastic stability of tapered I-beams. *Journal of the Structural Division, ASCE*, **98** (ST3), 713–28.
3. Galambos, T.V. (ed.) (1988) *Guide to Stability Design Criteria for Metal Structures*, 4th edn, John Wiley, New York.
4. Bradford, M.A. and Cuk, P.E. (1988) Elastic buckling of tapered monosymmetric I-beams. *Journal of Structural Engineering, ASCE*, **114** (5), 977–96.
5. Standards Australia (1990) *AS4100-1990 Steel Structures*, Standards Australia, Sydney.
6. Nethercot, D.A. (1973) Lateral buckling of tapered beams. *Publications*, IABSE, **33-II**, 173–92.
7. Brown, T.G. (1981) Lateral-torsional buckling of tapered I-beams. *Journal of the Structural Division, ASCE*, **107** (ST4), 689–97.
8. Bradford, M.A. (1988) Stability of tapered I-beams, *Journal of Constructional Steel Research*, **9** (3), 195–216.
9. Lee, G.C., Morell, M.L. and Ketter, R.L. (1972) Design of tapered members. *WRC Bulletin No. 173.*
10. American Institute of Steel Construction, (1986) *Load and Resistance Factor Design Specification for Structural Steel Buildings*, AISC, Chicago.
11. Morell, M.L. and Lee, G.C. (1974) Allowable stress for web-tapered beams with lateral restraints. *WRC Bulletin No. 192.* February.
12. Kitipornchai, S. and Trahair, N.S. (1975) Elastic behaviour of tapered monosymmetric I-beams. *Journal of the Structural Division, ASCE*, **101** (ST8), 1661–78.
13. Horne, M.R., Shakir-Khalil, H. and Akhtar, S. (1979) The stability of tapered and haunched beams. *Proceedings, Institution of Civil Engineers*, **67** (2), 677–94.
14. Horne, M.R. Shakir-Khalil, H. and Akhtar, S. (1979) Tests on tapered and haunched beams. *Proceedings, Institution of Civil Engineers*, **67** (2), 845–50.
15. Bradford, M.A. (1988) Lateral stability of tapered beam-columns with elastic restraints. *The Structural Engineer*, **66** (22), 376–82.
16. Bradford, M.A. (1989) Inelastic buckling of tapered monosymmetric I-beams. *Engineering Structures*, **11** (2), 119–26.
17. Trahair, N.S. and Booker, J.R. (1970) Optimum elastic columns. *International Journal of Mechanical Sciences*, **12** (11), 973–83.
18. Wang, C.-M. Thevendran, V., Teo, K.L. and Kitipornchai, S. (1986) Optimal design of tapered beams for maximum buckling strength. *Engineering Structures*, **8** (4), 276–84.
19. Wang, C.-M. Kitipornchai, S. and Thevendran, V. (1990) Optimal designs of I-beams against lateral buckling. *Journal of Engineering Mechanics, ASCE*, **116** (9), 1902–23.
20. Bleich, F. (1952) *Buckling Strength of Metal Structures*, McGraw-Hill, New York.
21. Attard, M.M. and Lawther, R. (1988) Effect of secondary warping on lateral buckling, in *Proceedings*, 11th Australasian Conference on the Mechanics of Structures and Materials, University of Auckland, pp. 219–25.
22. Cherry, S. (1960) The stability of beams with buckled compression flanges. *The Structural Engineer*, **38** (9), 277–85.
23. Bradford, M.A. and Hancock, G.J. (1984) Elastic interaction of local and lateral buckling in beams. *Thin-Walled Structures*, **2**, 1–25.
24. Wang, S.T., Yost, M.I. and Tien, Y.L. (1977) Lateral buckling of locally buckled beams using finite element techniques. *International Journal of Computers and Structures*, **7** 469–75.
25. Menken, C.M., Groot, W.J. and Stallenberg, G.A.J. (1991) Interactive buckling of beams in bending. *Thin-Walled Structures*, **12**, 415–34.

26. Kubo, M. and Fukumoto, Y. (1988) Lateral torsional buckling of thin-walled I-beams. *Journal of Structural Engineering, ASCE,* 114 (4), 841–55.
27. Hancock, G.J. (1978) Local, distortional, and lateral buckling of I-beams. *Journal of the Structural Division, ASCE,* 104 (ST11), 1787–98.
28. Lau, S.C.W. and Hancock, G.J. (1990) Inelastic buckling of channel columns in the distortional mode. *Thin-Walled Structures,* 10, 59–84.
29. Bradford, M.A. and Trahair, N.S. (1981) Distortional buckling of I-beams. *Journal of Structural Engineering, ASCE,* 107 (2), 355–70.
30. Johnson, R.P. and Bradford, M.A. (1983) Distortional lateral buckling of unstiffened composite bridge girders, in *Proceedings,* Michael R. Horne Conference on the Instability and Plastic Collapse of Steel Structures, Granada, London, pp. 569–80.
31. Seah, L.K. and Khong, P.W. (1990) Lateral-torsional buckling of channel beams. *Journal of Constructional Steel Research,* 17 (4) 265–82.
32. Cheung, Y.K. (1976) *Finite Strip Method in Structural Analysis,* Pergamon Press, New York.
33. Hancock, G.J., Bradford, M.A. and Trahair, N.S. (1980) Web distortion and flexural torsional buckling. *Journal of Structural Engineering, ASCE,* 106 (7), 1551–71.
34. Bradford, M.A. and Trahair, N.S. (1982) Distortional buckling of thin-web beam-columns. *Engineering Structures,* 4 (1) 2–10.
35. Chin, C.K., Al-Bermani, F.G.A. and Kitipornchai, S. (1992) Stability of thin-walled members having arbitrary flange shape and flexible web. *Engineering Structures,* 14 (2), 121–32.
36. Bradford, M.A. (1992) Lateral-distortional buckling of steel I-section members. *Journal of Constuctional Steel Research,* 23 (1–3), 97–116.
37. Trahair, N.S. and Woolcock, S.T. (1973) Effect of major axis curvature on I-beam stability. *Journal of the Engineering Mechanics Division, ASCE,* 99 (EM1), 85–98.
38. Vacharajittiphan, P., Woolcock, S.T. and Trahair, N.S. (1974) Effect of in-plane deformation on lateral buckling. *Journal of Structural Mechanics,* 3 (1), 29–60.
39. Roberts, T.M. and Azizian, Z.G. (1983) Influence of pre-buckling displacements on the elastic critical loads of thin-walled bars of open cross-section. *International Journal of Mechanical Sciences,* 25 (2), 93–104.
40. Pi, Y.L. and Trahair, N.S. (1992) Prebuckling deflections and lateral buckling–theory. *Journal of Structural Engineering, ASCE,* 118 (11), 2949–66.
41. Pi, Y.L. and Trahair, N.S. (1992) Prebuckling deflections and lateral buckling–applications. *Journal of Structural Engineering, ASCE,* 118 (11), 2967–85.
42. Woolcock, S.T. and Trahair, N.S. (1974) Post-buckling behaviour of determinate beams. *Journal of the Engineering Mechanics Division, ASCE,* 100 (EM2), 151–71.
43. Timoshenko, S.P. and Gere, J.M. (1961) *Theory of Elastic Stability,* 2nd edn, McGraw-Hill, New York.
44. Masur, E.F. and Milbradt, K.P. (1957) Collapse strength of redundant beams after lateral buckling. *Journal of Applied Mechanics, ASME,* 24 (2), 283–8.
45. Woolcock, S.T. and Trahair, N.S. (1975) Post-buckling of redundant rectangular beams. *Journal of the Engineering Mechanics Division, ASCE,* 101 (EM4), 301–16.
46. Woolcock, S.T. and Trahair, N.S. (1976) Post-buckling of redundant I-beams. *Journal of the Engineering Mechanics Division, ASCE,* 102 (EM2), 293–312.
47. Flugge, W. (1967) *Viscoelasticity,* Blaisdell Publishing Co., Waltham, MA.
48. Booker, J.R. Frankham, B.S. and Trahair, N.S. (1974) Stability of visco-elastic structural members. *Civil Engineering Transactions,* Institution of Engineers, Australia, CE16 (1), 45–51.

49. Clough, R.W. and Penzien, J. (1975) *Dynamics of Structures*, McGraw-Hill, New York.

50. Attard, M.M. and Somervaille, I.J. (1987) Stability of thin-walled open beams under non-conservative loads. *Mechanics of Structures and Machines*, **15** (3), 395–412.

51. Dux, P.F. and Kitipornchai, S. (1990) Buckling of suspended I-beams. *Journal of Structural Engineering, ASCE*, **116** (7) 1877–91.

52. Peart, W.L., Rhomberg, E.G. and James, R.W. (1992) Buckling of suspended cambered girders. *Journal of Structural Engineering, ASCE*, **118** (2), 505–528.

53. Simitses, G.J. (1976) *An Introduction to the Elastic Stability of Structures*, Prentice-Hall, Englewood Cliffs, NJ.

54. Ings, N.L. and Trahair, N.S. (1987) Beam and column buckling under directed loading. *Journal of Structural Engineering, ASCE*, **113** (6), 1251–63.

55. Bolotin, V.V. (1963) *Nonconservative Problems of the Theory of Elastic Stability*, Pergamon Press, New York.

56. Ziegler, H. (1968) *Principles of Structural Stability*, Blaisdell Publishing Co., Waltham, MA.

57. Dabrowski, R. (1991) Two examples of instability under follower load. *Journal of Constructional Steel Research*, **19** (2), 153–61.

17 Appendices

17.1 In-plane bending

17.1.1 MEMBER AND LOADING

The member of length L shown in Figure 17.1 is of doubly symmetric cross-section. The x, y principal centroidal axes are defined by

$$\int_A x \, dA = \int_A y \, dA = \int_A xy \, dA = 0. \tag{17.1}$$

The following cross-section properties are defined:

$$\left. \begin{array}{l} A = \displaystyle\int_A dA, \\[3mm] I_n = \displaystyle\int_A y^2 \, dA. \end{array} \right\} \tag{17.2}$$

The member is acted on by end loads $Q_{y1}, Q_{y2}, Q_{z1}, Q_{z2}$, and distributed loads per unit length q_y, q_z which act at the centroid, and moments M_{x1}, M_{x2}. These actions move with the cross-section of the member, but remain parallel to their original directions and planes defined by the original Y, Z axes of the straight member.

17.1.2 DISPLACEMENTS, ROTATIONS, AND CURVATURES

The centroid of a cross-section undergoes small deflections v, w parallel to the Y, Z axes under the applied actions, as shown in Figure 17.2. The cross-section rotates $-v'$ approximately ($' \equiv d/dz$), and the z axis has an approximate curvature $-v''$.

The displacements v_P, w_P in the Y, Z directions of a point P in the cross-section which has coordinates x, y are given by

$$v_P \approx v - yv'^2/2 \tag{17.3}$$

$$w_P \approx w - yv'. \tag{17.4}$$

17.1.3 STRAINS

The longitudinal normal strain ε_P at P parallel to the z axis may be defined in terms of the rates of change of the deflections v_P, w_P along the element, as shown in

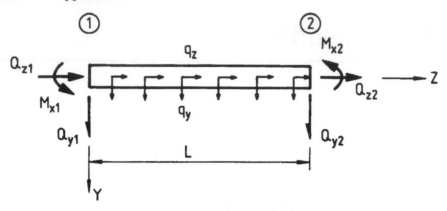

Figure 17.1 Member and loading.

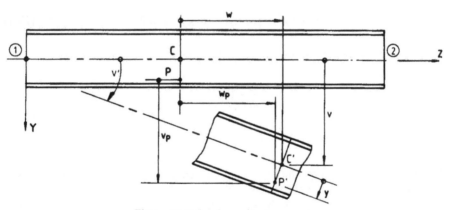

Figure 17.2 In-plane displacements.

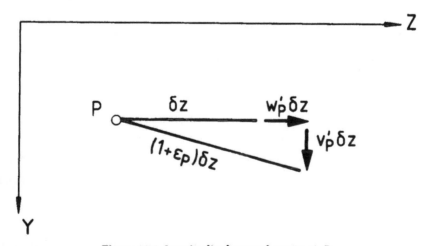

Figure 17.3 Longitudinal normal strain at P.

Figure 17.3. The new elemental length is given by

$$(1 + \varepsilon_p)\delta z = \sqrt{\{(\delta z + w_p'\delta z)^2 + (v_p'\delta z)^2\}}, \tag{17.5}$$

so that

$$\varepsilon_p \approx w_p' + v_p'^2/2, \tag{17.6}$$

whence

$$\varepsilon_p \approx (w' - yv'') + v'^2/2. \tag{17.7}$$

Any small shear strains caused by bending shear forces are neglected in this treatment.

17.1.4 STRESSES AND STRESS RESULTANTS

The longitudinal normal stress at the point P in the cross-section is given by

$$\sigma_p = E\varepsilon_p \tag{17.8}$$

in which E is the Young's modulus of elasticity.

The resultants $\{\sigma\}$ of the stresses σ_p may be defined by

$$\{\sigma\} = \left\{ \begin{matrix} \int_A \sigma_p \, dA \\ \int_A \sigma_p y \, dA \end{matrix} \right\} = \left\{ \begin{matrix} N \\ M_x \end{matrix} \right\} \tag{17.9}$$

in which N is the internal tension and M_x is the internal bending moment, whence

$$\left. \begin{matrix} N = EA(w' + v'^2/2) \\ M_x = -EI_x v'' \end{matrix} \right\} \tag{17.10}$$

after using equations 17.1, 17.2, 17.7, and 17.8. Thus

$$\sigma_p = N/A + M_x y/I_x. \tag{17.11}$$

17.1.5 TOTAL POTENTIAL

The total potential U_T (see section 2.3) of the strained length L of the member and its loading system is given by

$$U_T = U + V \tag{17.12}$$

in which U is the strain energy stored in the member, and V is the potential energy of the loading system measured from the straight position of the member.

The strain energy U (see section 2.2.3) may be expressed as

$$U = \frac{1}{2}\int_0^L \int_A \varepsilon_p \sigma_p \, dA \, dz \tag{17.13}$$

and when equations 17.7, 17.8, 17.10, and 17.11 are substituted, this becomes

$$U = \frac{1}{2}\int_0^L \{EA(w' + v'^2/2)^2 + EI_x v''^2\} \, dz \tag{17.14}$$

after using equations 17.1 and 17.2.

The potential energy V of the loading system (see section 2.2.5) may be expressed as

$$V = -\int_0^L (q_y v + q_z w)\,dz - \sum_{1,2}(Q_y v + Q_z w - M_x v').$$ (17.15)

17.1.6 VIRTUAL DISPLACEMENTS AND EQUILIBRIUM

Suppose the member undergoes a set of virtual displacements δv, δw from an equilibrium position v, w while under the action of constant loads and moments. For equilibrium of the position v, w, the principle of stationary total potential equivalent to the principle of virtual work (see section 2.4) requires that

$$\delta U_T = 0$$ (17.16)

for all sets of virtual displacements $\delta v, \delta w$.

Substituting equations 17.12, 17.14, and 17.15 leads to

$$\int_0^L [\delta v'\{EA(w' + v'^2/2)v'\} + \delta v''(EI_x v'') + \delta w'\{EA(w' + v'^2/2)\} - \delta v q_y - \delta w q_z]\,dz$$

$$- \sum_{1,2}(\delta v Q_y + \delta w Q_z - \delta v' M_x) = 0.$$ (17.17)

Integrating by parts leads to

$$\int_0^L [\delta v\{-(Nv')' - M_x'' - q_y\} + \delta w\{-N' - q_z\}]\,dz$$

$$+ [\delta v(Nv') - \delta v' M_x + \delta v M_x' + \delta w N]_0^L - \sum_{1,2}(\delta v Q_y + \delta w Q_z - \delta v' M_x) = 0$$ (17.18)

after substituting equation 17.10.

Since this must hold for all admissible sets of virtual displacements δv, δw, then

$$-M_x'' - (Nv')' = q_y,$$ (17.19)

$$-N' = q_z,$$ (17.20)

which are the differential equilibrium equations, and

$$\left.\begin{array}{l} (M_x' + Nv')_L = Q_{y2}, \\ -(M_x' + Nv')_0 = Q_{y1}, \\ -(M_x)_L = -M_{x2}, \\ (M_x)_0 = -M_{x1}, \\ (N)_L = Q_{z2}, \\ -(N)_0 = Q_{z1}, \end{array}\right\}$$ (17.21)

which are the boundary conditions for these equations.

17.1.7 NEUTRAL EQUILIBRIUM AND BUCKLING

For the stability of an equilibrium position defined by v, w to be neutral (see section 2.5.3)

$$\tfrac{1}{2}\delta^2 U_T = 0 \tag{17.22}$$

for any set of small buckling displacements $\{\delta v, \delta w\}$ which takes place from the equilibrium position under constant loads and moments. In this case, the adjacent buckled position $\{v + \delta v, w + \delta w\}$ is also one of equilibrium.

Using equation 17.15 and noting that the loads q_y, q_z, Q_y, Q_z and moments M_x remain constant and that the second variations of v, w, and v' vanish so that

$$\tfrac{1}{2}\delta^2 V = 0 \tag{17.23}$$

then equation 17.22 for neutral equilibrium leads to

$$\frac{1}{2}\int_0^L [EA(\delta w')^2 + 2EAv'\,\delta v'\,\delta w' + EI_x(\delta v'')^2 + \{N + EAv'^2\}(\delta v')^2]\,dz = 0 \tag{17.24}$$

or

$$\frac{1}{2}\int_0^L [EA(w_b')^2 + 2EAv'(v_b'\,w_b') + EI_x(v_b'')^2 + \{N + EAv'^2\}(v_b')^2]\,dz = 0, \tag{17.25}$$

in which $\{v_b, w_b\} \equiv \{\delta v, \delta w\}$ are the buckling displacements.

In the special case of inextensional buckling for which the stress resultants N remain constant so that

$$w_b' + v_b'^2/2 = 0 \tag{17.26}$$

and if the pre-buckling transverse displacements v are zero (i.e. Q_y, q_y, M_x are zero), then equation 17.25 becomes

$$\frac{1}{2}\int_0^L \{EI_x(v_b'')^2 + N(v_b')^2\}\,dz = 0 \tag{17.27}$$

for infinitesimal buckling deflections v_b. This is the energy equation for inextensional buckling. The sign convention for N in this equation is tension positive, and so N is negative for a member in compression.

17.1.8 EQUILIBRIUM OF THE BUCKLED POSITION

Because the buckled position $\{v + v_b, w + w_b\}$ is one of equilibrium, this position can be found by considering a set of virtual displacements δv_b, δw_b from it. Thus using equation 17.17

$$\int_0^L [\delta v_b'\{EA[w' + w_b' + (v' + v_b')^2/2](v' + v_b')\} + \delta v_b''EI_x(v'' + v_b'')$$
$$+ \delta w_b'\{EA(w' + w_b' + (v' + v_b')^2/2)\} - \delta v_b q_y - \delta w_b q_z]\,dz$$
$$- \sum_{1,2}(\delta v_b Q_y + \delta w_b Q_z - \delta v_b' M_x) = 0. \tag{17.28}$$

Equation 17.17 can also be applied to the prebuckled equilibrium position $\{v, w\}$, whence

$$\int_0^L [\delta v_b'\{EA(w' + v'^2/2)v'\} + \delta v_b'' EI_x v'' + \delta w_b' EA(w' + v'^2/2) - \delta v_b q_y - \delta w_b q_z]\,dz$$

$$- \sum_{1,2} (\delta v_b Q_y + \delta w_b Q_z - \delta v_b' M_x) = 0. \tag{17.29}$$

Subtracting equation 17.29 from equation 17.28 leads to

$$\int_0^L [\delta v_b' EA\{w' + v'^2/2\}v_b' + \{\delta w_b' + \delta v_b' v'\} EA\{w_b' + v'v_b'\} + \delta v_b'' EI_x v_b'']\,dz = 0 \tag{17.30}$$

after assuming that the buckling displacements $\{v_b, w_b\}$ are infinitesimal. Integrating by parts leads to

$$\int_0^L [\delta v_b\{ -(Nv_b')' + (EI_x v_b'')'' - (EA\{w_b' + v'v_b'\}v')'\} + \delta w_b\{ -(EA\{w_b' + v'v_b'\})'\}]\,dz$$

$$+ [\delta v_b(Nv_b') + \delta v_b(EA\{w_b' + v'v_b'\}v') + \delta v_b'(EI_x v_b'') - \delta v_b(EI_x v_b'')'$$

$$+ \delta w_b(EA\{w_b' + v'v_b'\})]_0^L = 0 \tag{17.31}$$

after substituting equation 17.10. Since this must hold for all admissible sets of virtual displacements $\{\delta v_b, \delta w_b\}$, then

$$(EI_x v_b'')'' - (Nv_b')' - \{EA(w_b' + v'v_b')v'\}' = 0 \tag{17.32}$$

$$- \{EA(w_b' + v'v_b')\}' = 0 \tag{17.33}$$

which are the buckling differential equilibrium equations, and

$$\left. \begin{array}{r} [-(EI_x v_b'')' + Nv_b' + EA(w_b' + v'v_b')v']_{0,L} = 0 \\ [EI_x v_b'']_{0,L} = 0 \\ [EA(w_b' + v'v_b')]_{0,L} = 0 \end{array} \right\} \tag{17.34}$$

which are the boundary conditions for these equations.

These buckling equilibrium equations can also be obtained from the calculus of variations, according to which the functions v_b, w_b which make

$$\frac{1}{2}\delta^2 U_T = \int_0^L F(z, v_b, v_b', v_b'', w_b, w_b', w_b'')\,dz \tag{17.35}$$

stationary satisfy the conditions

$$\frac{\partial F}{\partial v_b} - \frac{\mathrm{d}}{\mathrm{d}z}\left(\frac{\partial F}{\partial v_b'}\right) + \frac{\mathrm{d}^2}{\mathrm{d}z^2}\left(\frac{\partial F}{\partial v_b''}\right) = 0, \tag{17.36}$$

$$\frac{\partial F}{\partial w_b} - \frac{\mathrm{d}}{\mathrm{d}z}\left(\frac{\partial F}{\partial w_b'}\right) + \frac{\mathrm{d}^2}{\mathrm{d}z^2}\left(\frac{\partial F}{\partial w_b''}\right) = 0. \tag{17.37}$$

Thus using the energy equation of equation 17.25 in place of equation 17.35 leads also to equations 17.32 and 17.33.

In the special case for which the pre-buckling transverse displacements v remain zero, equation 17.32 simplifies to

$$(EI_x v_b'')'' - (Nv_b')' = 0 \tag{17.38}$$

which is the differential equilibrium equation for the buckled position. The first two of the boundary conditions of equation 17.34 reduce to

$$\left.\begin{array}{r}[-(EI_x v_b'')' + Nv_b']_{0,L} = 0, \\ [EI_x v_b'']_{0,L} = 0, \end{array}\right\} \tag{17.39}$$

which are the boundary conditions for equation 17.38.

17.1.9 ORTHOGONALITY RELATIONSHIPS

The differential equilibrium equation for the buckled position (equation 17.38) can be written as

$$(EI_x v_b'')'' - \lambda(Nv_b')' = 0 \tag{17.40}$$

in which N now represents an initial distribution of axial stress resultants and λ is a load factor. This equation has an infinite number of distinct solutions (eigenfunctions) v_{bn}, each of which is associated with a load factor (eigenvalue) λ_n. These eigenvalues may be arranged in ascending order so that

$$0 < \lambda_1 < \lambda_2 < \cdots < \lambda_n < \cdots.$$

The equality of equation 17.40 can be used to show that

$$\int_0^L [v_{bs}\{(EI_x v_{br}'')'' - \lambda_r(Nv_{br}')'\} - v_{br}\{(EI_x v_{bs}'')'' - \lambda_s(Nv_{bs}')'\}]\,dz = 0, \tag{17.41}$$

whence

$$\int_0^L [\{v_{bs}'' EI_x v_{br}'' \to \lambda_r v_{bs}' Nv_{br}'\} - \{v_{br}'' EI_x v_{bs}'' \to \lambda_s v_{br}' Nv_{bs}'\}]\,dz = 0, \tag{17.42}$$

after integration by parts and substitution of the boundary conditions of equation 17.39.

Thus

$$\int_0^L (\lambda_r - \lambda_s)Nv_{br}' v_{bs}'\,dz = 0 \tag{17.43}$$

and so

$$\int_0^L Nv_{br}' v_{bs}'\,dz = 0 \tag{17.44}$$

when $\lambda_r \neq \lambda_s$.

A similar argument starting from

$$\int_0^L [v_{bs}\{(EI_x v_{br}'')''/\lambda_r - (Nv_{br}')'\} - v_{br}\{(EI_x v_{bs}'')''/\lambda_s - (Nv_{bs}')'\}]\, dz = 0 \quad (17.45)$$

can be used to show that

$$\int_0^L EI_x v_{br}'' v_{bs}''\, dz = 0, \quad (17.46)$$

when $\lambda_r \neq \lambda_s$. Equations 17.44 and 17.46 are the orthogonality relationships associated with equation 17.38.

17.1.10 RAYLEIGH'S QUOTIENT
The energy equation for inextensional buckling (equation 17.27) can be written as

$$\frac{1}{2}\int_0^L \{EI_x(v_{bn}'')^2 + \lambda_n N(v_{bn}')^2\}\, dz = 0, \quad (17.47)$$

in which N is again an initial distribution of axial stress resultants and λ_n is one of the load factors at buckling. Thus

$$\lambda_n = \frac{-\displaystyle\int_0^L EI_x(v_{bn}'')^2\, dz}{\displaystyle\int_0^L N(v_{bn}')^2\, dz}. \quad (17.48)$$

By writing

$$f_n = \int_0^L N(v_{bn}')^2\, dz \quad (17.49)$$

then

$$-\int_0^L EI_x(v_{bn}'')^2\, dz = \lambda_n f_n. \quad (17.50)$$

An approximation

$$v_{ba} = \sum(a_n v_{bn}) \quad (17.51)$$

for the buckled shape v_{b1} associated with the lowest eigenvalue λ_1 may be formed from the eigenfunctions v_{bn} and a set of arbitrary multipliers a_n. The corresponding approximation λ_a for λ_1 is obtained from equation 17.48 as

$$\lambda_a = \frac{-\displaystyle\int_0^L EI_x\left[\sum(a_n v_{bn}'')\right]^2\, dz}{\displaystyle\int_0^L N\left[\sum(a_n v_{bn}')\right]^2\, dz}, \quad (17.52)$$

which simplifies to

$$\lambda_a = \frac{-\int_0^L EI_x \sum (a_n^2 v_{bn}''^2) \, dz}{\int_0^L N \sum (a_n^2 v_{bn}'^2) \, dz} \tag{17.53}$$

after using the orthogonality relationships of equations 17.44 and 17.46. Thus

$$\lambda_a = \frac{\lambda_1 a_1^2 f_1 + \lambda_2 a_2^2 f_2 + \cdots}{a_1^2 f_1 + a_2^2 f_2 + \cdots} \tag{17.54}$$

after substituting equations 17.49 and 17.50, and so

$$\lambda_a > \lambda_1 \tag{17.55}$$

since

$$\lambda_1 < \lambda_2 < \lambda_3 < \cdots.$$

This result shows that Rayleigh's quotient λ_a is never less than the lowest eigenvalue λ_1.

17.1.11 MATRIX FORMULATION

The in-plane bending behaviour discussed in sections 17.1.1 to 17.1.8 may also be presented in a matrix formulation. The normal strain ε_p of equation 17.7 can be expressed in terms of the generalized strains $\{\varepsilon\}$ by

$$\varepsilon_p = \{S\}^T \{\varepsilon\} \tag{17.56}$$

in which

$$\{\varepsilon\} = \left\{ \begin{matrix} w' \\ -v'' \end{matrix} \right\} + \left\{ \begin{matrix} v'^2/2 \\ 0 \end{matrix} \right\} \tag{17.57}$$

and

$$\{S\}^T = \{1, y\}. \tag{17.58}$$

Equation 17.57 can also be expressed as

$$\{\varepsilon\} = [B_L + B_Q]\{\Phi\} \tag{17.59}$$

in which

$$\{\Phi\} = \{v', v'', w'\}^T \tag{17.60}$$

$$[B_L] = \begin{bmatrix} 0 & 0 & 1 \\ 0 & -1 & 0 \end{bmatrix} \tag{17.61}$$

$$2[B_Q] = \begin{bmatrix} v' & 0 & 0 \\ 0 & 0 & 0 \end{bmatrix} \tag{17.62}$$

It can be shown that

$$[\delta B_Q]\{\Phi\} \equiv [B_Q]\{\delta\Phi\} \tag{17.63}$$

Generalized stress resultants $\{\sigma\}$ corresponding to the generalized strains $\{\varepsilon\}$ given by equation 17.57 may be defined by

$$\{\delta\varepsilon\}^{\mathrm{T}}\{\sigma\} = \int_A \delta\varepsilon_{\mathrm{p}}\sigma_{\mathrm{p}}\,\mathrm{d}A \tag{17.64}$$

Substituting equations 17.8 and 17.56 leads to

$$\{\sigma\} = [D]\{\varepsilon\} \tag{17.65}$$

in which

$$[D] = \int_A \{S\}^{\mathrm{T}}E\{S\}\,\mathrm{d}A \tag{17.66}$$

so that

$$[D] = \begin{bmatrix} EA & 0 \\ 0 & EI_x \end{bmatrix} \tag{17.67}$$

The generalized stresses $\{\sigma\}$ defined in this way are identical with those of equations 17.9 and 17.10.

The strain energy component U of the total potential U_{T} given by equation 17.13 can be expressed as

$$U = \frac{1}{2}\int_0^L \{\varepsilon\}^{\mathrm{T}}\{\sigma\}\,\mathrm{d}z \tag{17.68}$$

after using equations 17.56, 17.65, and 17.66.

The potential energy component V of the total potential U_{T} given by equation 17.15 can be expressed as

$$V = -\int_0^L \{v\}^{\mathrm{T}}\{q\}\,\mathrm{d}z - \sum_{1,2}\{v\}^{\mathrm{T}}\{Q\} \tag{17.69}$$

in which

$$\{q\} = \{q_y, q_z, 0\}^{\mathrm{T}} \tag{17.70}$$

$$\{Q\} = \{Q_y, Q_z, M_x\}^{\mathrm{T}} \tag{17.71}$$

and

$$\{v\} = \{v, w, -v'\}^{\mathrm{T}} \tag{17.72}$$

when q_y, Q_y, M_x act at the centroid. When q_y, Q_y act away from the centroid, then $\{v\}$ can be expressed in the form of

$$\{v\} = [A_{\mathrm{L}} = A_{\mathrm{Q}}]\{\theta\} \tag{17.73}$$

in which

$$\{\theta\} = \{v, v', w\}^{\mathrm{T}}. \tag{17.74}$$

It can be shown that

$$[\delta A_{\mathrm{Q}}]\{\theta\} \equiv [A_{\mathrm{Q}}]\{\delta\theta\} \tag{17.75}$$

which is similar to equation 17.63.

The virtual work equilibrium requirement of equation 17.16 can be expressed as

$$\int_0^L (\{\delta\varepsilon\}^T \{\sigma\} - \{\delta v\}^T \{q\})\,dz - \sum_{1,2} \{\delta v\}^T \{Q\} = 0. \tag{17.76}$$

In this

$$\{\delta\varepsilon\} = [B_L + 2B_Q]\{\delta\Phi\} \tag{17.77}$$

and

$$\{\delta v\} = [A_L + 2A_Q]\{\delta\theta\} \tag{17.78}$$

which make use of equations 17.59, 17.63, 17.73, and 17.75.

It is now assumed that the displacements v, w can be expressed in the form of

$$\{v, w\}^T = [N]\{\delta\} \tag{17.79}$$

in which

$$\{\delta\} = \{v_1, v_2, v_1', v_2', w_1, w_2,\}^T \tag{17.80}$$

and the elements of $[N]$ are functions of z. In this case, $\{\Phi\}$, $\{\theta\}$, and $\{\{\theta_1\}^T, \{\theta_2\}^T\}^T$ can be expressed as

$$\{\Phi\} = [N_\sigma]\{\delta\} \tag{17.81}$$

$$\{\theta\} = [N_q]\{\delta\} \tag{17.82}$$

and

$$\{\{\theta_1\}^T, \{\theta_2\}^T\}^T = [N_Q]\{\delta\}. \tag{17.83}$$

Substituting these relationships and equations 17.77 and 17.78 allows equation 17.76 to be expressed as

$$\{\delta\delta\}^T \Bigg(\int_0^L ([N_\sigma]^T[B_L + 2B_Q]^T\{\sigma\} - [N_q]^T[A_L + 2A_Q]^T\{q\})\,dz$$

$$- [N_Q]^T[A_{LQ} + 2A_{QQ}]^T\{\{Q_1\}^T, \{Q_2\}^T\}^T \Bigg) = 0. \tag{17.84}$$

Since this must hold for all sets of virtual deformations defined by $\{\delta\delta\}$, then

$$\int_0^L ([N_\sigma]^T[B_L + 2B_Q]^T\{\sigma\} - [N_q]^T[A_L + 2A_Q]^T\{q\})\,dz$$

$$- [N_Q]^T[A_{LQ} + 2A_{QQ}]^T\{Q_1^T, Q_2^T\}^T = \{0\} \tag{17.85}$$

which are the non-linear equilibrium equations.

The neutral equilibrium condition of equation 17.22 is equivalent to

$$\frac{1}{2}\int_0^L (\{\delta\varepsilon\}^T\{\delta\sigma\} + \{\delta^2\varepsilon\}^T\{\sigma\}) - \{\delta^2 v\}^T\{q\})\,dz - \frac{1}{2}\sum_{1,2}\{\delta^2 v\}^T\{Q\} = 0 \tag{17.86}$$

in which

$$\{\delta^2\varepsilon\} = [2\delta B_Q]\{\delta\Phi\} \tag{17.87}$$

and

$$\{\delta^2 v\} = \{0\}. \tag{17.88}$$

Using equation 17.81 with equations 17.63 and 17.65 leads to

$$\{\delta\delta\}^{\mathrm{T}}[K_{\mathrm{T}}]\{\delta\delta\} = 0 \tag{17.89}$$

or

$$\{\delta_b\}^{\mathrm{T}}[K_{\mathrm{T}}]\{\delta_b\} = 0 \tag{17.90}$$

in which

$$[K_{\mathrm{T}}] = \int_0^L [N_o]^{\mathrm{T}}([B_{\mathrm{L}} + 2B_{\mathrm{Q}}]^{\mathrm{T}}[D][B_{\mathrm{L}} + 2B_{\mathrm{Q}}] + [M_o])[N_o]\,\mathrm{d}z \tag{17.91}$$

which makes use of the identity

$$[2\delta B_{\mathrm{Q}}]^{\mathrm{T}}\{\sigma\} = [M_o]\{\delta\Phi\}, \tag{17.92}$$

in which

$$[M_o] = \begin{bmatrix} N & 0 & 0 \\ 0 & 0 & 0 \\ 0 & 0 & 0 \end{bmatrix}. \tag{17.93}$$

The neutral equilibrium condition of equation 17.90 is satisfied when

$$|K_{\mathrm{T}}| = 0 \tag{17.94}$$

and the load sets which satisfy this equation are the buckling load sets.

Because the adjacent buckled position is one of equilibrium, the condition for neutral equilibrium may also be obtained by using virtual work to analyse the equilibrium of the buckled position. Thus if δv, δw are virtual displacements from the buckled position $v + v_b$, $w + w_b$, then

$$\left.\begin{array}{l} \{\delta\varepsilon\} = [B_{\mathrm{L}} + 2B_{\mathrm{Q}} + 2B_{\mathrm{Qb}}][N_o]\{\delta\delta\} \\ \{\delta v\} = [A_{\mathrm{L}} + 2A_{\mathrm{Q}} + 2A_{\mathrm{Qb}}][N_q]\{\delta\delta\} \end{array}\right\} \tag{17.95}$$

and

$$\{\sigma\} + \{\sigma_b\} = \{\sigma\} + [D][B_{\mathrm{L}} + 2B_{\mathrm{Q}}][N_o]\{\delta_b\}. \tag{17.96}$$

Substituting into the virtual work equilibrium condition of equation 17.76 leads to

$$\begin{aligned} \{\delta\delta\}^{\mathrm{T}}\Bigg(&\int_0^L ([N_o]^{\mathrm{T}}[B_{\mathrm{L}} + 2B_{\mathrm{Q}}]^{\mathrm{T}}\{\sigma\} - [N_q]^{\mathrm{T}}[A_{\mathrm{L}} + 2A_{\mathrm{Q}}]^{\mathrm{T}}\{q\})\,\mathrm{d}z \\ &- [N_{\mathrm{Q}}]^{\mathrm{T}}[A_{\mathrm{LQ}} + 2A_{\mathrm{QQ}}]^{\mathrm{T}}\{\{Q_1\}^{\mathrm{T}}, \{Q_2\}^{\mathrm{T}}\}^{\mathrm{T}} \\ &+ \int_0^L ([N_o]^{\mathrm{T}}([2B_{\mathrm{Qb}}]^{\mathrm{T}}\{\sigma\} + [B_{\mathrm{L}} + 2B_{\mathrm{Q}}]^{\mathrm{T}}[D][B_{\mathrm{L}} + 2B_{\mathrm{Q}}][N_o]\{\delta_b\})\,\mathrm{d}z \\ &- \int_0^L [N_q]^{\mathrm{T}}[M_q][N_q]\{\delta_b\}\,\mathrm{d}z \Bigg) = 0 \end{aligned} \tag{17.97}$$

which uses the identity $[2A_{Qb}]^T\{q\} \equiv [M_q][N_q]\{\delta_b\}$, which is similar to equation 17.92. If equation 17.84 for the equilibrium of the pre-buckled position is substituted, then this simplifies for infinitesimal buckling displacements to

$$\{\delta\delta\}^T[K_T]\{\delta_b\} = 0,\qquad\qquad(17.98)$$

and since this must hold for all sets of virtual displacements $\{\delta\delta\}$, then

$$[K_T]\{\delta_b\} = \{0\}\qquad\qquad(17.99)$$

which are the buckling equilibrium equations.

17.2 Uniform torsion

17.2.1 NARROW RECTANGULAR SECTION MEMBER AND LOADING

The narrow rectangular section member shown in Figure 17.4 is of width b, thickness, t, and length L. The member is acted on by uniform end torques M_{u1}, M_{u2}, and distributed uniform torques per unit length m_u. Any warping torques M_{w1}, M_{w2} and warping torques per unit length m_w (see section 17.3) are assumed to be negligible.

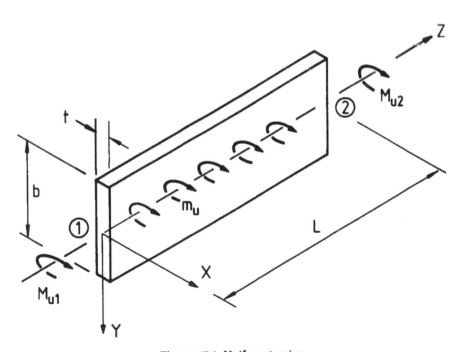

Figure 17.4 Uniform torsion.

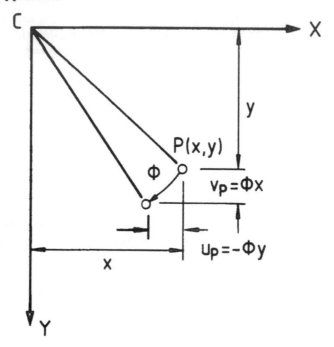

Figure 17.5 Twist rotation in uniform torsion.

17.2.2 TWIST ROTATIONS AND DEFLECTIONS

A member cross-section rotates through a small angle ϕ about the Z axis, as shown in Figure 17.5. When the distributed torque m_u is zero, the twist rotation varies linearly along the member, so that the twist $\phi' \equiv d\phi/dz$ is constant.

The small rotation ϕ causes small displacements u_P, v_P in the X, Y directions of a point P in the cross-section which has coordinates x, y. These displacements are given by

$$u_P = -\phi y, \tag{17.100}$$

$$v_P = \phi x. \tag{17.101}$$

The point P also displaces w_P in the Z direction, where w_P is a function of the coordinates x, y, so that

$$w_P = \phi'\Psi(x, y). \tag{17.102}$$

17.2.3 FIRST-ORDER SHEAR STRAINS

The first-order shear strains in the member are related to the displacements u_P, v_P, w_P through

$$\gamma_{xy} = \partial v_P/\partial x + \partial u_P/\partial y, \\ \gamma_{xz} = \partial w_P/\partial x + \partial u_P/\partial z, \\ \gamma_{yz} = \partial w_P/\partial y + \partial v_P/\partial z, \Bigg\}$$ (17.103)

so that

$$\gamma_{xy} = 0, \\ \gamma_{xz} = \phi'(\partial \Psi/\partial x - y), \\ \gamma_{yz} = \phi'(\partial \Psi/\partial y + x). \Bigg\}$$ (17.104)

17.2.4 SHEAR STRESSES AND UNIFORM TORQUE

The shear stresses are related to the shear strains through

$$\tau_{xz} = G\gamma_{xz}, \\ \tau_{yz} = G\gamma_{yz}, \Bigg\}$$ (17.105)

in which G is the shear modulus of elasticity, so that

$$\tau_{xz} = G\phi'(\partial \Psi/\partial x - y), \\ \tau_{yz} = G\phi'(\partial \Psi/\partial y + x). \Bigg\}$$ (17.106)

While the shear stress distributions may be obtained by solving the equilibrium equation [1]

$$\frac{\partial \tau_{xz}}{\partial x} + \frac{\partial \tau_{yz}}{\partial y} = 0$$ (17.107)

for the warping function $\Psi(x, y)$, these are more commonly found in terms of a stress function $\theta(x, y)$ [1] defined by

$$\tau_{xz} = \partial \theta/\partial y, \\ \tau_{yz} = -\partial \theta/\partial x. \Bigg\}$$ (17.108)

Contours of the maximum shear stress for a rectangular section are shown in Figure 17.6a. The tangents to the contours indicate the directions of the maximum shear stresses, while the spacing between the contours is inversely proportional to the maximum shear stress magnitude.

For a very narrow rectangular section, the stress function is approximated [2] by

$$\theta \approx G\phi'(t^2/4 - x^2).$$ (17.109)

The corresponding shear stresses are given by

$$\tau_{xz} = 0, \\ \tau_{yz} = G\phi'(2x), \Bigg\}$$ (17.110)

so that the shear stress contours become straight lines as shown in Figure 17.6b.

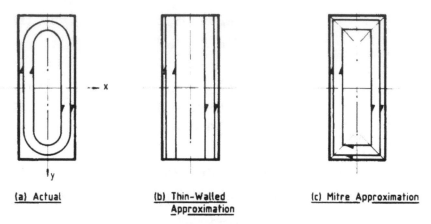

(a) Actual   (b) Thin-Walled   (c) Mitre Approximation
                              Approximation

Figure 17.6 Uniform torsion shear stress contours.

While this approximation for the stress function is quite accurate over most of the narrow rectangular section, it is in error near the ends of the section, where the shear stresses become parallel to the x axis, as shown in Figure 17.6a. The omission of these shear stresses leads to a serious underestimation of the uniform torsion stress resultant M_u.

A more accurate approximation is obtained from the 'mitred' contours shown in Figure 17.6c [3], which leads to shear stresses

$$\tau_{xz} = G\phi't\left(\frac{b}{t} - \frac{2y}{t} - 1\right) \tag{17.111}$$

in the end regions. The torque resultant of these and the shear stresses τ_{yz} is given by

$$M_u = GJ\phi' \tag{17.112}$$

in which J is the torsion section constant approximated by

$$J = \frac{bt^3}{3}\left(1 - \frac{t}{4b}\right). \tag{17.113}$$

The maximum shear stress

$$\tau_m = G\phi't \tag{17.114}$$

may also be expressed as

$$\tau_m = \frac{M_u t}{J}. \tag{17.115}$$

More accurate approximations may be obtained from the numerical solutions reported in [1] as

$$J = \frac{bt^3}{3}\left(1 - 0.674\frac{t}{b} + 0.096\frac{t^2}{b^2}\right) \tag{17.116}$$

and

$$\tau_m = \frac{M_u t}{(bt^3/3)(1 - 0.666t/b + 0.288t^2/b^2).} \tag{17.117}$$

17.2.5 OTHER THIN-WALLED OPEN SECTIONS

For very narrow rectangular sections, $t/b \to 0$, and so

$$J \approx bt^3/3. \tag{17.118}$$

Other very thin-walled open sections can be considered as being composed of a series of very narrow rectangular sections, so that

$$J \approx \sum bt^3/3. \tag{17.119}$$

The use of this relationship in equations 17.112 and 17.115 will lead to quite accurate solutions, except for the shear stresses at junctions, where re-entrant corners may cause high stress concentrations [1].

Corrections for finite thickness, end effects and junction effects for other thin-walled open sections can be obtained from [4].

17.2.6 TOTAL POTENTIAL, VIRTUAL ROTATIONS, AND EQUILIBRIUM

The total potential U_T (see section 2.3) of a twisted member and its loading system is given by

$$U_T = U + V \tag{17.120}$$

in which the strain energy stored in the member is given by

$$U = \frac{1}{2}\int_0^L \int_A (\gamma_{xz}\tau_{xz} + \gamma_{yz}\tau_{yz})\,dA\,dz \tag{17.121}$$

and the potential energy of the loading system measured from the untwisted position is given by

$$V = -\int_0^L m_u\phi\,dz - \sum_{1,2} M_u\phi. \tag{17.122}$$

Substituting equations 17.110 and 17.111 into equation 17.121 leads to

$$U = \frac{1}{2}\int_0^L GJ\phi'^2\,dz. \tag{17.123}$$

When the member undergoes a set of virtual rotations $\delta\phi$ from an equilibrium position ϕ while under the action of constant torques, the principle of virtual work requires that

$$\delta U_T = 0 \qquad (17.124)$$

for all sets of virtual rotations $\delta\phi$. Thus

$$\int_0^L \{\delta\phi'(GJ\phi') - \delta\phi m_u\}\,dz - \sum_{1,2}\delta\phi M_u = 0 \qquad (17.125)$$

and integrating by parts leads to

$$\int_0^L \delta\phi\{(-GJ\phi')' - m_u\}\,dz + [\delta\phi(GJ\phi')]_0^L - \sum_{1,2}\delta\phi M_u = 0. \qquad (17.126)$$

Since this must hold for all admissible sets of virtual rotations $\delta\phi$, then

$$-(GJ\phi')' = m_u \qquad (17.127)$$

and

$$\left.\begin{array}{l} (GJ\phi')_L = M_{u2}, \\ -(GJ\phi')_0 = M_{u1}. \end{array}\right\} \qquad (17.128)$$

Equation 17.127 is the first-order differential equilibrium equation for uniform torsion, and equations 17.128 are the boundary conditions.

17.3 Warping torsion

17.3.1 MEMBER AND LOADING

The member of length L shown in Figure 17.7a is of thin-walled open cross-section. The principal centroidal axes x, y of the cross-section are shown in Figure 17.7b. The positions of these axes are defined by equation 17.1.

The position of the shear centre $S(x_0, y_0)$ of the cross-section shown in Figure 17.7b is defined by the conditions

$$\int_A x\omega\,dA = \int_A y\omega\,dA = 0, \qquad (17.129)$$

in which ω is defined by

$$\omega = \frac{1}{A}\int_A\left\{\int_0^s \rho_0\,ds\right\}t\,ds - \int_0^s \rho_0\,ds \qquad (17.130)$$

in which t is the wall thickness, ρ_0 is the perpendicular distance from the shear centre to the mid-thickness tangent, and s is the distance around the mid-thickness line. Equation 17.130 satisfies

$$\int_A \omega\,dA = 0. \qquad (17.131)$$

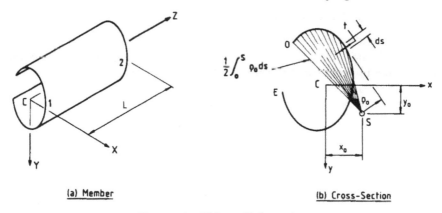

(a) Member   (b) Cross-Section

Figure 17.7 Thin-walled member.

The warping section constant of the cross-section is defined by

$$I_w = \int_A \omega^2 \, dA. \tag{17.132}$$

The member is acted on by end warping torques M_{w1}, M_{w2} and end bimoments B_1, B_2, and distributed warping torques per unit length m_w, as shown in Figure 17.8. It is assumed that the section is so thin-walled that any uniform torques (section 17.2) are negligible.

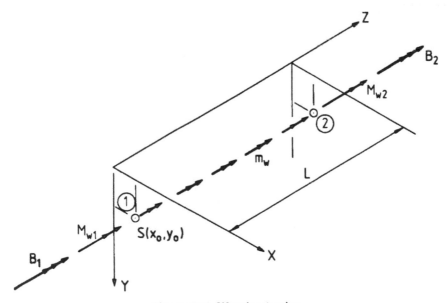

Figure 17.8 Warping torsion.

17.3.2 TWIST ROTATIONS AND DEFLECTIONS

A member cross-section rotates ϕ about the shear centre axis through $S(x_0, y_0)$ shown in Figure 17.8. It is assumed that any shear strains are small, so that warping due to shear can be neglected. In this case, the longitudinal warping displacements are due to twisting alone.

As a result of the rotation ϕ, a line through a point $P(x, y)$ at a distance

$$a_0 = \sqrt{\{(x - x_0)^2 + (y - y_0)^2\}} \tag{17.133}$$

from the shear centre rotates through an angle $a_0 \, d\phi/dz$ as shown in Figure 17.9, and an element of the wall $\delta z \times \delta s$ warps

$$\delta w_P = -\rho_0 \frac{d\phi}{dz} \delta s \tag{17.134}$$

as shown in Figure 17.9. Integration leads to

$$w_P = w_{PO} - \frac{d\phi}{dz} \int_0^s \rho_0 \, ds. \tag{17.135}$$

If the arbitrary constant of integration w_{PO} is chosen so that the average warping displacement is zero, then equation 17.135 becomes

$$w_P = \omega \frac{d\phi}{dz} \tag{17.136}$$

in which ω is the warping function given by equation 17.130 which satisfies equation 17.131.

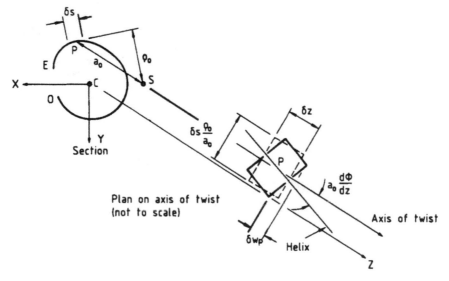

Figure 17.9 Warping of an element $\delta z \times \delta s \times t$ due to twisting.

17.3.3 FIRST-ORDER STRAINS

The first-order longitudinal normal strains at the point P resulting from the warping displacements w_P are

$$\varepsilon_P = w_P' = \omega \phi''. \tag{17.137}$$

All shear strains are neglected.

17.3.4 STRESSES AND STRESS RESULTANTS

The longitudinal normal stress at the point P in the cross-section is given by

$$\sigma_P = E\varepsilon_P. \tag{17.138}$$

Because of equations 17.129 and 17.131,

$$\left.\begin{array}{c} \displaystyle\int_A \sigma_P \, dA = 0, \\[2mm] \displaystyle\int_A \sigma_P x \, dA = 0, \\[2mm] \displaystyle\int_A \sigma_P y \, dA = 0, \end{array}\right\} \tag{17.139}$$

and so the axial force and bending moment stress resultants of σ_P are all zero. The non-zero stress resultant of σ_P is the bimoment

$$B = \int_A \sigma_P \omega \, dA = EI_w \phi''. \tag{17.140}$$

Although the shear strains have been assumed to be so small that they can be neglected, they are generally not zero, and correspond to warping shear stresses τ_P which are induced by variations of the longitudinal normal stresses σ_P along the length of the member, in the same way that bending shear stresses are induced by variations of the bending normal stresses along the member. Thus

$$\tau_P t = -\int_0^s \frac{d\sigma_P}{dz} t \, ds = -E\phi''' \int_0^s \omega t \, ds. \tag{17.141}$$

The warping shear stresses τ_P have a warping torque stress resultant

$$M_w = \int_A \rho_0 \tau_P t \, ds \tag{17.142}$$

which can be expressed as [2]

$$M_w = -(EI_w \phi'')'. \tag{17.143}$$

Substituting equation 17.140 leads to

$$M_w = -B'. \tag{17.144}$$

17.3.5 TOTAL POTENTIAL, VIRTUAL ROTATIONS AND EQUILIBRIUM

The total potential U_T (see section 2.3) of the twisted length of the member and its loading system is given by

$$U_T = U + V \tag{17.145}$$

in which the strain energy stored in the member is given by

$$U = \frac{1}{2}\int_0^L \int_A \varepsilon_p \sigma_p \mathrm{d}A\,\mathrm{d}z \tag{17.146}$$

and the potential energy of the loading system measured from the untwisted position is given by

$$V = -\int_0^L m_w \phi \mathrm{d}z - \sum_{1,2} M_w \phi - \sum_{1,2} B\phi'. \tag{17.147}$$

Substituting equations 17.132, 17.137, and 17.138 into equation 17.146 leads to

$$U = \frac{1}{2}\int_0^L EI_w \phi''^2 \mathrm{d}z. \tag{17.148}$$

When the member undergoes a set of virtual rotations $\delta\phi$ from an equilibrium position ϕ while under the action of constant torques and bimoments, the principle of virtual work requires that

$$\delta U_T = 0 \tag{17.149}$$

for all sets of virtual rotations $\delta\phi$. Thus

$$\int_0^L \{\delta\phi'' EI_w \phi'' - \delta\phi m_w\}\mathrm{d}z - \sum_{1,2}\{\delta\phi M_w + \delta\phi' B\} = 0 \tag{17.150}$$

and integrating by parts leads to

$$\int_0^L \delta\phi\{(EI_w\phi'')'' - m_w\}\mathrm{d}z + [\delta\phi' EI_w\phi'' - \delta\phi(EI_w\phi'')']_0^L$$
$$- \sum_{1,2}\{\delta\phi M_w + \delta\phi' B\} = 0. \tag{17.151}$$

Since this must hold for all admissible sets of virtual rotations $\delta\phi$, then

$$(EI_w\phi'')'' = m_w \tag{17.152}$$

and

$$\left.\begin{aligned}
-(EI_w\phi'')'_L &= M_{w2}, \\
(EI_w\phi'')'_0 &= M_{w1}, \\
(EI_w\phi'')_L &= B_2, \\
-(EI_w\phi'')_0 &= B_1.
\end{aligned}\right\} \tag{17.153}$$

Equation 17.152 is the first-order differential equilibrium equation for warping torsion, and equations 17.153 are the boundary conditions.

17.4 Energy equations for flexural-torsional buckling

17.4.1 MEMBER AND LOADING

The member of length L shown in Figure 17.7a is of thin-walled open cross-section. The positions of the principal centroidal axes x, y of the cross-section

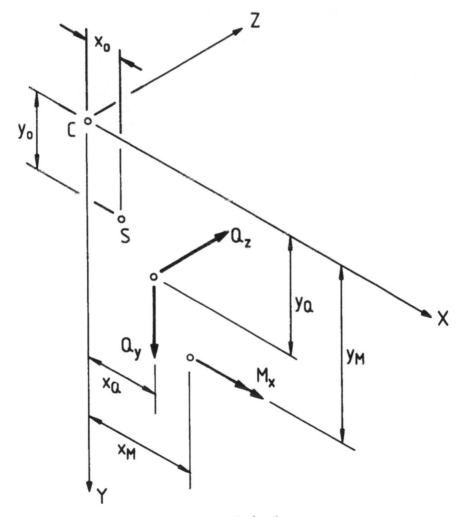

Figure 17.10 End actions.

shown in Figure 17.7b are defined by equation 17.1. The section properties are defined by equation 17.2 and

$$
\left.
\begin{aligned}
I_y &= \int_A x^2 \, \mathrm{d}A, \\[6pt]
I_P &= \int_A (x^2 + y^2) \, \mathrm{d}A, \\[6pt]
I_{Px} &= \int_A y(x^2 + y^2) \, \mathrm{d}A.
\end{aligned}
\right\}
\qquad (17.154)
$$

The position of the shear centre $S(x_0, y_0)$ of the cross-section shown in Figure 17.7b is defined by equation 17.129. The torsion section properties are defined by equations 17.116 or 17.119 (see also [4]) and 17.132.

The member is acted on by end loads $Q_{y1}, Q_{z1}, Q_{y2}, Q_{z2}$, end moments M_{x1}, M_{x2} (Figure 17.10), and distributed loads per unit length q_y. The loads act through points defined by x_Q, y_Q, or x_q, y_q, and the moments in planes defined by x_M, y_M. These actions move with the member but remain parallel to their original directions and the planes defined by the X, Y, Z axes of the straight, untwisted member.

17.4.2 DISPLACEMENTS AND ROTATIONS

A member cross-section undergoes shear centre displacements u, v, w_s parallel to the original X, Y, Z axes, and twist rotations ϕ about the shear centre axis, as shown in Figure 17.11. The corresponding displacements of a point $P(x, y)$ in the cross-section are given by

$$u_P \approx u - (y - y_0)\phi, \qquad (17.155)$$

$$v_P \approx v + (x - x_0)\phi, \qquad (17.156)$$

$$w_P \approx (w - xu' - yv' + \omega\phi') + (-xv'\phi + yu'\phi). \qquad (17.157)$$

The displacements u_P, v_P are shown diagrammatically in Figure 17.12c.

The longitudinal displacement w_P given by equation 17.157 is obtained by considering

$$w_P = w_s - u'\{(x - x_0) - (y - y_0)\phi\} - v'\{(y - y_0) + (x - x_0)\phi\} + \omega\phi'. \quad (17.158)$$

The first component of this is the shear centre displacement, while the last is the warping displacement caused by the twist ϕ' (see section 17.3.2). The second component is the result of the rotation u' of the line SP as shown in Figure 17.12b, and the third component is the result of the rotation v' as shown in Figure 17.12a. Substituting $x = 0$, $y = 0$, into equation 17.158 leads to

$$w_c = w_s + x_0(u' + \phi v') + y_0(v' - \phi u') + \omega_c \phi' \qquad (17.159)$$

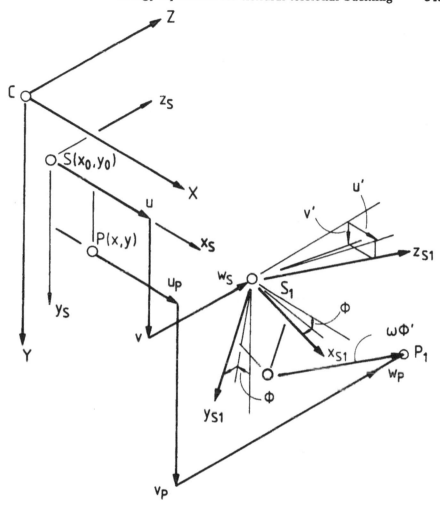

Figure 17.11 Displacements and rotations.

and defining

$$w = w_c - \omega_c \phi' \qquad (17.160)$$

and substituting into equation 17.158 leads to equation 17.157.

17.4.3 STRAINS

The longitudinal normal strain ε_P at P parallel to the Z axis may be defined in terms of the rates of change of the deflections u_P, v_P, w_P along the element, as shown in Figure 17.13. The new elemental length is given by

$$(1 + \varepsilon_P)\delta z = \sqrt{\{(\delta z + w'_p \delta z)^2 + (u'_p \delta z)^2 + v'_p \delta z)^2\}} \qquad (17.161)$$

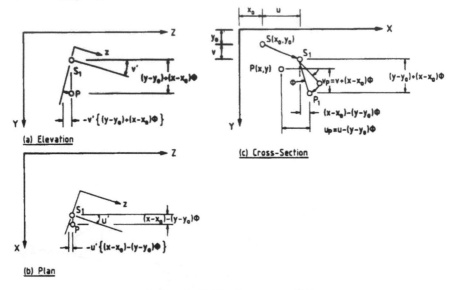

Figure 17.12 Displacements of P.

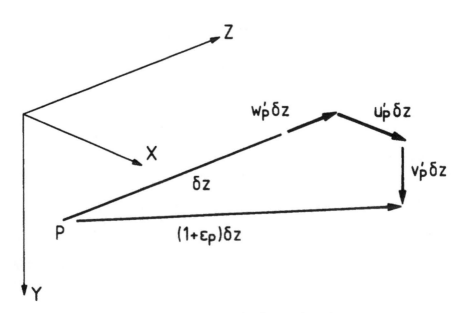

Figure 17.13 Longitudinal normal strain.

so that

$$\varepsilon_P \approx w'_P + \tfrac{1}{2}(u'^2_P + v'^2_P). \tag{17.162}$$

Substituting equations 17.155–17.157 leads to

$$\varepsilon_P = \{w' - xu'' - yv'' + \omega\phi''\} + \{\tfrac{1}{2}[u'^2 + v'^2 + (x_0^2 + y_0^2)\phi'^2] - x_0v'\phi' + y_0u'\phi'$$
$$+ x[-x_0\phi'^2 - \phi v''] + y[-y_0\phi'^2 + \phi u''] + \tfrac{1}{2}[x^2 + y^2]\phi'^2\}. \tag{17.163}$$

The shear strain γ_P at P resulting from twisting ϕ' of a thin-walled open section member is approximated by (see section 17.2)

$$\gamma_P = 2t_P\phi' \tag{17.164}$$

in which t_P is the perpendicular distance of P from the mid-thickness line of the cross-section.

Shear strains due to bending and warping shears are neglected.

17.4.4 STRESSES AND STRESS RESULTANTS

The longitudinal normal stress at the point P in the cross-section is given by

$$\sigma_P = E\varepsilon_P \tag{17.165}$$

and for small deformations, this is approximated by

$$\sigma_P \approx E(w' - xu'' - yv'' + \omega\phi''). \tag{17.166}$$

This can be written as

$$\sigma_P \approx \frac{N}{A} + \frac{M_x y}{I_x} - \frac{M_y x}{I_y} + \frac{B\omega}{I_w} \tag{17.167}$$

in terms of the stress resultants

$$\left.\begin{array}{c} N = \displaystyle\int_A \sigma_P dA = EAw', \\[2mm] M_x = \displaystyle\int_A \sigma_P y dA = -EI_x v'', \\[2mm] -M_y = \displaystyle\int_A \sigma_P x dA = -EI_y u'', \\[2mm] B = \displaystyle\int_A \sigma_P \omega dA = EI_w \phi''. \end{array}\right\} \tag{17.168}$$

These equations make use of the section properties of equations 17.2 and 17.132 and the conditions of equations 17.1 and 17.129.

The uniform torsion shear stress at P is given by

$$\tau_P = G\gamma_P \tag{17.169}$$

which can be written as

$$\tau_P = \frac{2M_u t_P}{J} \qquad (17.170)$$

in which

$$M_u = 2 \int_A \tau_P t_P dA = GJ\phi' \qquad (17.171)$$

in which the factor 2 compensates for the effects of the transverse shear stresses at the ends of the elements of the thin-walled section (see section 17.2.4).

17.4.5 TOTAL POTENTIAL, EQUILIBRIUM AND BUCKLING

The total potential U_T (see section 2.3) of the strained length of the member and its loading system is given by

$$U_T = U + V \qquad (17.172)$$

in which U is the strain energy stored in the member, and V is the potential energy of the loading system. The strain energy U may be expressed as

$$U = \frac{1}{2} \int_0^L \int_A (\varepsilon_P \sigma_P + \gamma_P \tau_P) dA\, dz \qquad (17.173)$$

and the potential energy V as

$$V = -\int_0^L (q_y v_q + q_z w_q) dz - \sum (Q_y v_Q + Q_z w_Q - M_x v_M'). \qquad (17.174)$$

For equilibrium of a deformed position, the principle of virtual work requires that

$$\delta U_T = 0 \qquad (17.175)$$

for all sets of virtual deformations δu, δv, δw, $\delta \phi$.

For neutral equilibrium (buckling)

$$\tfrac{1}{2}\delta^2 U_T = 0 \qquad (17.176)$$

for any set of buckling deformations δu, δv, δw, $\delta \phi$.

Now

$$\frac{1}{2}\delta^2 U = \frac{1}{2} \int_0^L \int_A (\delta \varepsilon_P \delta \sigma_P + \delta \gamma_P \delta \tau_P + \delta^2 \varepsilon_P \sigma_P + \delta^2 \gamma_P \tau_P) dA\, dz \qquad (17.177)$$

in which

$$\left. \begin{aligned}
\delta \varepsilon_P &= \delta w' - x \delta u'' - y \delta v'' + \omega \delta \phi'', \\
\delta \gamma_P &= 2t_P \delta \phi', \\
\delta \sigma_P &= E \delta \varepsilon_P, \\
\delta \tau_P &= G \delta \gamma_P,
\end{aligned} \right\} \qquad (17.178)$$

and

$$
\left.
\begin{aligned}
\delta^2 \varepsilon_P = & \{\delta u'^2 + \delta v'^2 + (x_0^2 + y_0^2)\delta \phi'^2 - 2x_0 \delta v' \delta \phi' + 2y_0 \delta u' \delta \phi'\} \\
& - x\{2x_0 \delta \phi'^2 + 2\delta \phi \delta v''\} - y\{2y_0 \delta \phi'^2 - 2\delta \phi \delta u''\} + (x^2 + y^2)\delta \phi'^2, \\
\delta^2 \gamma_P = & \, 0, \\
\sigma_P = & \, N/A + M_x y/I_x, \\
\tau_P = & \, 0.
\end{aligned}
\right\}
$$

$$(17.179)$$

Also

$$
\tfrac{1}{2}\delta^2 V = -\int_0^L (q_y \delta^2 v_q + q_z \delta^2 w_q)\mathrm{d}z - \sum (Q_y \delta^2 v_Q + Q_z \delta^2 w_Q - M_x \delta^2 v'_M). \quad (17.180)
$$

17.4.6 BUCKLING OF AN AXIALLY LOADED COLUMN

For an axially loaded column, $M_x = q_y = Q_y = 0$ and $\delta^2 w_q = \delta^2 w_Q = 0$. It is assumed that during buckling, the centroidal strain remains zero (inextensional buckling), so that the column buckles under constant force N. (In the extensional buckling of a latticed column, for example, the longitudinal members strain axially w' during buckling). For this inextensional buckling, equation 17.163 requires δw to be of small order compared with the other buckling deformations $\delta u, \delta v,$ and $\delta \phi$ and a term $\tfrac{1}{2}\int_L EA w'^2 \mathrm{d}z$ arising from equation 17.177 can be ignored.

If the buckling deformations $\{\delta u, \delta v, \delta w, \delta \phi\}$ are rewritten as $\{u, v, w, \phi\}$, then equation 17.176 becomes

$$
\frac{1}{2}\int_L \{EI_y u''^2 + EI_x v''^2 + EI_w \phi''^2 + GJ \phi'^2\}\mathrm{d}z
$$

$$
+ \frac{1}{2}\int_L N\{u'^2 + v'^2 + (r_0^2 + x_0^2 + y_0^2)\phi'^2 - 2x_0 v' \phi' + 2y_0 u' \phi'\}\mathrm{d}z = 0 \quad (17.181)
$$

in which

$$
r_0^2 = (I_x + I_y)/A = I_P/A. \quad (17.182)
$$

The sign convention for N in equation 17.181 is tension positive, and so N is negative for a member in compression.

17.4.7 BUCKLING OF A MONOSYMMETRIC BEAM-COLUMN

For a monosymmetric beam-column, $x_0 = 0$. For axial loading $\delta^2 w_q = \delta^2 w_Q = \delta^2 v'_M = 0$, while

$$
\left.
\begin{aligned}
\delta^2 v_q = -\tfrac{1}{2}(y_q - y_0)\delta \phi^2, \\
\delta^2 v_Q = -\tfrac{1}{2}(y_Q - y_0)\delta \phi^2,
\end{aligned}
\right\}
$$

$$(17.183)$$

as shown in Figure 17.14, in which y_q, y_Q are the distances the centroidal axis of the points of application of the loads q_y, Q_y.

It is assumed that during buckling, the centroidal strain and the curvature in the principal yz plane remain zero, so that the beam-column buckles under constant axial force N and bending moment M_x (this is the beam-column equivalent of the inextensional buckling discussed in section 17.4.6). In this case, the buckling displacements $\delta v, \delta w$ are of a small order compared with the other buckling deformations $\delta u, \delta\phi$ in equation 17.178, and after rewriting the buckling deformations $\{\delta u, \delta v, \delta w, \delta\phi\}$ as $\{u, v, w, \phi\}$, equation 17.176 becomes

$$\frac{1}{2}\int_L \{EI_y u''^2 + EI_w \phi''^2 + GJ\phi'^2\}dz + \frac{1}{2}\int_L N\{u'^2 + (r_0^2 + y_0^2)\phi'^2 + 2y_0 u'\phi'\}dz$$

$$+\frac{1}{2}\int_L M_x\{2\phi u'' + \beta_x\phi'^2\}dz + \frac{1}{2}\int_L q_y(y_q - y_0)\phi^2\, dz + \frac{1}{2}\sum Q_y(y_Q - y_0)\phi^2 = 0$$

(17.184)

in which

$$\beta_x = \int_A y(x^2 + y^2)\, dA/I_x - 2y_0 = I_{Px}/I_x - 2y_0 \qquad (17.185)$$

and N is negative for a beam-column in compression.

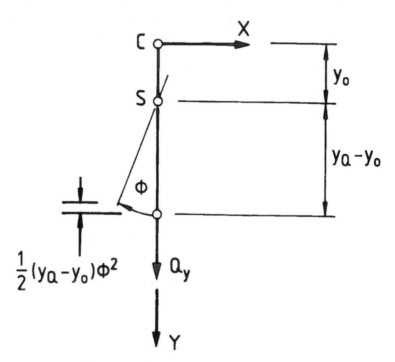

Figure 17.14 Second-order load displacement.

17.5 Differential equilibrium equations for the buckled position

17.5.1 BUCKLING OF AN AXIALLY LOADED COLUMN

The buckling equilibrium equations for an axially loaded column can be obtained by using the calculus of variations, according to which the functions u, v, ϕ which make

$$\frac{1}{2}\delta^2 U_T = \int_0^L F(z, u', u'', v', v'', \phi, \phi', \phi'')\,dz \qquad (17.186)$$

stationary satisfy the conditions

$$\left.\begin{aligned}
-\frac{d}{dz}\left(\frac{\partial F}{\partial u'}\right) + \frac{d^2}{dz^2}\left(\frac{\partial F}{\partial u''}\right) &= 0, \\[1mm]
-\frac{d}{dz}\left(\frac{\partial F}{\partial v'}\right) + \frac{d^2}{dz^2}\left(\frac{\partial F}{\partial v''}\right) &= 0, \\[1mm]
\frac{\partial F}{\partial \phi} - \frac{d}{dz}\left(\frac{\partial F}{\partial \phi'}\right) + \frac{d^2}{dz^2}\left(\frac{\partial F}{\partial \phi''}\right) &= 0.
\end{aligned}\right\} \qquad (17.187)$$

Substituting equation 17.181 into equation 17.186 leads to

$$(EI_y u'')'' = (N\{u' + y_0\phi'\})', \qquad (17.188)$$

$$(EI_x v'')'' = (N\{v' - x_0\phi'\}), \qquad (17.189)$$

$$(EI_w \phi'')'' - (GJ\phi')' = (N\{r_0^2 + x_0^2 + y_0^2\}\phi')' - (Nx_0 v')' + (Ny_0 u')', \quad (17.190)$$

which are the differential equilibrium equations for the buckled position u, v, ϕ for bending about the x, y axes and torsion. The sign convention for N in equations 17.188–17.190 is tension positive, and so N is negative for a member in compression.

17.5.2 BUCKLING OF A MONOSYMMETRIC BEAM-COLUMN

The buckling equilibrium equations for a monosymmetric beam-column with shear centre loading ($y_q = y_Q = 0$) can also be obtained by using the calculus of variations, this time by substituting equation 17.184 into equation 17.186, which leads to

$$(EI_y u'')'' = (N\{u' + y_0\phi'\})' - (M_x\phi)'', \qquad (17.191)$$

$$(EI_w \phi'')'' - (GJ\phi')' = (N\{r_0^2 + y_0^2\}\phi')' + (Ny_0 u')' - (M_x u'') + (M_x\beta_x\phi'), \qquad (17.192)$$

which are the differential equilibrium equations for the buckled position u, ϕ for bending about the y axis and torsion.

17.6 References

1. Timoshenko, S.P. and Goodier, J.N. (1970) *Theory of Elasticity*, 3rd edn, McGraw-Hill, New York.
2. Vlasov, V.Z. (1961) *Thin Walled Elastic Beams*, 2nd edn, Israel Program for Scientific Translations, Jerusalem.
3. Billinghurst, A., Williams, J.R.L., Chen, G. and Trahair, N.S. (1992) Inelastic uniform torsion of steel members. *Computers and Structures*, 22 (6), 887–94.
4. El Darwish, I.A. and Johnston, B.G. (1965) Torsion of structural shapes. *Journal of the Structural Division, ASCE*, 91 (ST1), 203–27.

Index